W9-ABY-797

American Entomologists

By the Same Author

Handbook of Pest Control: The Behavior, Life History and Control of Household Pests (5th ed., 1969)

American Entomologists

Arnold Mallis

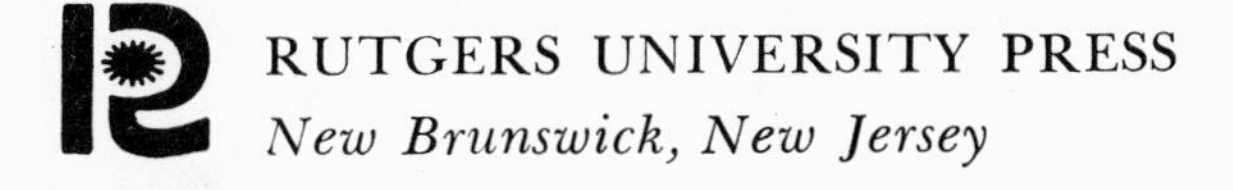
RUTGERS UNIVERSITY PRESS
New Brunswick, New Jersey

Permission to quote has been kindly granted by the following:

Doubleday and Co., Inc.: from *archy and mehitabel,* by Don Marquis.

Estate of Curtis Hidden Page: from *Japanese Poetry,* by Curtis Hidden Page.

George Allen and Unwin, Ltd.: from "The Child in School," a Chinese folk song, in *Flower Shadows,* by Alan S. Lee.

John Murray, Ltd.: from "Autumn Wind," in *Wisdom of the East,* tr. Lewis Mackenzie.

University of California Press: from *The Year of My Life,* tr. Nobuyuki Yuasa. Originally published by the University of California Press; reprinted by permission of The Regents of the University of California.

Library of Congress Catalogue Card Number: 78-152316
ISBN: 0-8135-0686-7
Manufactured in the United States of America
By Quinn & Boden Company, Inc., Rahway, N.J.

To
Dr. Stuart W. Frost
and
To the memory
of Dr. E. C. Van Dyke and Professor J. J. Davis

Contents

List of Illustrations

Preface

The man who loves living things, who is attuned to the world of plants and animals and the countryside around him, who harkens to the sound of insects on a sunny day or a starry night, and who derives pleasure from the glistening wings of the dragonfly or the lazy flight of a butterfly, will, I think, enjoy meeting the American entomologists who appear in this book.

I am, of course, aware of the fine works on American entomology and entomologists by E. O. Essig, L. O. Howard, H. B. Weiss, and H. Osborn, but I have found that these authors, for the most part, dwelled on the accomplishments of the entomologists and told us little about the men themselves. The thought occurred to me that one reason for this neglect was that there was little of interest to say about them. But, as I delved deeper and deeper into the old volumes in museum libraries, I found their lives more and more interesting.

It was then that I decided to try my hand at writing the stories of a number of America's outstanding entomologists, describing, of course, some of their entomological accomplishments, but wherever possible dwelling on the men themselves, fleshing them out a bit, so that they would appear to be something more than a list of publications. If I have emphasized some at the expense of others, it is chiefly because of the necessity for keeping the book within reasonable bounds.

For a variety of reasons, I have often failed in presenting the portrait I desired; in many instances there simply was a dearth of published information about the subject of the sketch. Nevertheless, I have diligently dug and probed, breathed the dust of little-used volumes, and written to many old-timers (my wife says I have the oldest correspondents in the country) to come up with a letter or some bit of unpublished information that lent zest to the chase.

A work like this can proliferate like aphids on a rose, and for this reason it was necessary to resort to certain restrictions. This work is limited to North American entomologists, all of whom are deceased. European entomologists, such as Linnaeus, Fabricius, Olivier, and many more, who contributed so much to the naming of American insects, are *not* the subject of this book; whereas J. Abbot, C. R. Osten-Sacken, B. D. Walsh, H. A. Hagen, and numerous others, who went to

America and became part of entomological history there, play an important part in these sketches.

Other historians of entomology have definitely influenced my choice of the older entomologists, but I was pretty much on my own in assembling some of the outstanding workers in the first half of this century. I can only hope that I have used good judgment in my selection, and, where I have failed, possibly I can make amends at some future date.

This collection of biographical material was made possible through the facilities of many libraries. I am especially indebted to the Museum Library in Carnegie Institute as well as Carnegie Library in Pittsburgh; I was also fortunate in being permitted to use the library facilities of the Academy of Natural Sciences of Philadelphia, American Museum of Natural History, Boston Museum of Science, Congressional Library, Harvard Museum of Comparative Zoology, and Yale University Library. Finally, I must acknowledge my debt to the very helpful *Bibliography of Biographies of Entomologists* by Mathilde M. Carpenter.

Many individuals aided me with information or photographs and I wish to acknowledge the help of the following: J. B. Adams, C. P. Alexander, S. F. Bailey, T. L. Bissell, E. P. Breakey, F. M. Carpenter, E. N. Cory, I. S. Creelman, P. J. Darlington, Jr., P. DeBach, A. B. Gurney, E. J. Hansens, D. E. Howell, the late O. E. Jennings, F. H. Lathrop, H. B. Leech, F. W. Lechleitner, M. D. Leonard, J. H. Lilly, C. D. Michener, C. E. Mickel, A. C. Miller, C. F. W. Muesebeck, J. R. Parker, C. A. Peters, V. T. Phillips, D. Pimentel, L. F. Pinkus, E. S. Ross, G. W. Simpson, C. F. Smith, T. E. Snyder, and G. E. Wallace. I am especially indebted to S. W. Frost, E. I. McDaniel, W. Dwight Pierce, and J. J. Davis.

R. H. Nelson kindly permitted me to use photographs in the collection of the Entomological Society of America. I understand that many of these were collected by J. S. Wade. D. E. Garvey made copies of many of the photographs used herein.

I also wish to accredit the following sources for the photographs and illustrations in this book: *Annals of the Entomological Society of America,* Boston Society of Natural History, *Bulletin of the Brooklyn Entomological Society,* California Mosquito Control Association, *Canada Entomology Division News Letter, Canadian Entomologist,* Carnegie Museum, Columbia University Press, Constable Publishers, Cornell University, Cornell University Press, *Entomological News,* Entomological Society of America, Harvard University, Illinois Natural History Survey, *Journal of Economic Entomology, Journal of the Kansas Entomological Society, Journal of the New York Entomological Society,* Lancaster County Historical Society, *Lepidopterists' News,* New York State Museum, Pacific Coast Entomological Society, *Pan-Pacific Entomol-*

ogist, Philadelphia Academy of Natural Sciences, *Proceedings of the Entomological Society of Washington, Psyche;* Rutgers State University, *Scientific Monthly,* Smithsonian Institute, *Transactions of the American Entomological Society, Transactions of the Connecticut Academy of Arts and Sciences,* United States Department of Agriculture, University of Kansas Press, University of Massachusetts, University of Maine.

Finally, I wish to acknowledge the invaluable help of Margaret A. McGregor, who translated my cryptic handwritten notes into legible typescript.

Arnold Mallis

American Entomologists

Children, if you meet a cricket,
Please remember not to kick it;
Not a youth whose nature's sweet'll
Strike a ladybug or beetle.

If a daddy longlegs passes
Do not slay it (as alas! is
Often done by wicked urchins
Who deserve the soundest birchin's).

When a gnat—that lively hummer,
Which you'll hear when it is summer,
Comes a-buzzing round your hat, it
Is wrong to throw a pebble at it.

Do not think me sentimental
When I ask you to be gentle
With the insect population
Of our free and glorious nation.

If you're kind in this partic'lar
They will buzz in your auric'lar;
Every child, of course, can see
What a pleasure that will be.

London Globe and
Entomological News, 1906

1 Pioneer Entomologists

A good world it is, indeed!
When the beetle rings his little bell
And the hawk pirouettes!

Issa

One of the treasures of the New World was the hordes of unknown insects eagerly sought by such European students of entomology as C. Linnaeus, J. C. Fabricius, G. A. Olivier, P. F. M. A. Dejean, P. A. Latreille, C. DeGeer, J. W. Meigen, W. Kirby and many others. Men such as Mark Catesby and John Abbot went to Colonial America, were fascinated by the insects they captured, and drew or painted them. The first American entomologists sent the insects or drawings of them to Europe for identification. Later a few, courageously using the European literature, began to name their insect gems. This chapter tells the story of some of these notable pioneer entomologists.

JOHN ABBOT (1751–1840?)

Dr. A. N. Avinoff, an eminent lepidopterist, artist, orientalist, linguist, Russian nobleman, émigré, and former Director of the Carnegie Museum, wrote to Dr. Frank Morton Jones in a letter dated November 17, 1938, about a volume of plates in his possession:

"At the time I was leaving Russia in the summer of 1917 I took this book with me to America, as I considered it the most valuable of my whole library of some eight thousand volumes on entomology. I could carry only a suitcase with me, so you will understand that I had to make a selection of something truly valuable . . ." This unpublished volume by John Abbot contained ninety-seven plates of insects, mostly Lepidoptera, dated from June 1767 to January 1773.

Abbot was one of the first naturalists in the New World to show a serious interest in insects. An accomplished artist, he specialized in water colors of plants, insects, and birds. Although he was not a trained

Figure 1. John Abbot. Self portrait. From C. L. Remington, "Notes on My Life (John Abbot)" in *Lepidopterists' News* (1948), 2:28–30

scientist, he reared many of the insects he painted and forwarded his notes on their life histories to serious students of entomology in Europe. The only portrait of this remarkable man is a candid study he made of himself, which is in one of his volumes in the British Museum.

Although Abbot was the first artist in the New World to produce an extensive series of insect drawings, other artists such as John White and Mark Catesby had illustrated American insects long before him. Holland (1929) notes that John White, a commander of Sir Walter Raleigh's third expedition in 1587 to Virginia, painted the tiger swallowtail, *Papilio turnus.* A woodcut of this painting appeared in Thomas Moufet's book, *Insectorum Sive Minimorum Animalium Theatrum* (1634).

Almost one hundred years later Mark Catesby's illustrations of insects appeared in print. His two-volume work, *Natural History of Carolina, Florida and the Bahaman Islands,* illustrates plants and birds as well as twenty-seven insects and one centipede. Here we recognize such common insects as the Luna moth, the Cecropia moth, the Polyphemus moth, the monarch butterfly, and the American cockroach.

Dow (1914) reports that Francillon, a silversmith of the Strand, London, offered for sale some time before 1797 pinned and well-spread specimens from the "Province of Georgia" in North America, rare larvae that were cleverly inflated, as well as beautiful drawings.

The drawings were not only of adult insects but also of the immature stages and their food plants and came with manuscript notes about the insects and plants. The drawings were excellent and also remarkably cheap. "It was business for Francillon, for he bought at 3d. and did not propose his wares should be secured by others from first hand to his undoing."

Two folio volumes, which are classics of their kind, were published by Sir James Edward Smith under the title *The Natural History of the Rarer Lepidopterous Insects of Georgia Collected from the Observations of John Abbot, with the Plants on Which They Feed.* These volumes contain 104 plates which Sir James Smith purchased from Francillon. Abbot only became aware of the book long after it was published. Similar volumes by J. A. Boisduval and J. E. LeConte did not accredit Abbot for his drawings.

We know that Abbot made well over three thousand drawings. At

Figure 2. A stag beetle. From A. N. Avinoff's collection of plates of John Abbot's drawings. Courtesy of the Carnegie Museum

times he employed assistants to help him, and he often retouched their work. Over the years many of these drawings disappeared. The British Museum of Natural History has a seventeen-volume work on American insects by Abbot, by far the largest collection extant. The Boston Society of Natural History has several hundred of his drawings, of both insects and birds, and Avinoff's volume of insects, mentioned previously, is now part of the collection of the Carnegie Museum.

George Marx (1891) tells us that John Abbot anticipated Marx's own interest in American spiders in his catalogue of over five hundred illustrations of previously unknown spiders, largely from Colonial Georgia. Unfortunately, Baron C. A. Walckenaer used these plates to prepare vague and inadequate descriptions of spiders so that few of them can be recognized today.

In view of Avinoff's report of bringing his volume of Abbot's drawings of insects to America, it is interesting to note that Edward Doubleday, an English correspondent of T. W. Harris', sent Harris in 1839 the volume of insect drawings now in the possession of The Boston Society of Natural History. Doubleday (1869), in a letter to Harris dated June 4, 1839, says:

"A few days since I found at a bookseller's, eighty-four drawings by Abbot, containing one hundred and fifty figures of Georgian Coleoptera, and about three hundred and fifty of Lepidoptera. They are bound in a small folio volume, and did not belong to Swainson. As many of the things figured are new to me, I thought they might not be known to you either, and so gave seven guineas for them, and brought them away, determining to send them as a trifling present to you in my next parcel. I hope they may contain something new to you."

Until recently we had very little information about Abbot's early years. C. L. Remington (1948) discovered in the locked files of Harvard's Museum of Comparative Zoology Abbot's *Notes on My Life,* excerpts of which are reproduced here.

I was born in the Year 1751, the first of June Old Stile at the West end of the Town London, in Bennet Street St. James, my father was an Attorney at Law, I was his 2d. Son my brother dying before I was born, at the time of my leaving England I had 2 Sisters, Elizabeth & Charlotte, and a brother Thomas then 7 years old. I had a very early love for Books laying out my pocket money for little Story books, and an early taste for drawing, which might be much increased by my father having a large & valueble collection of prints, of some of the best Masters, he had also many good paintings

My peculiar liking for Insects was long before I was acquainted with any method of catching or keeping them I remember knocking down a Libella & pining it, when I was told it wou'd sting, as bad as a Rattle Snake bite

My Father had a Country house, at Turnham Green 5 miles from London at £25 a Year rent, at the early part of my Life, I remember breeding some there, when I had no method of keeping them after I had done it when the Lease expired my father gave it up, as the grounds & house was divided, between the heirs of it I have a drawing of the house, which I will send you some time hence

In one of my Walks after Insects I met with a Mr Van Dest the famous flower painter, he invited me to come & see him, he had been a small Collector, showed me a pattern of the large Net, & gave me some rare Insects, I got me immediately a Net made & begun to understand keeping them better

My father got a Mr Boneau an Engraver, & Drawing Master, to give me lessons of Drawing at our own house, he was acquainted with a Mr Rice a Teacher of Grammer, who had likewise been a Collector of Insects, Mr Boneau, did not paint in Water Colours, he only understood the Rules of Drawing & perspective, he praised my Drawings of Insects, & got me through Mr Rice introduced to Mr Drury who had been president of the Linnean Society & who then was allowed to have the best Collection of Insects both English & foreign of any one.

I leave You to Judge my pleasure & astonishment at the sight of his Cabinets the first I had ever seen of the kind he very politely offered to lend me Insects to Draw, & we immediately became well acquainted That hour may be said to have given a new turn to my future life I had immediately a Mahogany Cabinet made of 26 Draws, covered with sliding tops of Glass, it cost me 6 Guineas, & began to collect with an unceasing Industry I met with soon after & purchased a parcel of beautiful Insects from Surinam I soon begun to have a respectable Collection but not Satisfied with it I craved more . . .

Abbot now began to think of going abroad to collect insects. To prepare himself as a naturalist, he obtained George Edward's *Birds* in four volumes, and Mark Catesby's *Natural History of Carolina.*

I was Articled to my Father as his Clerk for 5 Years to be an Attorney, but Deeds, Conveyances & Wills &c. was but little to my liking when my thoughts was ingrosed by Natural-history . . .

In July 1773 Abbot set out for the Colonies and sailed into the mouth of the James River six weeks later. By means of another boat he reached "old James Town, the first settlement in Virginia."

Soon after my arrival at Mr Goodalls it became very sickly in the neighbourhood, with fevers & fluxes, one of Mr Goodalls Negroes died of the Flux, & many of the neighbours; in one family 22 died in 2 years white & black leaving only a little Girl heir to the Estate.

I was very fortunate in not being sick at all during my 2 Years stay in Virginia, and escaping a seasoning to the Country, & it was not till the 2d year in Georgia before I had the Ague & fever.

Abbot almost lost his taste for collecting when an insect collection from Virginia sank with a ship off the English coast, and another collection gathered the following summer went down with a ship in a September storm. In later years he became so discouraged at times that he would temporarily abandon his activities as collector, artist, and naturalist.

Shortly after the Colonies declared their independence of England, Abbot left Virginia for Georgia with his friends, the Goodalls. Later he served in the Revolutionary War as a private in the 3rd Georgia Continental Battalion.

A. S. Basset (1938), who studied Abbot's career because he was also a pioneer American ornithologist, wrote: "The earliest Georgia records of him point to Burke County as his place of residence. He married Penelope Warren and his son, John Abbot, Jr., was born in 1779, the years that the terrors of the Revolutionary War first struck upper Georgia." Afterwards he received a bounty of several hundred acres of Georgia land for his services in the war.

"John Abbot and his wife probably started out simply—busy young people, she with her housework and young son, John with his work as planter, schoolmaster and naturalist. After the invention of the cotton gin, 1794–1795, slave labor became more common in upper Georgia. John Abbot became prosperous, owned several slaves and had a comfortable and commodious home. . . . It seems safe to refer to the twenty odd years of John Abbot's life spent in Burke County as the happy years of his life.

"Just what happened or when it happened is not recorded but something did happen to change all this. Most likely it was the death of his wife. At any rate we find John Abbot, Sr., and John Abbot, Jr., in Savannah where they are listed as taxpayers in 1806."

American entomology and American ornithology owe a great deal to Abbot. He made at least 700 bird paintings for English collectors and sent abroad innumerable stuffed birds. Alexander Wilson, the pioneer American ornithologist, met Abbot when Abbot was fifty-seven. Abbot assisted Wilson freely and supplied him with much useful information. Wilson often credited Abbot for his help on his nine-volume work, *American Ornithology,* published from 1808–1814, two volumes appearing posthumously.

In his old age Abbot was deaf, and it was necessary to use a slate to converse with him. He had become quite corpulent and had to hire boys to catch insects and birds for him to paint. He seemed to be well supplied with funds to pay for his collections and for shipping them to London. Aaron Cone, born in 1810, who collected for Abbot as a boy thereby making a good deal of pocket money, says Abbot lived the life of a hermit in a little log hut. It is believed that Abbot lived in a small house on the property of his friend W. E. McElveen, and when he died about 1840 he was buried in a family grave there. Oemler, a friend of both Abbot's and Dr. T. W. Harris', went to Abbot's house shortly after his death to save his drawings, but they had already been destroyed by children playing with his effects.

Although Abbot had never mastered entomology as a science, he was an exceptional artist and a student of the life histories of many of the insects he collected, and he is held in great esteem as a pioneer illustrator and collector. Dow pays him this tribute: "He was an untutored optimist, with a constitutional smile, who looked forward only to the day's reward, who had talent with the brush, who had the assiduity to rear every insect species he could find for over fifty years. May the earth lie lightly upon him. No man has done better."

FREDERICK VALENTINE MELSHEIMER (1749–1814) AND HIS SONS

Frederick Valentine Melsheimer was called the "Father of American Entomology," following Thomas Say, in his *American Insects,* who referred to him as the "Parent of American Entomology." This title later was applied to Say himself. The high regard Say had for Melsheimer, the first serious American entomologist, was shared by Harris and other early entomologists.

Melsheimer made the first important collection of insects and wrote the first important entomological work in the United States. This book had a powerful influence on the pioneer entomologists. The title page of the work reads *"A Catalogue of Insects of Pennsylvania,* by Fred Val. Melsheimer, Minister of the Gospel. Part First, Hanover, York County: Printed for the Author, by W. D. Lepper, 1806." The next time the reader browses through secondhand entomological literature, let him keep an eye out for this sixty-page catalogue: there are only eight known copies extant.

Melsheimer's publication was the only one available on insects alone for about twenty-five years, until Thomas Say's *American Entomology* appeared. Frost (1937) notes of Melsheimer's catalogue, "It was the first separate publication on insects to appear in America. Before this time American insects were described by European workers such as DeGeer, Linnaeus, and Fabricius. This little book of sixty pages was a pioneer work in the field of entomology. It deals only with beetles but Melsheimer intended to publish a second volume to cover other insects.

Figure 3. F. V. Melsheimer. Portrait by an unknown artist. Courtesy of the Lancaster County Historical Society

Sickness in his last years prevented him from accomplishing his purpose. The book contains the classification of 1,363 species of American insects, of which about 400 are recognized today. One of the merits of the book is the interest and attention he paid to the habits of insects. He frequently mentions the food plants and the life histories of insects. He gives the oldest description of a coleopterous larva published in America. Occasionally he makes reference to the destructiveness of certain species and on two occasions recommends methods of control. Dr. E. A. Schwarz doubts that the recommendations did much good as they were printed in Latin and the book had a very limited circulation. Nevertheless, the introduction of control measures was a new adventure and there is evidence that later workers followed him."

The Reverend F. V. Melsheimer not only studied entomology but is also believed to have contributed to other phases of natural history, as well as mineralogy and astronomy. His ability was recognized by fellow scientists, and in 1795 he was elected a member of the American Philosophical Society in Philadelphia.

Melsheimer, himself, sketched a brief and modest account of his life. He says (Heisey, 1937) that he was born on the twenty-fifth of September 1749 in Regenborn, a village in the Duchy of Brunswick. His father was the ducal forester. Frederick Valentine studied in a ducal cloister school and then entered the University of Helmstadt in 1772, which he attended until 1776. He was ordained shortly after leaving the University, when he became chaplain to a regiment of Hessian dragoons.

February 10, 1776, I preached my first sermon to the Dragoons. As this regiment of Dragoons had been selected to go to America, we embarked and left the Vaterland on the 22nd of February, 1776, and on the 1st of June the same year we cast anchor at Quebec. We remained at Quebec until May 1777, when our regiments were ordered to the northern provinces. On the 16th of August 1777 our regiment, myself included, was taken prisoners at the battle of Bennington, by the American troops. We were taken to Brimfield in the province of Massachusetts. After an imprisonment of 14 months at the latter place, I was on the 29th of September 1778, paroled and sent to New York. In the following year all paroled prisoners of war were recalled, and so it happened, that on the 3rd of March 1779, I, with the officers of the regiment were again prisoners and we were sent to Bethlehem in the province of Pennsylvania.

On account of some difficulties I had with my brother officers, I resigned my commission as chaplain, and assumed charge of several congregations in Lancaster County, where on the 13th day of May I preached the first sermon. On June 3rd, 1779 I was married to Maria Agnes Man. The birthplace of my wife was Bethlehem, Pennsylvania. Her father's name was Samuel Man and her mother's name Anna Catharina.

November 12th, 1784, I removed to Manheim, where I took charge of four congregations.

October 25th,* I removed to New Holland, where I served four congregations.

July 25th, 1787, I removed to Lancaster and on the 19th of August, 1789, to McAllister,† where I served five congregations.

Heisey says that the Rev. F. V. Melsheimer was one of the "Ablest of the early Lutheran clergymen in America" and that he "exerted a large and beneficial influence among the German settlers of Pennsylvania." Shortly after his resignation in 1779 as chaplain, he began to preach in Lancaster County, Pennsylvania. From 1784 to 1789 he preached in such towns as Manheim, New Holland, and Strasburg, as well as others. He faithfully and ably served the congregation of Hanover from 1789 until his death.

The Reverend Mr. Melsheimer helped found German-American schools in a number of communities where he preached, and he became Professor of German, Latin, and Greek at the founding in 1787 of Franklin College (now Franklin and Marshall) and was president of the College in its second year.

As can be imagined, the Reverend's activities in pursuit of bug, beetle, or butterfly were a source of amusement to his parishioners, who considered it a harmless eccentricity on the part of their pastor. Melsheimer died June 30, 1814, from the consequences of a lung disease of thirty years' duration. In 1937, there was unveiled a marker in his honor in

* Probably 1785.

† Now Hanover, York County, Pa.

the borough of New Holland in Lancaster County, Pennsylvania. The bronze tablet affixed to the boulder reads: "Memorial to Frederick Valentine Melsheimer 1749–1814 known to science as Father of American Entomology to religion as Lutheran Pastor here and elsewhere to education as Second President of Lincoln College and Founder of a Common School one block east of this site in 1787. The Lancaster County Historical Society 1937."

Of Melsheimer's eleven children, two sons shared his interest in entomology, the Reverend John F. Melsheimer and Dr. Frederick Ernst Melsheimer.

According to Hagen (1884), the famous Harvard entomologist, a "Dr. Zimmerman called on Mrs. Melsheimer, and was told by her and her daughter that after his [Melsheimer's] death his eldest son, Johann Friedrich Melsheimer, succeeded his father as minister, whose love for natural history he had inherited, together with his collection and library. This J. F. Melsheimer is the entomologist quoted so often by the Th. Say." The Rev. J. F. Melsheimer, like his father, was an active collector of insects. He frequently met and corresponded with Thomas Say. J. F. Melsheimer was an "eloquent" preacher both in English and German.

When the eldest son died, the collection and library were inherited by Dr. F. E. Melsheimer, a country physician who lived fourteen miles from Hanover in Lancaster County. Zimmerman describes his visit to the home of the latter as follows:

"The house, rudely put together with boards, painted red, stood all alone in the middle of a forest, and looked more like a hut. His wife was at the spinning wheel. The reception was indeed, very cordial, and when he heard that his father's book was well known and mentioned in German, English and French works, which he had never dreamt of, he became animated and talked with great interest on entomological matters and books."

Zimmerman noted that the family lived on "plain but good fare" and that many little comforts were lacking but not missed. "Indeed there was no drinking glass in the house; cans or dippers served for the purpose. The cordial hospitality made one forget the lack of comfort."

Zimmerman considered Dr. F. E. Melsheimer to be a first-rate American entomologist and wondered how he was able to keep up in the natural sciences despite his isolation. Melsheimer kept the collection in good order, retaining his father's labels on the original specimens. The collection consisted largely of Coleoptera and Lepidoptera, apparently because his father's books on classification were limited to these orders.

According to Dow (1913), "Dr. Franz Ernst Melsheimer was a man of no mean attainments. He kept up the catalogue for many years and in

1842 was elected first president of The Entomological Society of Pennsylvania. In 1848 he offered his catalogue to the Smithsonian Institution of Washington for publication. The mss. was submitted to Dr. S. S. Haldeman of the University of Pennsylvania, and Dr. J. L. LeConte. Both authorities not only advised its publication but volunteered to bring it up to date (1851) and to see it safely through the press."

In 1853 the Smithsonian Institution published Dr. F. E. Melsheimer's *Catalogue of the Described Coleoptera of the United States,* as revised by Haldeman and LeConte. From 1846 to 1848 Melsheimer contributed seven papers on beetles to the Proceedings of the Academy of Natural Sciences of Philadelphia; and in 1853 he was president of The American Entomological Society.

After some correspondence with Dr. Melsheimer in 1860, Professor Louis Agassiz of Harvard University visited Davidsburg to inspect his collection, and in 1864 the collection was purchased for Harvard for $150. Despite the fact that the collection had been started more than eighty years ago by his father, it was still in a fine state of preservation. The collection contained more than 14,000 specimens of about 5,000 species. Many of Say's species can be recognized only from the authentic types in this collection.

Melsheimer sold this collection when he was eighty-two. Nine years later, on March 10, 1873, he died in Davidsburg, York County, Pennsylvania.

WILLIAM DANDRIDGE PECK (1763–1822)

It is said that William Dandridge Peck became interested in natural history after reading a copy of the *Systema Naturae* by Linnaeus, which he picked up from a shipwreck near his home. He was one of the first American biology students to devote himself to the life history and control of harmful insects in the United States. He studied the cankerworm, the slugworm, and the pests of pear trees. He also worked with oak, cherry, and locust tree pests. His reports appeared in *The Massachusetts Society for Promoting Agriculture Papers, The Massachusetts Agricultural Repository and Journal* and similar periodicals.

Peck was born in Boston on May 8, 1763. His mother died when he was seven, and it is possible that the loss of his mother at this early age may have been in part responsible for his mental illness later in life. His father, John Peck, was a noted naval architect and a man with ability in science and mechanics, talents which his son inherited.

There is in the files of Harvard University a copy of a letter written

Figure 4. W. D. Peck. Portrait by an unknown artist. From *A History of Applied Entomology,* by L. O. Howard. *Smithsonian Miscellaneous Collection,* Vol. 84 (1930). Courtesy of the Smithsonian Institution Press

by William Dandridge Peck to his father, possibly in 1779 or 1780, wherein he strongly expresses his desire to become a student of medicine, a desire that was apparently never gratified. The letter reads as follows:

That Time swift approaches Honord Pappa when the longest Act of Life begins, and tis proper I should determine the Part which I shall exhibit; and to You my dear Pappa tis proper I should signify my Inclinations. Since that Time when first I began to think of Science; Nature seem'd to point me in the Path of Physic, and in all my Considerations, my Prejudice has been in favour of that Branch; & can't but think I shall be more successful in it than in any other occupation. The little Course of Studies which I have lately followed, and other detached observations which I have made in my Connection at Harvard; have been in some measure preparatory, and the ease with which I presume I can acquire a considerable degree of Knowledge in this Profession rather than in any other, have confirmed my desire to pursue it.

Whatever the Reason may be honor'd Parent, I have an Idea that You would wish me to engage in the Law.

I do not wish to speak against it, But Pappa when You consider that to me there is nothing in it amusing; nothing to employ my most assiduous attention but abstruse disquisition You will I doubt not justify my Choice.

Something tells me my dear Pappa, I feel it impress'd upon me; That to force the Inclinations of Your Children in these Matters, is abhorrent to Your Nature. That is the Truth, I am convinced by those endearing Indulgencies, which I have so often experienced. . . .

A degree of Felicity attends Curiosity gratified and I find in Physic that which may satisfy the inquisitive Mind.

These Pappa are my Sentiments & the wait Your Approbation which if they obtain; twill be necessary I should study under some able Phactitioner. . . .
I wish my honor'd Parent, that in some Leisure Moment You would think of these Things, for on these Circumstances depends the future happiness of
Your duteous son . . .

An anonymous writer of 1843 notes that Peck received his Bachelor of Arts from Harvard in 1782, where he had been an apt student, especially interested in the classics and natural history. Upon graduating from college he entered a counting house.

Although Peck's father was an outstanding naval architect and was employed by the U.S. Government in designing naval vessels, his financial returns were so small that, in disgust, he retired to a small farm in Kittery, Maine. His son, who was also discontented, left the mercantile business shortly thereafter to join him on the farm.

There Peck remained for twenty years, only occasionally leaving the life of a recluse for visits to friends in Boston and Portsmouth. However, these twenty years of isolation were by no means wasted, for despite difficulties in obtaining books and papers, he constantly applied himself to the study of the sciences and became quite skilled in botany, entomology, ornithology, ichthyology, and related sciences. In time he amassed fine collections of insects, aquatic plants, and fish, and studied these specimens, it is believed, with a microscope of his own manufacture.

Peck began to write on insects in 1795, and in 1796 he won fifty dollars and a gold medal from the Massachusetts Society for Promoting Agriculture for a paper on the natural history of the cankerworm. Other papers appeared at irregular intervals until 1819. These reports were often illustrated with sketches that demonstrated a talent for drawing.

His studies in natural history won him some measure of fame, and on March 27, 1805, he was appointed the first professor of natural history at Harvard. He accepted this position with some reluctance because of "the hermit life he had led." As a professor he was given the opportunity to study in scientific institutions in England, France, and Sweden for three years, and he must have profited tremendously from the stimulating associations they afforded. Among his students was T. W. Harris, a pioneer economic entomologist.

Harvard University has an original handwritten copy of Peck's course on entomology. This was undoubtedly the first course on the subject in the United States, and it was part of the instruction in natural history. We do not know the exact date it was presented, but it must have been some time after Peck began his association with Harvard. As an example of Peck's teaching, we will quote his remarks about *Cimex*. The footnotes are actually his marginal notes:

> One species of *Cimex* is destitute of wings.* It is the *Cimex lectularius,* or common bed-bug, of a uniform rust-colour. This scourge of slovenly indolence is preyed upon by some other species of the same genus, but these are not numerous, and it can only be effectually expelled by neatness and mercurial applications. The *Cimex membranaceus* is one of the most curious of this genus. This has the thighs of the hinder pair of feet dentated, and the tibia membranaceous. It is from Surinam. The species which destroys the bed-bug is called *C. personatus,* or the masked cimex, because in the larva state it is always covered or masked with a thick coat of dust. It is wholly black, and is found in this country.
>
> Most species have an abominably offensive odour. †

Peck's periods of mental depression must have incapacitated him at times. He was very studious and had an "uncommon fear of exciting publick attention." Besides natural history and the classics, he was interested in painting, sculpture, and architecture. One of his favorite amusements was working with the lathe, "and he has left some fine specimens of turning, executed by him after he had wholly lost the use of one of his hands." Before he died, on October 3, 1822, he had suffered a series of strokes, and the final one left him unable to talk, so he wrote with his pencil "no funeral, no eulogy."

THOMAS SAY (1787–1834)

Admirers of Thomas Say called him the "Father of American Entomology." Others, as we saw, claimed the same title for Frederick Valentine Melsheimer. As we well know, two fathers of the same child lead to complications. Because both of these pioneer entomologists are worthy of the honors bestowed on them, possibly we can do justice to both by calling F. V. Melsheimer the "Father of American Entomology" and Thomas Say the "Father of American Descriptive Entomology," as suggested by H. B. Weiss. In any event, Thomas Say was the first American entomologist to compare favorably with such European entomologists as Fabricius, DeGeer, Dejean and others. Weiss and Ziegler (1931) made the most comprehensive study of Say, and the present author acknowledges his debt to them.

Say was of French Protestant stock. His grandfather, also Thomas

* The bed-bug is sometimes though seldom found with wings.

† Compared by some to that of over-ripe fruit.

Figure 5. Thomas Say. Portrait by Charles Willson Peale. Courtesy of the Academy of Natural Sciences

Say, was in the saddlery business in Philadelphia, but eventually became an apothecary and chemist and was very prominent in the affairs of the city. Benjamin Say, Thomas Say's father, who was both a physician and an apothecary, served in the Pennsylvania legislature and as a representative from Pennsylvania in the United States Congress. At the time of his death he was one of the wealthiest men in Philadelphia.

Thomas Say, the entomologist, was born June 27, 1787. His mother, Ann Bonsall Say, died October 15, 1793, two days after the death of a daughter, probably during an epidemic of yellow fever, which was the scourge of Philadelphia in those days. Mrs. Say left three children, Thomas, six years of age; Benjamin, about three; and Rebekah Ann, two months old. Say's father remarried again and had three children by the second marriage.

When Say was twelve years old he was sent to a Friends' academy near Philadelphia. He did not like his teachers or his studies, and George Ord (1869), a friend of his, observes that country schools often left pupils with a "distaste for letters" for the remainder of their lives.

"Of the name or character of the teachers, to whom the education of Thomas Say was confided, I have not been informed; but there is reason to infer, from his deficiency in elementary learning, on his arriving at manhood, and his indifference to polite literature, that his teacher had been either grossly negligent or incompetent."

Although Say lacked the literary talents of George Ord, a reading of Say's works shows that Ord was somewhat harsh in his criticism of Say's education.

Say's formal schooling in the Quaker school did not last more than three years, for he entered when he was twelve and left at fifteen, the maximum permissible age in the school. He helped his father in his apothecary shop, did a good deal of reading, and collected beetles and butterflies, in which he was encouraged by his great-uncle, William Bartram, who was an ardent collector, too.

Fortunately for Say, Philadelphia in 1812 already had such botanists as the Bartrams, Muhlenbergs, and Bartons; the ornithologist, Alexander Wilson; and the Peales and their museum; as well as other naturalists. These men started an association of scientists which soon became the Academy of Natural Sciences. Thomas Say was one of the founders. Because it played so important a part in Say's life, George Ord's comments about the Academy should be of interest: "Of the origin of this highly respectable and useful institution, I shall at this time merely assert, that its founders had any thing in view but the advancement of science. Strange as this may appear, it is nevertheless true, that the club of humorists, which subsequently dignified the association under the imposing title of Academy, held its weekly meetings merely for the purpose of amusement; and consequently, confined itself to those objects which it was thought would be most conducive to that end.

"But in the process of time, when it was found that mere colloquial recreation soon loses its charms, a higher object was suggested to the attention of the association, one which it was thought would tend to waken public curiosity, and thereby procure an accession of members, and, consequently an accession of means: this object was the collecting and preserving of natural curiosities. At the date of Mr. Say's joining the Society, this plan had been recently adopted; but how great was his surprise, on being inducted into the temple of science, to find that the whole collection consisted of some half a dozen common insects, a few madrepores and shells, a dried fish and a stuffed monkey: a display of objects of science calculated rather to excite merriment than to procure respect, but which in the end, proved to be the nucleus of one of the most beautiful and valuable collections in the United States."

Although young Thomas was primarily interested in the study of natural history, he acquiesced in his father's desire that he continue in the drug business. When he had acquired a knowledge of the drug business, his father helped him to form a partnership with John Speakman, an apothecary who had an interest in the natural sciences. According to Weiss and Ziegler (1931) "Speakman very generously assumed the active management of the shop so that Say's time could be devoted to natural history. It is not known when the partnership was formed, although it did not last very long. Speakman and Say endorsed notes for friends and when they became due the firm had to stand the losses which were so great that the business was wiped out. Even the loose change in Say's pockets, so one account states, was used in satisfying creditors." Say was twenty-five when he failed in business, and he never again ventured into business for himself. From this time on he devoted his energies almost entirely to natural history and solved his monetary problems by living in the Academy on the plainest foods, such as bread and milk, with an occasional chop or egg. He barely took time to eat and is believed to have slept beneath the skeleton of a horse in the Academy.

Once again we are indebted to George Ord for an insight into Say's problems: "The readiness with which Mr. Say attended to the wants of others, his liberality in communicating his knowledge to those who sought it, together with his urbanity and companionable qualities, were the occasion of such repeated interruptions, that he felt constrained to appropriate those hours of his private studies, which ought to have been devoted to rest; hence to him the season of midnight was the hour of prime, it was the time of stillness and tranquillity; and so greatly did he enjoy these vigils, that he not infrequently prolonged them, even during the summer, until the approach of day. Of this injudicious application to study, he soon became sensible, by the derangement of his digestive organs, which resulted in dysenteric affection, that, probably, were the remote cause of the illness which carried him to the grave."

Say was "six feet in height, slender, with a slight lisp to his articulation, which gave to his naturally gentle voice a musical softness." He had black hair and a dark complexion and was "apparently endowed, before his health was injured by repeated illness, with considerable strength." People found him to be "very amicable" and it was "impossible to quarrel with him." George Ord (1869) said "the name of Thomas Say was synonymous with honour, and his word the expression of truth."

Ord describes him as modest, likeable but with little confidence in himself; he refused a professorship in natural history in a university because of his belief that he could not "lecture in an acceptable manner." Weiss and Ziegler (1931) are of the opinion that "his refusal to take advantage of certain offers which might have increased his worldly possessions may have been due to his dislike or lack of interest, and he was too honest to feign interest where it did not exist. Certainly Say's work does not exhibit any lack of confidence in his ability in his chosen field, and his standards of usefulness or advancement may have differed from those of his friends. . . .

"During the early years of the Academy Say was always in the rooms, always ready to help others, always ready to make specimens and books available, and always agreeable and obliging."

In 1814 Say's studies were interrupted by his service as a private in the First City Troop of Philadelphia, although he did not engage in any active campaign during the War of 1812.

Encouraged by his friend Alexander Wilson, the ornithologist, Say began his work on American entomology in 1816 and actually published a small section in 1817. This work he had to suspend for lack of finances. He also began to publish papers in the newly-founded *Journal of the Academy of Natural Sciences,* and the first number contained his paper on "Descriptions of Seven Species of American Fresh Water and Land Shells, Not Noticed in the Systems," and the second number had his paper "Descriptions of Several New Species of North American Insects." Say was a steady contributor to the *Journal* for the next ten years.

Say corresponded with F. V. Melsheimer and later with John F. Melsheimer. He wrote the latter on November 6, 1817, "you will see in the 'Journal' that I have been describing the Crustacea of our waters; but my dear sir, I assure you that Shells and Crustacea are but secondary things with me, INSECTS are the great objects of my attention, I hope to be able to renounce everything else & attend to them only—"

In the fall of 1817 Say, in the company of William Maclure, George Ord, and Titian R. Peale, left Philadelphia for an expedition to the sea islands and nearby coastal areas of Georgia and eastern Florida to collect and study animal and insect specimens. This trip was shortened because of the reaction of the Florida Indians to United States attacks on them. Say's thoughts about the Government's conduct appear in a letter of June 10, 1818, to J. F. Melsheimer: "Thus, in consequence of this most cruel & inhuman war that our government is unrighteously & unconstitutionally waging against these poor wretches whom we call savages, our voyage of discovery was rendered abortive as we were not in Florida at the season we wished, the Spring, we therefore obtained but very few Insects."

In 1819 Say was appointed as a zoologist on Major Stephen H. Long's expedition to the Rocky Mountains. This was both a military and scientific expedition, and Major Long was in charge of the scientific aspect. The scientists included a mathematician, a botanist-physician-surgeon, a geologist, and two artists, including Say's friend T. R. Peale, as well as several others. The aim of the expedition was to explore the Mississippi and the Missouri and the Missouri's major branches, including the Red and Arkansas Rivers. Life was very strenuous for these pioneer scientists, and Say was in ill health throughout much of the trip. In a letter dated October 10, 1819, he wrote to his friend, Jacob Gilliams, "I will merely observe that during our last pedestrian excursion, we met with a large Pawnee War Party, who robbed us of our pack-horses & sundry small articles amongst which was your double barrelled pistols; We had however the good fortune to escape personal injury."

"N.B. We have just now held a council with the Pawnees in which most of the articles taken from us were restored & amongst others your pistols."

This expedition reached the present site of Denver and several of the men climbed Pike's Peak. Later in the year three deserting soldiers robbed the expedition, and among the stolen articles were Say's five notebooks which, among other things, contained descriptions of previously undescribed animals and their habits and vocabularies of mountain Indians. Apparently these fruits of months of effort were thrown to the winds. On this trip Say had the opportunity to study the "Otto," "Konza," "Omawhaw," Sioux, "Minnetare," Pawnee, and Cherokee Indians, and as a result he and Mr. John Daugherty prepared a section in Major Long's report on *Vocabularies of Indian Languages*. He also described in this report "wolves, serpents, birds, toads, rats, the prairie dog and other animals, as well as shells." The insects collected on this trip were not mentioned in the report.

Shortly after his return to Philadelphia at the end of 1820 he began to describe and publish papers on insects and living and fossil shells, many of which were collected in his Southern and Western expeditions. In 1821 he became curator of the American Philosophical Society, an office he held until 1827. He was appointed professor of natural history in the University of Pennsylvania in 1822, but it is believed he was not too active in this capacity because he left for another Western trip a year after his appointment.

Say's second Western trip was once again an expedition with Major Long. This time Say was both official zoologist and antiquary.

In the course of his journey, he visited Chicago but was not greatly taken with it: "The village presents no cheering prospect, as notwithstanding its antiquity, it consists of but few huts, inhabited by a miserable race of men, scarcely equal to the Indians from whom they are descended. Their log or bark houses are low, filthy and disgusting, displaying not the least trace of comfort. Chicago is perhaps one of the oldest settlements in the Indian country, its name, derived from the Potawatomi language, signified either a skunk, or a wild onion."

A printed report of this expedition appeared in 1824, and Say was largely responsible for the information on natural history and the Indians.

Among Say's entomological correspondents, besides the Melsheimers, were T. W. Harris of Boston and Charles Pickering of Salem, Massachusetts. Say had many correspondents other than entomologists; much of his time was spent answering questions on natural history. From 1821 to 1825 Say also lectured occasionally on zoology in Peale's Museum of art and natural history.

In 1825 Say left Philadelphia permanently, though without relinquishing his curatorship, to participate in the altruistic, socialist community at New Harmony, Indiana, on the Wabash River. Two idealistic men were primarily responsible for this community: Robert Owen, a Scotch social reformer and manufacturer, who tried to remedy the unhealthy social conditions resulting from the Industrial Revolution, and William Maclure, a wealthy landowner, geologist, and principal patron of the Philadelphia Academy of Natural Sciences.

Webster (1895) says that Say and other scientists left Pittsburgh in the winter of 1825–1826 on a craft resembling a Western flat boat, which, because of its "cerebral" passengers, was known as the "Boat Load of Knowledge." These passengers landed at Mt. Vernon, Indiana, and were conveyed overland to New Harmony. In this community Say was the Superintendent of Literature, Science, and Education. Here, in January of 1827, Say married Lucy Way Sistare.

Greed, jealousy, and the usual shortcomings of the human species wrecked New Harmony shortly after its founding. Owen and Maclure lived in anything but harmony and soon came to legal blows, which were settled by a compromise. "One returned to his native country to concoct new measures of the reformation of domestic policy; and the other retired in disgust to the republic of Mexico, to brood over his misfortunes." Say accompanied Maclure, but they returned from Mexico in 1828; only Say went back to New Harmony.

Maclure provided Say with a fine library for those times, but unfortunately Say became more and more involved in business matters, which took him away from his descriptive work. Madame Fretageot, a Frenchwoman who taught in Philadelphia and New Harmony, wrote Maclure, "He does not like to be troubled with money and dislikes business as much as I like it.

"But Mr. Say had become involved for life," says Ord. "He had married; he had accepted the agency of the property [at New Harmony], the duties of which compelled him to a residence there; he had no other means of support but what the bounty of his patron, Mr. Maclure, offered him; he, therefore, set himself down with his usual composure, to await the turn of events, appropriating all his moments of leisure to his favorite pursuits; and not allowing a thought of the future to disturb the equanimity of his mind."

At times the worthy Mr. Ord appears to have been unduly harsh with Say, and E. A. Schwarz's (1890) remarks may give us a clue to this severity: "What induced Say to print between the years 1830 to 1832, four important descriptive papers in the *New Harmony Disseminator,* the most obscure village paper that could be found in this country, even in those early times, has not yet been explained; but to anyone who understands how to read between the lines of Ord's biography of Say it becomes apparent that just about that time there was a great coolness between Say and his former friends in the East, and that Say in his isolated position in the West, had probably no other place for publication. Among these papers is that on the North American Heteroptera, of which according to [Samuel] Scudder, only one copy is known to exist in this country, viz: that which came into the possession of the Boston Society of Natural History from Harris' library. It was finally republished by Fitch in 1857, and LeConte's second republication in 1859 is a reprint of Fitch."

When Say first went to New Harmony, he was already troubled with stomach disorders and suffered from repeated attacks of fever and dysentery. Undoubtedly he was paying the piper for the irregular eating and sleeping habits of his youth. Despite repeated requests from kin and friends for him to return to Philadelphia, Say felt obliged to stay in New Harmony. After suffering from poor health for a long time, Say died in the Maclure House in New Harmony on October 10, 1834, at the age of forty-seven. Say's ashes rest in a tomb in the rear of the house. An inscribed monument of white marble was erected by Alexander Maclure at the request of his brother, William.

Mrs. Say is worthy of more than a passing phrase. Webster says that she "is remembered as a very amiable lady, scrupulously neat in all that pertained to herself or her household, though somewhat given to complaining." Miss Lucy Way Sistare came to New Harmony with two younger sisters on the same "Boat Load of Knowledge" that brought Thomas Say. In New Harmony she was in charge of the school children and the household. She was interested in shells, and made some excellent drawings for Say's *American Conchology*. After the initial failure of New Harmony she wanted Say to return to the East, but he felt obligated to remain at New Harmony. After Say's death she moved to New York to live in the home of a sister. She corresponded with naturalists about shells and various animals, and with Harris and Haldeman and other entomologists about insects, although she admitted she had no knowledge of the latter. In 1841 she was elected the first woman member of the Philadelphia Academy of Natural Sciences. William Maclure died in 1840 and left her an annuity of $300. Mrs. Say died in Lexington, Massachusetts, in 1886, when she was almost eighty-six years of age.

Say's reputation is solidly based because of such publications as *American Entomology* and *American Conchology*. Both of these important

Figure 6. Frontispiece drawn by Charles Lesueur. From Thomas Say's *American Entomology* (1824)

pioneer works in taxonomy are dedicated to William Maclure, his patron and lifelong friend.

The full title of Say's work on insects is *American Entomology or Descriptions of the Insects of North America.* The three-volume work appeared in 1824, 1825, and 1828 and contained fifty-four plates prepared by T. R. Peale, C. A. Lesueur, W. W. Wood, and H. B. Bridport. Dr. J. L. LeConte, a noted entomologist, collected all of Say's entomological writings and had them published in 1869. It is this publication that contains George Ord's important memoir of Say. Although Say has been criticized by some for being too brief in his descriptions of insects, LeConte apparently is not of this opinion, as we can see from his comment in his 1869 edition of *American Entomology*. "The entire destruction of his original specimens would be the subject of much greater regret, were it not for the fact that his descriptions are so clear as to leave scarcely a doubt regarding the object designated. I am thus enabled to assign to nearly all of his Coleoptera their proper place in the modern system."

Say is also spoken of as the founder of American conchology, because of his six papers on the subject printed and published in New Harmony. A seventh paper was almost complete at the time of his death, and this was also published. These had sixty-six well-drawn plates by Mrs. Say, and two by Charles Lesueur. The plates were colored by Mrs. Say with the help of some of her pupils. *American Conchology,* which contained Say's complete writings on the subject, was edited by W. G. Binney and published by Baillière Brothers, New York, in 1859. Say also wrote on fossil shells, birds, reptiles, mammals and other animals, as well as on the linguistics and ethnology of Western Indians.

Weiss and Ziegler (1931) note that Say "was the first efficient and extensive describer of North American insects, especially Coleoptera, and the first to demonstrate to Europe that America contained entomologists whose ability equaled their own."

Say has been criticized as being a "closet" entomologist, because he seemed to be interested only in naming specimens and apparently spent little of his time determining their relationships, biology, or behavior. This criticism is not altogether valid, because Say was dealing with myriads of undescribed insects, and the first step in making order out of chaos was to name the hitherto unknown species.

Upon Say's death, his collection and books were given to the Academy of Natural Sciences in Philadelphia. In 1836 Say's collection was sent to Dr. T. W. Harris in Cambridge, Massachusetts. By the time it came into Harris's hands the collection had been destroyed by the ravages of dermestids and the transportation by stagecoach. Fortunately F. E. Melsheimer's collection contained many authentic types of Say's species. Moreover, as mentioned previously, Dr. LeConte found most of the descriptions to be so accurate as to leave relatively few difficulties in recognizing the species described by Say. But the collection returned to the Academy in 1842 was a far cry from the superb collection that Say had so feverishly accumulated in his lifetime.

THADDEUS WILLIAM HARRIS (1795–1856)

The high regard entomologists have for T. W. Harris has not diminished with time. His profile appeared on every cover of the *Journal of Economic Entomology* from 1917 until 1953, when the American Association of Economic Entomologists and the Entomological Society of America were united and a seal was designed for the new organization.

Figure 7 (A).

Figure 7 (B). (A) and (B). T. W. Harris. (A) "Entomological Correspondence of Thaddeus William Harris, M.D." *Occasional Papers of the Boston Society of Natural History*. Courtesy of the Museum of Science, Boston, Mass. (B) Courtesy of the Academy of Natural Sciences

Many of the older entomologists miss the familiar profile of Dr. Harris on the *Journal*'s cover.

T. W. Higginson (1869), Harris' student and friend, and E. D. Harris' (1882), wrote sympathetic accounts of the life and work of T. W. Harris.

Harris was born November 12, 1795, in Dorchester, Massachusetts, the eldest child of Mary (Dix) Harris and the Reverend Thaddeus Mason Harris of the First Congregational Church in Dorchester. The Reverend Mr. Harris was a student of nature and the author of the *Natural History of the Bible.* Possibly T. W. helped in its preparation. Mrs. Harris, who apparently had a fondness for nature, for many years raised silkworms and thereby obtained sewing silk for her own use.

A classmate of Harris' at Harvard described him as a "timid, and sensitive, rather a nervous and recluse youth," but E. D. Harris notes "that this reputation was due more to his habits of retirement than to timidity. Certainly later in life this trait was not prominent in his character." It is believed that T. W. became interested in natural history only after graduating from Harvard in 1815, having entered in 1811 at the age of sixteen. Harris continued at Harvard until he obtained a medical degree in 1820. During this time he attended special lectures on natural history by Professor W. D. Peck, and Harris later said, "It was this early and much esteemed friend who first developed my taste for entomology, and stimulated me to cultivate it." As early as 1819 he studied botany and entomology in his course on *Materia Medica.*

Dr. Harris became associated in medicine with Dr. Amos Holbrook, a well-known practitioner in Milton, Massachusetts, and in 1824 married his youngest daughter, Catherine. Several years later he moved to Dorchester Village near Boston. He apparently was not too successful as a country doctor, and in 1829 he wrote to Professor Nicholas Marcellus Hentz, a favorite correspondent of his, about his desire for a change in employment. It is believed that it was during the years he was in Milton and Dorchester that he did a great many of his field studies in entomology. It was there that he began to prepare articles for the *New England Farmer* and other publications, all of which offered him some compensation for his efforts.

In 1831 Harris was appointed librarian of Harvard University, a position held by his father from 1791 to 1793. Although the income from this job was small, it was at least steady, a matter of prime concern to the head of a family who ultimately fathered twelve children. He held the position until the time of his death. However, if Harris believed that this post would permit him leisure to pursue his studies of natural history, he was gravely mistaken. He was an excellent librarian, zealous in performing his duties and very helpful to visitors. Besides his knowledge of the natural sciences, he had a keen appreciation of fine arts; he was also interested in geography, history, classics, and mathematics.

From 1837 until 1842 he lectured twice a week at Harvard on natural history. Higginson writes: "Even these scanty lessons were, if I rightly remember, a voluntary affair; we had no 'marks' for attendance, and no demerits for absence, and they were thus to a merely ambitious student a waste of time, so far as college rank was concerned. Still they proved so interesting that Dr. Harris formed, in addition, a private class in entomology, to which I also belonged. It included about a dozen young men from different college classes, who met one evening of every week at the room where our teacher kept his cabinet, in Massachusetts Hall. These were very delightful exercises, according to my recollection, though we never got beyond the Coleoptera. Dr. Harris was so simple and eager, his tall spare form and thin face took on such a glow and freshness, he dwelt so lovingly on antennae and tarsi, and handled so fondly his little insect-martyrs, that it was enough to make one love this study for life beyond all branches of Natural Science, and I am sure that it had that effect on me."

"If this sketch of Dr. Harris as he stood in the lecture-room is truthful," writes E. D. Harris, "it is equally so of him in his outings. Indelibly fixed on the writer's memory are the recollections of the bright, sunny days spent with him in rambling over the beautiful country about Belmont and Waltham; of the sudden rushes after some flying Buprestis, or the wary chase of some shy Cynthian; of the bark-stripping in search of the Curculio larva, and the search in the meadow pools for the Dytiscus. And sometimes net would be dropped and stone wall scaled that an old, mossy grave-stone in a forgotten burying-ground could be cleaned and its epitaph transcribed in the note-book. On such occasions every passing insect and every way-side plant furnished a text for such lessons as only he could give."

Although Dr. Harris was a silent, reserved man, according to E. D. Harris, "For the science to which he devoted so large a portion of his thoughts he possessed an enthusiasm that seemed almost at variance with his nature. Nothing else could arouse it, if his love for natural scenery is excepted. In the fields or on the mountains he was like another man."

Some of his students attempted to arrange and classify the collection of insects in the Harvard Natural History Society. They referred to Dr. Harris' cabinet in this endeavor, and Higginson says, "the more we used this ready assistance, the more profound became the wonder how Dr. Harris himself had identified them. There were no manuals, no descriptions, no figures accessible to us; even in the college library there were only a few books on tropical insects, and a few vast encyclopedias, which appeared to hold everything but what was wanted. It seemed as if a special flight of insects must have come to Dr. Harris from the skies, all ready pinned and labelled. Older heads than ours were equally perplexed, and the mystery was never fairly solved until after the death of our dear preceptor, and the transfer of his cabinet and papers to the Boston Society of Natural History.

"It was then apparent by what vast labor Dr. Harris had compiled for himself of literary apparatus of his scientific study. A mass of manuscript books, systemized with French method, but written in the clearest of English handwritings, show how he opened his way through the mighty maze of authorities. First comes, for instance, a complete systematic index to the butterflies described by Godart and Latreille, in the *Encylopédie Méthodique*. Every genus or species is noted, with authority, reference and synonyms; the notes being then rearranged alphabetically and pasted into a volume—perhaps three thousand titles in all. This was done in 1835."

Studies similar to the above were made by Dr. Harris on the works of G. A. Olivier, P. Cramer, C. Stoll, J. Hübner, D. Drury, J. O. Westwood, and other entomologists and naturalists, and most of it was done in so neat a hand that it is as legible as print.

Although entomology was flourishing in Europe, the serious student of entomology in the United States was indeed an isolated and solitary figure, often an object of laughter. In a letter dated September 23, 1829, Dr. Harris expresses his loneliness to Mr. E. Doubleday, an entomologist friend in England:

> You have never, and can never know what it is to be alone in your pursuits, to want the sympathy and the aid and counsel of kindred spirits; you are not compelled to pursue science as it were by stealth, and to feel all the time, while so employed, that you are exposing yourself, if discovered, to the ridicule, perhaps, at least to the contempt, of those who cannot perceive in such pursuits any practical and useful results. But such has been my lot,—and you can therefore form some idea how grateful to my feelings must be the privilege of an interchange of views and communication with the more favored votaries of science in another land.

Although, at first, native entomologists were a rarity, there were a few noteworthy ones with whom Dr. Harris carried on a regular correspondence, including N. M. Hentz, naturalist and specialist in spiders, Thomas Say, F. E. Melsheimer, W. LeBaron, and J. L. LeConte.

If Dr. Harris was a lonely entomologist, lacking in books and reference collection, he had certain compensations, which Higginson points out: "The conditions of a new country, implying these drawbacks, imply also a great wealth of material. In older countries it is rare to discover a new species; it is something to detect even a new *habitat*. But these lonely American entomologists seem as one reads their correspondence, like so many scientific Robinson Crusoes, each with the insect-wealth of a new island at his disposal. They are monarchs of all they survey."

Harris was cognizant of his opportunities and was preparing to take full advantage of them, as he indicates in this letter to Hentz, dated November 19, 1828:

> I am aware that the "New England Farmer" is not likely to be much circulated among men of science, and therefore will not be considered the best authority; but it is a convenient vehicle at present; and, such is the ambition of European entomologists to anticipate Americans, that I willingly yield to the solicitations of several friends in publishing what may possibly contain many new species; and in doing so, I am not actuated so much by personal considerations as by a desire to aid several young entomologists in this vicinity, and by the wish to promote American science in general: *pro patria.* The "Farmer" is taken at New Harmony, and will therefore come under the eye of Prof. Say: it is my intention, after these descriptions shall have undergone his rigid scrutiny, to republish them, either by themselves, or in some respectable scientific journal.

A month later, he notes in a letter to Hentz that he intends to prepare a work entitled *Insectorum Faunula Bostoniensis,* with plates, which will describe the common species in the vicinity.

But this ambitious undertaking was never to see the light of day in its contemplated form because, as the unknown species in his collection multiplied, the gap between the insects and his small library grew ever wider and deeper, and his unfamiliarity with certain orders of insects such as the Diptera and Hymenoptera did not help matters. Moreover, his labors in the library, his lectures and similar activities, and the demands of his growing family, all competed for the spare hours he might have devoted to such a book. Nevertheless, his plans for the *Insectorum Faunula Bostoniensis* being sidetracked, he was considering the preparation of a synopsis of American insects, using Say's species as a basis in the preparation of a decriptive catalogue.

In a letter dated November 2, 1836, to Dr. D. H. Storer of Boston, Harris writes:

> Hereupon Dr. Pickering obtained leave of the Academy of Natural Sciences to send me the whole of Say's collections, only stipulating that I should put them in good order, and return them in a condition to be preserved after I had examined and arranged them. They arrived about the middle of July; but on examination were found to be in a deplorable condition, most of the pins having become loose, the labels detached, and the insects themselves without heads, antennae and legs, or devoured by destructive larvae, and ground to powder by the perilous shakings which they had received in their transportation from New Harmony. This irremediable destruction has in great measure

defeated my expectation of deriving benefit from examining the specimens and comparing them with those in my own collection, and that of Prof. Hentz.

However, Dr. Harris had corresponded with Thomas Say since 1822 and Dow (1913) records that at Say's suggestion Harris shipped his collection to Philadelphia in 1825 for Say to label as many of the specimens as he could, and that all these labels were intact, which adds immeasurable value to Harris' collection in view of the destruction of most of the Say collection. Harris' valuable collection, carefully labeled and arranged, consisting of almost ten thousand specimens with 140 types, is now safely in the museum of the Boston Society of Natural History.

Harris was not only interested in insect taxonomy but also was an effective student of insect biology. Whenever time was available he studied and drew the various life stages of many common insects. According to Higginson (1869), "His excursions, too, though rare, were effectual; he had the quick step, the roving eye and the prompt fingers of a born naturalist; he could convert his umbrella into a net, and his hat into a collecting box; he prolonged his quest into the night with a lantern, and into November by searching beneath the bark of trees. Every great discovery was an occasion for enthusiasm . . ."

In the 1833 issue of Hitchcock's *Report on the Geology, Mineralogy, Botany and Żoology of Massachusetts,* he published his *Catalogue of Insects.* This enumerated and classified 2,300 species. He received one copy of this report for his labors. In 1841 he completed his *Report on Insects Injurious to Vegetation,* for which the State paid him $175. The following year he republished it as a *Treatise,* and a second revised edition appeared in 1852. However, the edition usually seen was published shortly after his death and contains colored plates and beautiful woodcuts never seen by the author. This painstaking, comprehensive, and lucidly written work established Dr. Harris' reputation, was a standard source of information on American insects for many years, and is useful to this day. Both J. H. Comstock and L. O. Howard as boys were so fascinated by the book that they took up the study of insects as their life work.

Dr. Harris became an economic entomologist, probably because he was the only reliable source of information in his area, and people with problems, especially farmers, would inquire regarding means to curb important economic pests. Higginson reports that "many an old farmer who had travelled miles to bring him a newly discovered cabbage-pest, or a strange wheat-fly, was sent away from his house delighted with the story of insect life, and the practical hints that he received from the reserved but courteous gentleman who had welcomed him with the dignity and politeness of a by-gone age." For the most part the methods of control in his work consisted of cultural methods, pruning, hand-picking, and the use of simple dusts and washes.

Howard (1930) says, "There is no doubt that this 'Treatise on Insects Injurious to Vegetation' helped many farmers and gardeners and that it helped many entomologists. More than that, I feel sure that it made many entomologists. I shall never forget my delight when one Christmas Day, 1871, I was given a copy of the new edition with the wonderful illustrations by Sonrel and Burckhardt. They were the best illustrations on insects that had been published and nothing that has been issued since is better than some of them."

Although entomology was Harris' first interest, he was involved in many facets of science, especially botany and zoology, and he wrote a a very carefully researched monograph on the Cucurbitaceae. E. D. Harris recalls that "For several seasons his garden—next to his collections the delight of his life—was filled with squashes, pumpkins, and gourds of every conceivable shape, size, and color. Seeds came to him from all quarters of the globe, and their products formed a collection as unique as it was interesting." He was also a distinguished bibliographer and archaeologist as well as librarian.

Higginson (1869) says that he became less and less productive in the last fifteen years of his life. "Genius works many miracles, but it cannot secure leisure for science to a man who has twelve children, no private means, and the public library of a university to administer," and in time he had to abandon the hope of being both a librarian and naturalist. However, he was not an unhappy man, even if he regretted that he could not spend more time on natural history.

The portrait of Dr. Harris used on the seal of the American Association of Economic Entomologists displays the profile of a sensitive, intelligent, and somewhat emaciated man. Higginson, in discussing this portrait, says, "There is a certain rigidity about it which belonged to his face, perhaps, when in repose,—the result partly of overwork and partly of the frequent headaches which, in his own words, 'kept him always thin.' But the moment he spoke, his face had the kindest smile and such a play of sensitive expression that I cannot possibly associate with it anything like sternness." According to E. D. Harris, he was a tall man "measuring full six feet in height, but his spare, thin frame gave him the appearance of even greater stature." Dr. Harris "would give whole golden days of his scanty summer vacations to arranging and labelling the collections of younger entomologists. And it roused all the wrath of which his soul was capable when even a rival was wronged, as when [P. F. M. A.] Dejean ignored Say's descriptions, because he had not learned English enough to read them."

Though Harris was constantly overworked and at times suffered from nervous exhaustion and severe headaches, yet he was in relatively good health until the last two years of his life, when he began to suffer from pleurisy. In November 1855 he was confined to the house. He died on January 16, 1856, at the age of sixty.

Harris' library, cabinets, and manuscripts were purchased by the Boston Society of Natural History largely through the efforts of a few patrons of the Society. His library contained rare works by Say, John Abbot, and European entomologists such as Fabricius and Dejean.

At his death, Professor Louis Agassiz said he had few equals, and A. R. Grote (1889) noted that Dr. Harris "ran the first furrow, and his successors have but widened the field of practical and economic entomology."

SAMUEL STEHMAN HALDEMAN (1812–1880)

Haldeman was born August 12, 1812, at Locust Grove, Pennsylvania, "a beautifully situated country-seat on the east bank of the Susquehanna River, twenty miles below Harrisburg," says D. G. Brinton (1882), in what is now known as the Cumberland Valley. According to the obitu-

Figure 8. S. S. Haldeman. Courtesy of the Academy of Natural Sciences

aries, the Haldemans were of Swiss heritage, and Samuel's father, Henry, was a well-to-do, prominent businessman of keen intellect, who loved books and had an excellent library. Samuel was the eldest of seven sons, with an innate interest in natural history, which was encouraged by his father. His mother, an accomplished musician, died when Samuel was twelve years old, and he must have inherited from her his love of music and his uncanny ear for sound, which in later life made him such an outstanding phoneticist. As a boy he read the books in his father's library, collected shells, prepared skeletons of small animals, stuffed birds, and used his own keen senses to read the story of the fields and forests around him. In 1826 he went to a "Classical Academy" in Harrisburg, Pennsylvania, where he studied natural history under Professor H. D. Rogers, who later became a well-known geologist. He attended Dickinson College at Carlisle for two years, and then left because college routine was not for him. He wanted to pursue his studies by himself and occupy himself, in the words of J. P. Lesley (1886), with "his cabinets of minerals, plants, shells, and insects, and his library of scientific and philological books." He always relied on his own observations, did not copy notes from other students, and regarded information from others "as interesting *if true.*"

Because the industrious Pennsylvania Dutch of 1830 had a poor opinion of collectors, classifying them as "good for nothing," his father set him up to run a sawmill, a business which he found more irksome than college. According to Brinton, Haldeman said, "I developed a taste for rainy weather and impassable roads; then I could remain undisturbed in the perusal of my books, a supply of which I kept in a back office, where I retired as soon as the sky looked threatening." This taste for rainy weather remained with him all his life.

During the winter term of 1833–1834 he attended a course of lectures in the Medical Department of the University of Pennsylvania, not to undertake the study of medicine but to prepare himself for the natural sciences.

The next year he married Miss Mary A. Hough, an understanding woman who encouraged him in his interests and produced four children. After his marriage he moved to a spacious mansion built for him by his father at Chickies (short for the Indian name, Chicquesalunga), near Harrisburg, Pennsylvania, where he lived for the rest of his life. He also became a silent partner in an iron business managed by two of his brothers. His knowledge of chemistry was most helpful to them, and in time he became an authority on the smelting of iron.

Brinton writes: "Here books and cabinets accumulated under his laborious hands, only to be scattered again and give place to others when his insatiable appetite for knowledge led him into new fields of investigation. For forty-five years he spent most of his time in his library, where in his vigorous manhood he worked sixteen hours a day. For though he accepted several professorships and delivered a number of courses of lectures, he did so with reluctance, preferring to be master of his time and to spend it in the quiet of home.

"In person, Mr. Haldeman was of middle height, with small, well formed hands and feet, a large and remarkably round head, giving great breadth across from ear to ear; high forehead, Roman nose, full lips, black eyes, and in youth a quantity of black hair, which at his death was of snowy whiteness. Long before it was usual in America, he wore a moustache and a beard, not for adornment but for convenience. In speaking, he had a clear enunciation, penetrating voice, and much readiness at repartee. His movements were rapid, his disposition cheerful, his general health excellent, and his interest in science unflagging to the end of his life . . ."

In 1836 he was an assistant in the New Jersey Geological Survey, and in 1837 in the Pennsylvania Geological Survey. He was Professor of Zoology at Franklin Institute, Philadelphia, in 1842 and a Professor of Natural History at the University of Pennsylvania from 1850 to 1853. He held the same position from 1855 to 1858 at Delaware College (the University of Delaware) and at some period was also a Professor of Geology and Chemistry at the State Agricultural College. He was the first Professor of Comparative Philology appointed by the University of Pennsylvania in 1869, when the department was established. An honorary Doctor of Laws degree was conferred on him by the University of Pennsylvania in 1876. He had such a diversity of talents and interests that it was not possible to limit his contributions to any one field. As he himself said, according to Brinton, "I never pursue one branch of science more than ten years, but lay it aside and go into new fields." Entomology was just one of the sciences that profited by his interest. "In his earlier years natural history was his passion, while in his later life linguistics and archaeology occupied most of his attention." Failing eyesight prevented him from pursuing his studies in Zoology.

As a boy Haldeman had collected shells on the banks of the Susquehanna, and his interest in this specialty resulted in his first important paper, the *Freshwater Univalve Mullusca of the United States,* which was issued in nine parts, from 1840 to 1866. He also wrote on geology, conchology, entomology, philology, and several branches of zoology, on coal, the English and Chinese languages, American Indian languages, Latin pronunciation, chess, poetry, etymology, orthography, archaeology, arrowheads, lava, the Pennsylvania Dutch dialect, and many other subjects.

His entomological works include one of his first papers on insects, which appeared in Volume I of *The Proceedings of the Academy of Natural Sciences,* 1843, under the title, "Catalogue of the Carabideous Coleoptera of South Eastern Pennsylvania." In the same volume he wrote another article describing some newly discovered beetles. For fifteen years he turned out papers on the systematics of Coleoptera, as well as occasional articles on Hemiptera, Hymenoptera, and Orthoptera. He also wrote on the beetle, *Agrilus ruficollis,* as a pest of cultivated strawberries; on the transformation and development of that remarkable aquatic insect, the dobson fly or hellgrammite, *Corydalus cornutus* (Linn.); on beetles in ant nests; on sound production and organs of sound in moths; and other aspects of insect life. In 1853 the Smithsonian Institute published the *Catalogue of the Described Coleoptera of the United States,* by Friedrich Ernst Melsheimer, M.D., revised by S. S. Haldeman and J. L. LeConte. Haldeman's last paper on insects appeared in 1858, after which he devoted himself to new avenues of learning.

He could speak most of the European languages and many rare tongues and dialects, and was one of the founders of the American Philological Society in 1869. One anecdote, cited by C. H. Hart (1881), he would tell with great relish: at "a *bal d'opéra* in Paris, under a mask he talked with a Russian savant in all the principal European languages. His interlocutor, in vain attempting to guess his nationality, at last informed him that he must be à Russian, but with sarcastic incredulity, whereupon, Dr. Haldeman repeated a verse in *Russ,* that made the other gasp in wonder when he was told that he was conversing with an American."

As a teacher, he was sometimes difficult to follow, because he would depart from his subject to speak on whatever came to his mind, whether or not it was in logical sequence. We are told that once he was supposed to speak on the subject of fish "but the hour passed in philological statements, anecdotes of stupid blunders made by dictionary men, and illustrations of the unpronounceable sounds of savages, without any allusion to fish until he arrived at the last sentence." Nevertheless, his books and papers were well written and in good order.

Haldeman traveled extensively throughout the United States and Europe, where he attended many scientific meetings. He died suddenly at his home of heart failure on September 20, 1880, upon returning from a meeting of the American Association for the Advancement of Science in Boston.

2 Early State Entomologists

So, naturalists observe a flea
Has smaller fleas that on him prey;
And these have smaller still to bite 'em
And so proceed ad infinitum.
Thus every poet in his kind
Is bit by him that comes behind.
Jonathan Swift

Until the middle of the nineteenth century there were no entomologists in the United States who made a living from their knowledge of insects; entomology was an avocation pursued purely for the pleasure of learning and knowing. In 1854 Dr. Asa Fitch, an accomplished student of entomology, was authorized by the New York State Legislature to study the insects of the state, and he thus became the first professional entomologist ever appointed by a state.

ASA FITCH (1809–1879)

A mystery that as yet remains unsolved is how two outstanding pioneer economic entomologists such as Dr. Thaddeus W. Harris and Dr. Asa Fitch, in what were then otherwise unexplored fields of interest, apparently never wrote to each other or had anything to do with one another. Possibly this lack of communication can be explained by Dr. Fitch's appearing on the center of the stage about the time that Dr. Harris was making his exit into the wings. Thus, Dr. Fitch's first publication, "Insects of the Genus *Cecidomyia,* Including the Hessian and Wheat Fly," appeared in 1845, about the time that Dr. Harris was beginning to abandon his insect studies for his library and family affairs.

According to Howard (1930), "Much has been written about Harris and his charming personality. Comparatively little has been written about Fitch. Harris had the advantage of priority. He had the advantage possibly of a better literary style. He also had the advantage that his work was brought together in a compact whole and published with

Figure 9. Asa Fitch, Courtesy of the New York State Museum

beautiful illustrations. [C. V.] Riley, coming before the public at a time when Fitch was finishing his work, had that advantage, and, filled with later ideas and a positive genius for illustration work, received infinitely more contemporary fame than did Fitch. And yet, as Fitch's work is studied it is evident that he was quite as sound an entomologist as either his perhaps more famous predecessor or his undoubted more famous successor, and there can be no doubt that many of his studies were as finished and as valuable as anything that Riley ever wrote and were more complete than anything that Harris did."

The New York State Legislature appropriated $1,000 for a study of insects, especially those injurious to vegetation, and in 1854, at the age of forty-five, Dr. Fitch began his entomological investigations. These resulted in fourteen reports on the noxious, beneficial, and other insects of the State of New York, and they established his reputation as an economic entomologist. Although not officially designated the State Entomologist of New York, "he was always given that title by courtesy."

In his first report, published in 1855, he explains that "it is the primary object of this report to diffuse information upon an important topic with which very few are at present conversant, I have throughout endeavored to treat the subject in a plain, familiar manner, avoiding any unnecessary resort to technical language, and using no terms but such as will be found clearly defined in dictionaries which are in every school district in our State."

In the third report, published in 1856, he notes: "Many things which are most interesting and remarkable are brought to my view, in the researches in which I am occupied, and I sometimes think there is no kind of mischief going on in the world of nature around us but that there is some insect at the bottom of it. Certain it is that these little creatures, seemingly so insignificant and powerless as to be unworthy of a moment's notice from anybody but the curious, occupy a most important rank in the scale of creation, and on every side of us their performances are producing most important results, tending probably in an equal degree to our benefit in one direction as to our detriment in another."

Dr. Fitch's fourteen reports were published in the *Transactions of The New York State Agricultural Society* from 1855 through 1872; eleven of them also appeared separately. The first report was devoted to insects infesting fruit trees, and the first insect considered was the woolly aphid, "The Apple-Root Blight," *Pemphigus pyri*. Some of the other insects described by Dr. Fitch, often in great detail, were the lepidopterous enemies of fruit trees, pests of forest trees, and the insect enemies of grain and field crops, the garden, and domestic animals. Some of the more important agricultural pests of that period were the Hessian fly, the chinch bug, the wheat midge, the Angoumois grain moth, the asparagus beetle, the tobacco hornworm, the onion fly, the cabbage aphis, and others. Dr. Fitch's cultural methods for control were quite helpful but his chemical methods were shortly superceded by newly developed washes and dusts.

Asa Fitch was the second son of Asa Fitch, an M.D., and he was born February 24, 1809, at "Fitch's Point," Salem, in Washington County, New York. He attended school in Salem and applied his keen and observant mind to the study of nature, actively collecting and identifying plants. When he was seventeen he entered the Rensselaer School (now Rensselaer Polytechnic Institute) at Troy in 1826 and graduated with honors in 1827. While there he came under the influence of a famous pioneer teacher of science, Professor Amos Eaton, who taught courses in chemistry, physics, geology, botany, and other natural sciences. It was professor Eaton who chiefly spurred Asa's interest in botany, geology, conchology, and especially entomology. Asa accompanied Professor Eaton and his students on several collecting expeditions to Lake Erie. They traveled much of the way on the Erie Canal via boats towed by tandem teams of horses. This slow and leisurely mode of locomotion enabled them to jump ashore from time to time to collect plants or net insects in the adjacent fields.

This one year at the Rensselaer School served merely as an appetizer to Asa's insatiable appetite for knowledge. He was not only introduced to the sciences, but also to classical languages and philosophy. We can observe Asa's intellectual development from his diaries, now at Yale University (Mallis, 1963), which he maintained on and off from the age of twelve till shortly before his death in 1879. Samuel Rezneck (1961) speaks of Asa as a "religious youth in a religious age," and the depth of his faith can be seen in the pages of his diary.

Although Asa had a "passion" for studying insects at the completion of his year at Rensselaer, there were then neither courses nor opportunities for making a living in the pursuit of entomology. So at his father's request, he entered the Vermont Academy of Medicine at Castleton in 1827. There he received his M.D. degree in 1829.

He then attended lectures at Rutgers Medical College in New York City and continued his preparation for the medical profession in the office of a Dr. March of Albany. Apparently, he spent more time studying entomology than the arts of medicine, for Thurston (1879) remarks, "While thus engaged he made illustrious use of the libraries of that city [Albany] so far as they could aid in advancing his knowledge of entomology. Being unable to purchase the books he needed, and determined to possess all the information they contained about the insects of this country, he copied with great accuracy and rapidity from the various entomological works in both the State and academy libraries, all that had been written on American insects." Thurston reports that Professor Eaton considered Fitch at the age of twenty-one one of the most accomplished entomologists in the country.

In 1832 Fitch married Elizabeth McNeil of Stillwater, New York, and moved to Stillwater to practice. Apparently he was an unenthusiastic practitioner, for in 1838 he abandoned medicine to return to Salem and manage his father's business. Thurston notes that thereafter he spent more and more time on agricultural pursuits as well as insects. "It is related that he would frequently be seen after a shower, on his hands and knees, searching about for insects and all manner of 'creeping things,' and would finally return to the house with his tall old hat completely covered inside and out with the writhing victims of his scientific greed. He was nicknamed 'The Bug-Catcher' by his neighbors, and so eager became his quest for curious specimens in wood, field, and stream, that many thought him demented, while others declared that he destroyed more grain than his scientific investigations were worth."

In 1845 the *American Quarterly Journal of Agriculture* published his first paper, "Insects of the Genus *Cecidomyia*," describing a newly discovered willow gallfly. Two years later he was employed by the State of New York to collect and identify insects for the State Cabinet of Natural History.

Collins (1954) notes that Asa Fitch, as a student, found that he could get along on five hours sleep a night or as little as two or three, and this ability to apply himself tirelessly and wholeheartedly to his studies was characteristic of his seventeen years' work on the insect life of the State of New York.

Figure 10. Asa Fitch's "bughouse" in Salem. From D. L. Collins, "The 'Bug Catcher of Salem,'" in *Bulletin of the Schools* (New York, 1954), Vol. 40

According to Thurston "The little office a few yards from his residence became his workshop, and night and day sent forth light to the world. So close was the watch he kept at the hatching-time of the various larvae collected, that for a week together he would catch his sleep in an armchair, waking at intervals to note the wonderful changes taking place in the insect-life before him. At such times, his meals, and an extra hour after tea to read the news, was all the recreation he allowed himself, and even then his pocket-net was always within reach, to capture unwary moth or curious beetle whose love of light attracted it to the room. Dr. Fitch was a most devout Christian, and reading the Scriptures and prayer with his family was a daily habit of his life. But even when thus engaged it was not safe for an attractive insect to come in his way. A daughter, the one to whom he was indebted for many of the beautiful drawings which illustrate his writings, relates that on one such occasion when he had the Bible in his hands, and was about to begin reading, a moth of peculiar appearance alighted on the book before him. The ruling passion was too strong for time or circumstance; glancing about, as if conscious of the incongruity of the proceedings, he quickly seized his net, bagged the curious specimen, and with a half-guilty look proceeded with the reading. The capture was an important one, as the moth proved to be new to science."

Riley (1880) gives this sketch of Fitch: "In company with Mr. P. R. Uhler we enjoyed a brief visit to Fitch's Point and a day's delightful communion with the object of our sketch, in the fall of 1870. It was the first and last time we had the pleasure of meeting him. He had then been suffering for some time from illness and was very much bowed down. A strong and very tall man, he had become quite round-shouldered from the force of stooping in pursuit of his studies, while the constant usage of the microscope and lens had produced a noticeable contrast in the appearance of his left as compared with his right eye. Genial, enthusiastic, unassuming, he made a most favorable impression." From 1871 on, Fitch ceased to gather information for publication. He died on April 8, 1879.

Dr. J. A. Lintner, who succeeded Dr. Asa Fitch as State entomologist of New York, examined his collection and library at the family's request. He found 106 boxes of insects in all orders. According to Weiss (1936), in addition to the above, there were two large cases containing duplicate specimens consisting of about one hundred thousand Coleoptera and twenty thousand insects in other orders. One hundred and forty-eight notebooks "contain an exhaustive descriptive catalogue of the collection, each specimen with its date of capture, locality, etc., being numbered, beginning about the year 1833, and a brief diagnosis followed by a fuller description and remarks, accompanying the majority of the New York species. The number of specimens referred to in the note-books reach as before stated fifty-five thousand (*circa*), although doubtless many specimens have either been exchanged or destroyed." Dr. Fitch himself set a value of $5,000 on the collection, the fruit of untiring forty-five years of labor.

Lintner and Sanborn (1879) add that "A large and valuable library of works on entomology in various languages, containing many rare and curious volumes, and a valuable microscope by Nachét were likewise stored in the small wooden building or 'office' a few metres back of the dwelling house, the latter within a few months of its centennial anniversary.

"The faithful shepard dog that for some years past accompanied the good doctor in his walks, sleeps nightly on the floor of the little office porch guarding these treasures of science in their frail receptacle."

Fitch's collection was broken up, parts were sold to various collectors, and the balance purchased by the U.S. National Museum. However, there are still some parts of his collection and some types in the New York State Museum at Albany, New York.

Henry Ulke, who was shown the collection by Fitch, found many of the Coleoptera misnamed. Fitch explained that he had written to a number of coleopterists and repeatedly to Dr. LeConte and received no reply, and therefore he had to name them as well as he could without the help of the experts.

Fitch's original work and detailed observations gave him an international reputation and he corresponded with such European entomologists as J. O. Westwood and J. Curtis of London, V. A. Signoret of Paris, Gerstacker of Berlin, and C. R. Osten-Sacken of St. Petersburg.

BENJAMIN DANN WALSH (1808–1869)

Asa Fitch, Joseph Albert Lintner, and Benjamin Walsh, all made their most serious studies of insects long after their youth and all achieved fame in their chosen profession. C. V. Riley, who was closely associated with Walsh in the editing and illustrating of the *American Entomologist,* wrote an informative obituary of him in 1870, and fifty years later Mrs. E. A. Tucker (1920) prepared an excellent profile, which, with Riley's obituary, provides much of the information that follows.

"We are not stepping beyond the bounds of truth," says Riley, "in asserting that Mr. Walsh was one of the ablest and most thorough Entomologists of our time; and when we consider his isolation from any of the large libraries of the country, and the many other disadvantages under which he labored, we are the more astonished at the work he accomplished. He was essentially original and *sui generis;* everything about him was Walshian, and though he had some of those eccentricities which frequently belong to true genius, and though he made many enemies by his bold, outspoken manner, and his hatred of all forms of charlatanism; yet those best acquainted with him know

Figure 11. B. D. Walsh. Courtesy of the Illinois Natural History Survey

what a deep feeling, tender and generous heart lay hidden beneath the rough and uncouth exterior."

To appreciate Walsh and his remarkable knowledge of insects, you have to read the *Practical Entomologist* and the *American Entomologist.* Walsh heartily detested the mountebanks who sold the farmer worthless nostrums, and you could hear their howls of pain when he applied his cudgel. He wrote in 1869:

> We have received several circulars from the "Union Fertilizer Company of New York," crying up the merits of a miraculous panacea of theirs, which they kindly offer to the public at the low price of $45 per ton, packed in barrels ready for shipment. The Secretary of this company rejoices in the very appropriate and suggestive name of A. S. Quackenbosh, and he assures us that the article which he offers for sale, besides being an excellent Fertilizer, is "sure death and extermination to the Cankerworm, the Curculio, the Apple Moth, the Potato Bug, the Cotton Worm, the Tobacco Worm, the Hop Louse, the Army Worm, the Current Bug, and all descriptions of insect and vermicular life which infest and devastate the Orchard, the Garden, or the Farm." . . . The trouble with all such panaceas as this vaunted New York "Exterminator" is, that we hear nothing of the ninety and nine cases where the Universal Remedy was applied and found to do no good, while in the one case where the medicine worked well, or was supposed to work well, the happy experimenter lauds it to the skies in a flaming advertisement. In the words of the veteran sportsman, when his juvenile companions were bragging of their achievements with the fowling piece
>
> What is hit is history
> But what is missed is mystery. . . .
>
> Nothing is more certain than that there is no Royal Road to the destruction of the Bad Bugs; and the only way in which we can fight them satisfactorily, is by carefully studying out the habits of each species, and adopting mode of attack to the peculiarities of the fortification, which we are about to besiege.

Walsh was born in Frome, East Somerset, England, on September 21, 1808. He graduated from Trinity College, Cambridge, with a Master's degree at twenty-five and was a classmate of Charles Darwin's and according to an anonymous source (1894), "though he took up the latter's work on the Origin of Species with great prejudices against the development hypothesis, yet he became a thorough convert to Darwinism after he had once studied it." His parents wished him to enter the ministry; however, the conduct of some of the students for the ministry and his strong hatred of certain hypocrisies prejudiced him against entering the church. He published a pamphlet on university reforms, wrote articles for newspapers and periodicals, and in 1837 published a volume in London entitled *Walsh's Comedies of Aristophanes,* a translation of the Greek in corresponding English meter.

Mrs. Tucker reports that, in 1838, "After marrying the woman of his choice if not the choice of his people he came to America at the age of thirty landing at New York and expecting to settle in Chicago. It is not known why he came to choose a spot so far removed from his birthplace and settle down contentedly—as he did—so early in life. Possibly the best explanation one can glean from his tendencies was his love of unqualified freedom and his philosophical trend of mind. When he saw the few houses and the low, swampy ground of Chicago, he decided he did not want to live there, so journeyed on west by ox-team to the Red Oak neighborhood in Henry County, near Cambridge, Illinois. This especial part of the country, rich in verdure, may have appealed to his aesthetic tastes and reminded him of the beautiful flower-strewn English lanes. Here he built a mud-plastered log cabin, with a fireplace that took a log that had to be rolled in with oxen. A farm of 300 acres was purchased and tilled for about twelve years . . ." Riley says Walsh was so fond of his new country he never returned to England. He led a secluded life, associating very little with his neighbors, with whom he had little in common. "Yet he was thoroughly Democratic in his ideas, and had no false pride whatsoever: he did, as far as possible, all his own work, even making his own shoes and mending his own harness."

He remained in Cambridge until a Swedish colony settled in the vicinity and dammed the water nearby, thus causing the area to become malarial. We have no information to show that Walsh connected mosquitoes with malaria. Because his health was being undermined he moved, in 1850, to Rock Island, Illinois, where he opened a lumber yard in 1851. He was in this business for seven years. He was quite active in politics as a Radical Republican; he hated slavery and all forms of oppression and frequently expressed his strong views in print. In 1858 he retired from the lumber business after building a row of brick tenements, well constructed and reasonably priced, from which he derived some income.

Although he had made a small collection of insects in England, he had not devoted himself to entomology in this country, but at this point he began to pursue entomology in earnest. Riley (1870) writes, "Thus his Entomological career dates back scarcely a dozen years, but how faithfully and perseveringly he labored, the record of those years abundantly testifies. The first published account that we can find of Mr. Walsh as an Entomologist is in the report of a lecture which he delivered before the Illinois State Horticultural Society at the Bloomington Convention, in January, 1860. He there spoke extempore for two hours, displaying that rare faculty which he possessed of communicating his ideas in such a manner as to please and hold the popular ear. The reporter of this lecture, whom we take to be Mr. C. D. Bragdon, at the present time one of the editors of *Moore's Rural New Yorker,* states that he became so intensely interested, that his hand refused to move his pencil." Later he wrote articles on insects for a variety of farm newspapers, wherein he attempted to acquaint the farmers with a knowledge of their insect foes. From 1862 to 1866 he published about a dozen scientific papers in the *Proceedings of the Boston Society of Natural History* and the *Philadelphia Entomological Society.*

Mrs. Tucker says "He was a familiar figure about the streets and fields near Rock Island and was often a welcome visitor at the Holt greenhouse (located at 6th Avenue and 31st Street), where he gathered butterflies and insects that the flowers attracted. He wore a long cloak and a high peaked hat, cork lined, that gave him a quaint appearance. When he captured a bug or beetle that he wished to study, he would stick a pin through its body and attach it to the cork lining of his hat and in that manner bring it home. He was healthy and vigorous from a life passed in the green fields or along the brooks. The tap of his staff was well known upon our sidewalks, his rapid walk bearing testimony to the activity that tired not by age nor constant employment. He had the rude but genial simplicity of the scholar whose life is spent in the open air." However, according to the entomologist William LeBaron, Walsh applied himself so industriously to his studies that he had no time for the usual social amenities, and on occasion this affected his health and spirits.

In 1865 the Entomological Society of Philadelphia founded a monthly bulletin, the *Practical Entomologist,* which had on its editorial committee E. T. Cresson, A. R. Grote, and J. W. McAllister. Shortly thereafter Walsh became Associate Editor from the West, and finally sole editor of the second and last volume, the bulletin being discontinued in 1867.

Walsh had done so excellent a job of publicizing the dangers from insect pests that the Illinois State Legislature in the biennial session of 1866–1867 authorized the appointment of a State Entomologist at a salary of $2,000 per annum. Although Walsh was only confirmed in this office during the 1868–1869 biennial session, at the request of the farmers and horticulturists he began his work as State Entomologist in 1867 and, says Riley, "trusted to the future liberality of the Legislature to reimburse him for his work. As Acting State Entomologist he issued his First Annual Report for 1867, which was published as an appendix to the State Horticultural Transactions for that year." And in September 1868, C. V. Riley and Walsh started the *American Entomologist,* an outstanding popular periodical on insects. It continued for three volumes through 1880.

Of Walsh's unfortunate end, Osten-Sacken (1903) writes, "It is a strange coincidence that Walsh, Riley, and Walsh's successor as State Entomologist, LeBaron, all met with a more or less unnatural death: Walsh, from a railway accident, Riley from a fall with a bicycle, and LeBaron from the consequences of a sunstroke." Riley (1870) reports that on November 12, 1869, Walsh was returning from the post office, and as usual walked on the track of the Chicago and Rock Island Railroad. So engrossed was he in reading his mail he disregarded the ringing of the bell of a slow moving, oncoming train and only at the last moment flung himself clear of it. Unfortunately, his left foot was caught and terribly mangled. It was found necessary to amputate the foot above the ankle. According to Riley, "Mr. Walsh bore the amputation remarkably well and soon became quite cheerful, displaying his facetiousness by declaring in the most philosophical spirit, that nothing more fortunate could have happened to him. 'Why,' he would say to his grieving wife, 'don't you see what an advantage a cork foot will be to me when I am hunting bugs in the woods: I can make an excellent pincushion of it, and if perchance I lose the cork from one my bottles, I shall simply have to cut another one out of my foot.' "

On the day of the amputation he penciled the last letter Riley ever received from him. "It commenced with: 'I have been fool enough to get my left foot smashed,' and after dwelling at length on matters pertaining to the illustrating of his next State Report, concluded with: 'Adieu, Yours ever, the 99th part of a man!' "

Although he seemed to be getting better after the amputation, several days later his health began to deteriorate, possibly due to internal injuries sustained when he threw himself so violently aside in his attempt to escape the engine. "He lingered but a few days, and finally expired on Thanksgiving Day, the 18th of November. His mind was remarkably clear up to within a few hours of his death, and when the physicians informed him that his life was rapidly drawing to a close, he became perfectly calm and resigned to his fate."

Just prior to his death, Walsh and Riley were planning an exhaustive popular work on entomology, which was to be illustrated by Riley.

Speaking of a portrait of Walsh (possibly the one used here) Riley says, "We are much pleased with the plate, for it is a good likeness. In it the wonted humor twinkles from those eyes which are now closed forever in the great rest of the grave, and the facetious smile lurks around those lips which are nevermore to utter word again."

Walsh left a fine collection of ten thousand species. According to Mrs. Tucker, who for some unknown reason often referred to Riley as "General," "As soon as the death of Walsh was made known, General Riley came post haste from St. Louis and tried to gain possession of the collection, claiming that Walsh would have left it to him if he had time to arrange his affairs and that it was absolutely necessary to have it for the continuation of the *Entomologist,* as he needed it for reference. It is alleged he succeeded in boxing all the books and papers and nearly got away with the collection before being stopped." The State of Illinois purchased the collection and William LeBaron, Walsh's successor as State Entomologist, who was responsible for the purchase, had the collection removed from Rock Island to Springfield, Illinois. Later it was transferred to the Academy of Science in Chicago because of inadequate facilities at the State capitol. However, this valuable collection had as unlucky a fate as Mr. Walsh himself, for in 1871, when Mrs. O'Leary's cow kicked over a lamp, the collection, along with countless free-roaming bedbugs and cockroaches, were consumed in the Chicago holocaust (Anon., 1871).

A small part of his collection, which had been set aside for other purposes and was not in the Academy, is all that remains from the years of his painstaking toil. If not much of Mr. Walsh's collection survived him, he at least left a more permanent legacy, namely, his protégé, C. V. Riley, of whom we will hear more.

Mrs. Tucker concludes her biography of Mr. Walsh with some interesting information about Mrs. Walsh, who had her share of romance and tragedy. When she was eighteen, the young man she was to marry sailed to America to make his fortune and a home for her, but died aboard the ship and was buried at sea. She then, at the age of twenty, married Walsh. At first she was very homesick for England and quite unhappy about the early log cabins or "mud houses" she lived in. Cherished silver fell through holes in the floor and the corn meal became wormy. She mourned so much, says Mrs. Tucker, that Walsh told her that "if she could not endure it, he would take her to New York and put her on a ship for England. She never mentioned homesickness again. She was her husband's warmest appraiser, thoroughly in sympathy with his every idea, giving him leeway in his inclinations and worshipping his gods because they were his. He would become so deeply engrossed in some book that he would read at meals while the food grew cold. It was his custom to remain up till early hours correcting proof sheets and writing, oft times not retiring, simply smoking a pipe for rest." The Walshes were childless, and she lived until her eightieth year, dying in 1892.

One obituary on Walsh expressed the sentiment of many of his colleagues when it stated: "In the loss of Benjamin D. Walsh, State Entomologist of Illinois, there is gone one whose position may be filled, but his place never."

WILLIAM LeBARON (1814–1876)

William LeBaron, who succeeded Walsh as State Entomologist of Illinois, appears to be another entomologist sadly neglected by biographers.

He was born October 17, 1814, in North Andover, Massachusetts. His father and grandfather were doctors. As a boy he was greatly interested in natural history, moving successively from birds, to plants, to insects. Goding (1885) writes, "The neighboring cities soon attracted his less studious brothers, and upon William devolved the care of the aged mother and grandmother, both of whom he attended through long illnesses and [he] closed their eyes in death before he was twenty-five years old." He studied medicine under his uncle and practiced for a number of years in North Andover. He married in 1841 and moved to Geneva, Illinois, three years later. There he became "known far and near for his skill in diagnosis of disease and surgical operations." Later he resumed his medical studies and graduated from Harvard Medical College.

Around 1850 he began to write on economic entomology for farm newspapers and journals, and his account of the chinch bug was so highly esteemed by Dr. Asa Fitch that Fitch used it in his *Second New York Entomological Report,* noting that "Little requires to be added to this account." LeBaron never published his studies until they were very complete. He had a very practical viewpoint and carefully tested

Figure 12. William LeBaron. Courtesy of the Illinois Natural History Survey

the methods he recommended. In 1872 he suggested that fruit trees be sprayed with Paris green and water to control cankerworms and that insect parasites be transferred from one part of the country to the other.

In 1865 Dr. LeBaron became editor of the *Prairie Farmer* and in 1870, to his great surprise, he was appointed State Entomologist by the Governor of Illinois. He produced four annual reports; these and his papers on insects in various farm publications gave him an international reputation. His fourth and most extensive report, which specialized on beetles, provided a suitable and inexpensive text for students interested in Coleoptera. He was preparing a fifth report on Diptera when he became ill and subsequently, on October 14, 1876, died from sunstroke. His extensive collection augmented what remained of the Walsh collection. A classical scholar as well as a physician, he nevertheless preferred the study of insects to the practice of medicine, as did many medical men in the early days of entomology.

CYRUS THOMAS (1825–1910)

Thomas lived a long and varied life as lawyer, clergyman, teacher, entomologist, archaeologist, and ethnologist. Although the study of insects was only one of his interests, he became the third State Entomologist of Illinois in 1875 and a member of the United States Entomological Commission in 1876.

He was born July 27, 1825, in Kingsport, Tennessee, of Pennsylvania-German parents. Although he attended village schools and nearby

Figure 13. Cyrus Thomas. Courtesy of the Illinois Natural History Survey

academies, he never went to college. He learned science and mathematics on his own. In 1849 he began the study of law in Tennessee and then moved to Illinois, where, in 1851, he was admitted to the bar. He practiced in Murphysboro, Illinois, until 1864.

In 1850 Thomas married a sister of the noted Union general, John A. Logan. Shortly after she died in 1864 Thomas became a Lutheran minister, in which capacity he served for five years.

According to Goding (1888): "Early in his career Mr. Thomas sought for some branch of science in which he might carve a round in the ladder of fame. As a consequence the study of entomology was taken up, in 1856, deliberately, as being least expensive and with the materials closest at hand. After gaining a clear insight, by careful study and observation, into the intricacies of entomology in general, he made a specialty of the Orthoptera, eventually becoming our greatest American authority thereon. His views early took a practical turn, and his greatest delight was in contributing articles of economic importance to the agricultural press."

Thomas began to publish his entomological papers as early as 1859 while still practicing law, and in 1860, in the words of Forbes (1910), "he engaged with B. D. Walsh in a spirited controversy" on the life history of the army worm, which ran through several numbers of the *Prairie Farmer*. He also wrote on economic entomology for the *Prairie Farmer, Rural New Yorker, American Agriculturist, Farmer's Review*, and other papers and periodicals; and his more technical articles appeared in the scientific and entomological publications of the time.

Between 1869 and 1874 he served on the Hayden Geological Survey of the "Territories" in the West and Southwest. In 1870, he married one of his scientific assistants, a Miss Davis of Pennsylvania. He devoted his time to a study of agricultural resources and entomology and contributed a number of entomological reports, including a monograph on grasshoppers, which were then a scourge of both mountain and plain states.

In 1874 he was appointed Professor of Natural Sciences in the Southern Normal University (Southern Illinois University), Carbondale, Illinois, and severed his connection with the Survey. The following year he was appointed State Entomologist of Illinois, but retained his teaching post until 1876.

According to Forbes, "In the fifth volume of the Transactions of the State Agricultural Society (Illinois), printed in 1864, are three prize essays on entomology by Walsh, LeBaron and Thomas, who afterwards became respectively the first, second and third state entomologists of Illinois. Doctor Thomas became one of the leading specialists on the *Acrididae* of the United States, publishing a monograph of that family in the report of the Geological Survey of the Territories in 1873. Later he gave special attention to the *Aphididae,* and the third report as state entomologist is one of the most important descriptive publications on that family which has ever yet appeared." Of his six reports, the first is devoted to the Coleoptera, the second to the Lepidoptera, the third, as Forbes said, to the Aphididae, the fourth to cabbage insects and Acrididae of Illinois, the fifth to larvae of some Hymenoptera and Lepidoptera, and the sixth to noxious insects of that year.

Forbes adds, "In his economic work in Illinois he was severely handicapped by the penurious policy of the state with respect to the entomologist's office. In all his seven years' service he received nothing from the state treasury except his salary, and had no facilities for his work except those which he provided for himself and at his own expense. His official reports are consequently largely given to systematic articles, to resumes of pertinent matter already published, and to special articles based on his own personal observations and those of his assistants. Precise experimental work or systematic field operations of any sort were beyond his reach. His principal assistants, Professor G. H. French, Mr. D. W. Coquillet, and Miss Emily A. Smith, were engaged to enable him to share in the work of the United States Entomological Commission, appointed in consequence of destructive outbreaks of the western locust."

He was appointed to the U.S. Entomological Commission in 1877, along with such outstanding entomologists as C. V. Riley and A. S. Packard. This commission engaged in remarkable studies that resulted in the publication of two volumes on migratory grasshoppers and other volumes on cotton insects, forest insects, and a variety of other insects as well as bulletins on the cotton worm, Hessian fly, chinch bug, and additional agricultural pests.

In 1882, at the age of fifty-seven, he resigned his professorship at the Southern Normal University and his position as State Entomologist to join the Bureau of Ethnology in the Smithsonian Institution in Washington, D.C. There he applied himself to his new interests and in time became a great authority on archaeology, especially on Mayan inscriptions, the mound builders in the Mississippi Valley and the Gulf States, and the Cherokee and Shawnee Indians, among other tribes.

This remarkable self-taught entomologist died in Washington, D.C., on June 27, 1910, at the age of eighty-five.

JOSEPH ALBERT LINTNER (1822–1898)

Lintner is another example of an outstanding entomologist neglected by biographers. Felt (1899) prepared a 300-page listing of Dr. Lintner's publications, but unfortunately told little about Lintner himself.

Figure 14. J. A. Lintner

Felt reports that Lintner wrote 900 articles and thirteen reports, mostly on insects, during the last thirty-six years of his life, and that for twenty-five years he was the entomological editor of the *Country Gentleman.* "His reports, replete with valuable and practical information, are enduring monuments to their author. In simplicity of language, dignity of expression, conciseness and thoroughness of treatment, they are models."

Howard (1930) enlarges on Lintner's contribution to entomology:

"It is true that very many of these were newspaper articles, although published in journals of high class like the *Rural New Yorker* and the *Country Gentleman,* but many of them were important; and his larger publications, notably his *Entomological Contributions* published by the New York State Cabinet of Natural History, contained very many separate articles of high rank, and his twelve reports on the injurious and other insects of the State of New York are models. He had no laboratory facilities (all of his work was done in his office) and original investigations in applied entomology were almost impossible. So these reports are largely compilations and the results of correspondence and of such rather limited observations as he could make in the field. Nevertheless they are models. No other reports that have been published excel them in care of preparation, lucidity of style, bibliographical detail, fullness of indices, and general consultability. When Lintner wrote about an insect, it was certain that he overlooked nothing that had previously been done, and his papers were the latest words on that topic."

Lintner was born in Schoharie, New York, on February 8, 1822; he was of German parentage, and his father was the Rev. George Ames Lintner. The son graduated from the Schoharie Academy in 1837 at the age of fifteen. Although fascinated by the study of nature as a boy, a year after his graduation he entered commerce in New York City, to which he devoted himself for ten years. He moved back to Schoharie in 1848, and to Utica in 1860, where he was in the woollen-goods business.

Lintner began to collect insects seriously in 1853, when he aided Dr. Asa Fitch in an entomological survey for the State of New York. In 1862 Lintner published his first entomological article, and in 1868 he became a zoological assistant in the State Museum. He was placed in charge of entomological work in the Museum in 1874. Shortly after Dr. Fitch died, in 1880, he was appointed State Entomologist a position he held until the time of his death in 1898.

Dr. Howard, who knew him personally, writes: "He was a man of very fine appearance, an impressive speaker and lecturer, and of great personal charm, dignified, well dressed, in fact a man who by his personality helped much to dignify the profession. It always seemed to me that Harris must have been much of the same type. I said as much to Lintner once, and it embarrassed him greatly. He answered in very courteous words, but the idea I gained was that he thought me foolishly fulsome."

Felt says he was "modest, unassuming, gentle, yet never yielding to imposition, kind and loved by all." And Slingerland (1898) writes, "He was ever ready to impart from his vast fund of knowledge; and, being an impressive speaker, he always commanded the attention of scientific bodies which he was called upon to address. His frequent addresses before horticultural and agricultural societies in his own and other states, and farmer's meetings of all kinds, were always full of information."

In the winter of 1891, when Professor Henry Comstock was quite ill, Dr. Lintner wrote to Mrs. Comstock (1953), who observed: "He was a man of great reserve, but when he heard of Mr. Comstock's illness, he wrote me a sympathetic letter, full of solicitude for Harry's health and with kind words of praise for a proof which I made for him, from an engraving of a moth."

Although he did not begin a serious study of insects until he was thirty-one, by industrious and astute application he soon attained fame in his chosen field. In 1884 he received an honorary Ph.D. from the State University of New York, and in 1892 served as president of the American Association of Economic Entomologists.

Dr. Lintner had been granted, says Slingerland, "a well-earned six months' leave of absence, and was spending it in sunny Italy" when he died on May 5, 1898.

Figure 15(A).

Figure 15(B). (A) and (B). S. A. Forbes. Courtesy of the Illinois Natural History Survey

STEPHEN ALFRED FORBES (1844–1930)

L. O. Howard thought so highly of Forbes that he believed Forbes should have been Riley's successor rather than himself. The high regard Howard had for Forbes, the fourth and last State Entomologist of Illinois, is shared by other entomologists.

J. J. Davis, who, in the early years of this century, assisted Forbes, wrote me that he considered Forbes one of America's greatest entomologists. He was an accomplished scientist in many fields, a marvelous linguist, and a man with great competence in expressing himself in technical or popular writings. Davis said, "he was a great man and it was a privilege to know him," and he notes that, "Like most Civil War veterans he was a constant chewer of tobacco," which was the subject of "interesting incidents" he did not reveal to me.

Forbes was born on May 29, 1844, in Stephenson County, Illinois, in a log house on a small, primitive pioneer farm. His father, Isaac Forbes, lost this farm because he had endorsed a friend's bad note. His father began all over again on a new farm "in a house built of slabs." There were six children in the family. When Stephen was ten his father died. His older brother, Henry, who was twenty-one, took over as head of the family, and later sacrificed his own education to send Stephen to Beloit Academy.

Howard (1932) observes that Stephen attended the district school from school age to the age of fourteen, and from 1858 to 1860 he studied at home under his brother. In 1860 he attended Beloit Academy, but he had to give this up for lack of money. He was so poor he had to wheel his trunk to a railroad station in a wheelbarrow. Conditions at home were very difficult and he could barely afford a glass lamp to replace the candle light which was ruining his eyes. After the war he studied medicine at Rush College in Chicago, and almost completed his studies. However, lack of money and an aversion to undertaking surgical operations without the benefit of anaesthetics made him change his plans.

When the Civil War broke out, Henry borrowed money so that he could buy two horses to enable Stephen and himself to enter the cavalry service. Both brothers served with great distinction. At twenty Stephen was captain of his company.

Forbes (1930), in a short autobiographical sketch, tells a little about his service in the Federal Army:

> My army service, concerning which you ask particulars, was begun by my enlistment as a private in Company B, 7th Illinois Cavalry, in September, 1861, when I was seventeen years old, At eighteen I was made orderly sergeant; at nineteen, second lieutenant, and at twenty, captain of my company. Just after my eighteenth birthday, when sent to carry an important dispatch to a distant outpost near Corinth, Mississippi, I was put upon the wrong road and presently found myself inside the rebel lines as a prisoner of war. Telling my captors that I had a verbal message only, which I refused, and they did not compel me, to disclose, I availed myself of an opportunity to tear up my dispatch, secreting the fragments in the pistol holster of my saddle. At General Bragg's headquarters

I was threatened with hanging if I did not produce my dispatch and was thoroughly searched for it, as was also my saddle, but nothing was found. Later in the day I had a brief interview with General Beauregard, in command of the rebel army after the battle of Shiloh, and a much longer one with a major of his staff, who ended by wishing me good luck, and telling me to appeal to him if I got into trouble. I was in prison four months at Mobile, Alabama, Macon, Georgia, and Richmond, Virginia. Utilized my abundant leisure by studying Greek from books which I managed to buy at Mobile. When paroled and released, I was sent to a hospital for three months to recover from scurvy and malaria acquired in prison. Rejoined my regiment, reenlisted for the war, and was mustered out in November, 1865.

In 1863 the brothers participated in the famous Grierson cavalry raid from Lagrange in western Tennessee to Baton Rouge, Louisiana. Forbes (1907) wrote a very interesting account of this raid by a Federal force of about one thousand men behind the Rebel rear at Vicksburg. The story of the raid was based on his dead brother's notes. Professor Forbes wrote it partly in tribute to his brother, Captain Henry Clinton Forbes, later brevet colonel of volunteers. Henry's account of the last day of the raid, as reported by his brother, needs no comment:

Men by the score, and I think by fifties were riding sound asleep in their saddles. The horses excessively tired and hungry, would stray out of the road and thrust their noses to the earth in hopes of finding something to eat. The men, when addressed, would remain silent and motionless until a blow across the thigh or shoulder should awaken them, when it would be found that each supposed himself still riding with his company, which might perhaps be a mile ahead. We found several men who had either fallen from their horses or, dismounted and dropped on the ground, dead with sleep. Nothing short of a beating with the flat of a saber would awaken some of them. In several instances they begged to be allowed to sleep, saying that they would run all risk of capture on the morrow. Two or three did escape our vigilance, and were captured the next afternoon.

Ward (1930) says, "It was a source of great delight to hear the story of those days, when on rare occasions he [Forbes] could be persuaded to relate to younger friends some of his experiences in the field." And Dr. Forbes himself said, "To us war was not hell, but at the worst a kind of purgatory, from whose flames we emerged with much of the dross burned out of our characters, and with a fair chance still left to each of us to win his proper place in the life of the world."

We learn from Forbes' autobiographical sketch that upon leaving Rush College he became "infatuated" with the study of botany, studied natural history, and taught school for a living. When Major J. W.

Powell, the famous geologist, resigned as curator of the State Natural History Society of Normal, Illinois, in 1872, Forbes succeeded him. In 1877 he was appointed Director of the State Laboratory of Natural History, a position he held from that date until 1917, when he was made chief of the newly organized Illinois State Natural History Survey. From 1875 to 1878 he was a Professor of Zoology at the Illinois State Normal University, and in 1882 he was appointed State Entomologist of Illinois to succeed Cyrus Thomas.

In 1884, Forbes resigned as Professor of Zoology and Entomology, Director of the State Laboratory of Natural History, and State entomologist, to become Professor of Zoology, Professor of Entomology, and Dean of the College of Science at the University of Illinois. He notes that, in 1884, "Indiana University gave me the degree of doctor of philosophy 'on examination and thesis' entirely the product of private study, as I had taken no academic college course and had no bachelor's degree. In 1905, when I resigned as dean of the College of Science, the University of Illinois conferred on me, in recognition of my sixteen years' service in that capacity, the honorary degree of doctor of laws."

Metcalf (1930) wrote of Forbes' entomological interests: "His publications in natural history, begun in 1870 in the *American Entomologist* and *Botanist,* numbered at the time of his death over 500 titles, of which about 400 dealt with various phases of entomology. His most important papers are to be found in his eighteen *Reports on the Injurious and Beneficial Insects of Illinois,* in the *Bulletin of the State Laboratory of Natural History,* later the *Bulletin of the State Natural History Survey* and in his Final Reports on the Biology of Illinois. Some of the best known of his entomological publications deal with the insects of Indian corn, of strawberries, of sugar beets, the chinch bug, Hessian fly, white grubs, San Jose Scale, corn root aphid, army worm, codling moth, black flies and insect diseases. In all these subjects, and others, he made fundamental contributions to economic entomology and ecology which have endured and are today models of clarity, originality and completeness. His writings are characterized by their remarkably simple and lucid expression, their excellent illustrations, their intensely practical, economic nature, and reveal a deep appreciation of fundamental biological principles and of the importance of the interrelations of insects with their environment and with other living things."

Forbes also wrote on birds, fishes, crustacea, leeches, bacteria, rotifers, the parasites of swine, museum methods, and teaching. "His most intimate friends and associates," says Metcalf, marveled at his "interest in and depth of understanding of other fields of knowledge: history, music, art, politics, languages, literature, agriculture, horticulture, world affairs, the social sciences—he studied them all in order to relate his own work most effectively to the material and intellectual progress of his state."

According to his son Ernest B. Forbes (Howard, 1932), his father was a Republican and an enthusiastic backer of Abraham Lincoln. As a fourteen-year-old boy he had listened to the Lincoln-Douglas debate at Freeport, Illinois.

"Physically, my father was characteristically restless, active, and energetic. While at the height of his powers his course through the Natural History Building could be traced by the slamming of the doors behind him. The attitude of command, attained during his extensive military service, was habitual, and he expected action on the part of his subordinates." He was mentally quick, acute, and alert. He slept five hours a night on the average, and was a prodigious reader in several languages. He played the organ and was interested in the theater. In his early years he was easily bored and made few intimate friends. "There was, however, almost always some small and choice group of kindred spirits with whom he met for feasts of philosophy, and these men he enjoyed immensely." In later years he apparently was more sociable and widened his acquaintanceship.

Dr. C. A. Kofoid, the protozoologist, wrote to Dr. L. O. Howard, "He was at his best when with uplifted cigar in his hand he would discourse wittily, fluently and pointedly, till his cigar went cold, on subjects far afield from Hessian fly and corn aphids but with a familiarity and readiness as masterful as though he were in a group of his entomological colleagues."

Forbes was a handsome, scholarly appearing man about six feet in height. He exercised a great deal and claimed that bicycle riding added about ten years to his life. "Later he drove an automobile and became locally famous for a long series of minor mishaps—resulting from the fact that the automobile came late in the period of his physical adaptability and that he drove to the accompaniment of his intensely concentrated thinking, without knowing that he failed to give the job his whole attention." He was proud of the ticket he received for speeding on his eightieth birthday.

Forbes was the recipient of many honors. He was a member of the National Academy of Sciences, the American Philosophical Society, and other organizations. He was president of the American Association of Economic Entomologists in 1893 and 1908, being the only president other than Professor C. V. Riley to serve two terms, and of the Entomological Society of America in 1912. He also served as president of the Ecological Society of America and other organizations.

Forbes was intellectually active until nine days before his death on March 13, 1930, when he still held the position of Chief of the Illinois Natural History Survey. He was then almost eighty-six years old; his wife, the former Clara Show Gaston, whom he had married in 1873, died a few weeks before he did. They had had five children; and Mills (1964) paints a cheerful picture of the Forbes' home life.

3 Early Federal Entomologists

Pale tree-cricket with his bell
Ringing ceaselessly and well
Sounding silver to the brass
Of his cousin in the grass
Thomas Moore(?)

When Townend Glover was appointed in 1854 to the Bureau of Agriculture of the United States Patent Office to study seeds, fruits, and insects, he became the first Federal entomologist. He was followed by C. V. Riley, L. O. Howard, and many other notable workers who established entomology as an essential part of the economy. This chapter tells something about a number of early entomologists who gave Federal entomology its international reputation.

TOWNEND GLOVER (1813–1883)

The reputation of Townend Glover, the first U.S. entomologist, is somewhat obscured by his many singularities. We must not overlook the fact that he was one of the finest illustrators of insects in the United States. Unfortunately, his talents along this line are not well known because so few have seen his excellent plates. The library of the Carnegie Museum in Pittsburgh contains many original plates engraved and colored by him. Some of these illustrations are as good as any done today—and it is a pity that so few entomologists are even aware of their existence.

Glover was an individualist and a nonconformist whose behavior and mode of living were considered eccentric; he apparently had a different sense of values from that of most men. Howard calls him the prototype or caricature of all the entomologists who came after him.

Glover had a sympathetic biographer in his assistant, C. R. Dodge, and the acount given here is largely drawn from the biography he wrote in 1888.

Townend Glover was born in Rio de Janeiro February 20, 1813, where his father, Henry Glover, an Englishman, was in business. His mother,

Figure 16. Townend Glover

Mary Townend Glover, died when Townend was about six weeks old, and he was brought up by relatives in Leeds, England, where he was educated in a private school of good reputation. His father died in Rio when the child was six years old.

"By nature," Dodge writes, the boy Townend was of a reserved disposition, making few close friendships outside the immediate circle of his own family. He was, nevertheless, abounding in high animal spirits, possessed of a strong sense of fun and humor which always made him an agreeable and entertaining companion to those with whom he was wont to associate."

He was a good student, with a love for drawing and caricature; the latter at times got him into trouble. Even as a boy he had a predisposition for natural history, especially for the study of botany and the collection of insects.

Although his father had left him an ample inheritance, this all disappeared "save a portion which, unknown to Mr. Glover, until he had reached his majority, was reserved in trust by relatives in England. Mr. Glover not only thought that he had been cheated out of his patrimony [by the machinations of his father's partners], but on at least one occasion has intimated the suspicion that his father's death had occurred from other than natural causes."

At the age of twenty-one he left the wool business, in which he had been engaged for less than a year and for which he had no liking, and began to study German in preparation for the continuance of his schooling on the Continent. He kept a diary, about which Dodge writes, "To one who has known Mr. Glover intimately in later life his diary kept at this period is most interesting, as showing, even at the age of twenty-one, so many of those traits of character or individualisms, if the term may be used, which so strongly marked the mature man. Indifference to country or home, distrust of mankind and of the motives of people about him, self-reliance and a wish to be his own master, and at the same time frequent evidences of the good influences by which he had been surrounded in the family circle in which he was reared, appear on many pages."

Glover was soon on his way to Munich to study art, visiting a number of cities en route. In 1834 he began to study the painting of fruits and flowers; these along with insects, were his favorite models. Dodge reports that "When Margaret Fuller first saw some of the flower paintings she would hardly believe that they had not been done under the microscope, so delicate was the work. Whether or not his extreme shortsightedness made it difficult for him to paint in any other manner, it is impossible to say, however well adapted to the labors of his after life this special kind of work may have been. He could not have painted broadly had he desired to do so, for his almost microscopic vision saw everything in the minutest detail. This explains too, why his engravings of insects, particularly of *larvae* lack in action." After leaving Munich, he traveled through Europe and then returned to Leeds.

Dodge says a Mr. Oates provides this glimpse of Glover, the artist: "He would sit before his easel with a favorite lizard nestled in his breast, his coat pockets tenanted by snakes, and a blackbird perched upon his shoulder, whilst hanging on the walls of his apartment might be seen some tiny gauze cages, daintily constructed for the reception of tame spiders, which were periodically supplied with flies. There were also in the room a variety of other birds and such quadrupeds as mice, rats, and guinea pigs, all pets in a greater or lesser degree."

One of Glover's school friends visited him in his apartment and recalls one occasion "when he was painting a bunch of grapes, his blackbird as usual upon his shoulder. Glover had just completed the painting of the grapes, when the fancy seized him to add a fly, as though it had alighted on the fruit. This he did, and had scarcely withdrawn his hand from the work, when the blackbird darted from its master's shoulder and pecked lustily at the phantom fly." Glover also engraved on copperplate and stone and did wood carving.

In 1836 Glover sailed to America to visit some relatives. Then he roamed through the country at his leisure for a couple of years, particularly through the South. Dodge (1886) says that "in his earlier years he led a wild life." Sportsman that he was, he bore a scar on the side of his head made by the explosion of a gun barrel, which fractured his skull.

In 1838 he settled down in New Rochelle, New York, with gun, dog, horse, and a boat he had built. Then, in 1840, he married Miss Sarah T. Byrnes, whom he had met in Fishkill, New York, and whose father owned a large estate on the Hudson; he accordingly moved to that town. For the next five years he spent his time on floriculture, the study of natural history, taxidermy, fishing, hunting, and the painting of fruit and flowers. "Mr. Glover was a skillful taxidermist," Dodge reports, "and was a capital shot, not withstanding the peculiarity of his eye-sight; and as he tramped over the adjacent country, cane-gun in hand, using it also as a walking stick, he doubtless appeared more as a rural gentleman than the enthusiastic naturalist that he was."

In 1846 he purchased his father-in-law's estate. "In this beautiful place . . . he began in earnest the life of a country gentleman, busying himself with the planting and care of fruit and ornamental trees, and with his garden, which was noted for its fine flowers and vegetables. He also paid considerable attention to the cultivation of small fruits, all the leading varieties of which were tested by him." There he also began to develop a great interest in insects and to model fruits. The two thousand pomological specimens which he modeled in six years of "unremitting toil" cost him $3,000 and were exhibited in many places. The collection changed his mode of life from that of a gentleman farmer to that of a wanderer.

After exhibiting his specimens in Washington during the winter of 1853–1854, on June 14, 1854, he joined the staff of the Bureau of Agriculture, which had just been set up in the United States Patent Office, and displayed part of his collection in the one room devoted to the Bureau. His duties were to collect "statistics and other information on seeds, fruits and insects in the United States." As "the entomologist and special agent" of the Bureau, he traveled widely on agricultural missions, particularly in the South. In 1854 and 1855 he studied crop insects, including cotton and grape pests, in South Carolina and Florida. Dodge gives no clue as to what became of Glover's wife or his estate in the meantime.

Glover's Florida experiences inspired the following verse, which Dodge had to jot down from memory since Glover never would give anyone a copy of it:

> From red-bugs and bed-bugs, from
> sand-flies and land flies
>   Mosquitoes, gallinippers, and fleas,
> From hog-ticks and dog-ticks, from
> hen-lice and men-lice
>   We pray Thee, good Lord, give us ease:

And all the congregation shall *scratch*
and say Amen.

In 1856 and 1857 Glover visited British Guiana and Venezuela to obtain new plantings of sugar cane to replace the deteriorating stock on the Louisiana plantations. This expedition was quite successful.

Apparently he contracted fever (malaria?) in New Orleans in 1857 and was ill a great deal of the time while working on cotton plantations during the summer. Thus an entry in his journal for that year notes, "Quarrel between doctors, so I have to dismiss one, and the other says it is no use to attend. Saved my life by it."

He worked part of the following year on orange insects in Florida and there studied, among other things, the effect of alligator blood on scale insects. One of his entries in his journal notes, "Etching, itching, and scratching as usual from 8 to 4; scratching with pen from 8 till 12, and with finger nails continually."

Not only did he study insects, but also plant diseases, "soils and earths, vegetation, birds, animals, reptiles, Indian mounds, and even human nature." He made many original observations but delayed publishing these until he could incorporate them in one work. Some of the results of his industry can be gleaned in the reports of the Patent Office from 1854 to 1858.

In 1859 he severed his connection with the Patent Office because of friction with his superior, whom he considered a man of limited ability who thrived on the talents of his underlings. Glover caricatured his chief mercilessly and recited uncomplimentary doggerel, which he could make up "on the spur of the moment."

When Glover resided in Washington, his "den" was a single room, in which he "ate, wrote, sketched, engraved, and saw his few intimate friends. What with his engraving and writing tables, his book case (constructed from boxes), trunks, tool chest, and insect cases, in addition to the stove and regular bedroom furniture, there was little space to spare." Later he added more rooms, covering the walls with purchased pictures.

After leaving the Bureau of Agriculture in 1859, he joined the faculty of the Maryland Agricultural College (University of Maryland) as a professor of natural sciences, where he lectured, did field work, and made a collection of birds and insects. In that year, he began work on his proposed *Illustrations of North American Entomology,* which was to be "an illustrated encyclopedia of economic entomology."

According to Bissell (1960), while Glover was at the Maryland Agricultural College he was frequently "accompanied on his collecting trips by a child, Salome Lavinia Johns, daughter of Montgomery Johns, M.D.,

Professor of Agriculture, Chemistry, etc. Miss Johns later married Daniel C. Hopper and was the 'adopted daughter' referred to by Dodge, Glover's biographer. Glover gave the Hoppers their home in Baltimore and spent his years of retirement with them."

In 1862, a new Department of Agriculture was established and in April 1863 Glover was appointed United States Entomologist. His first reports in 1863 and 1864 were popular discussions of common insects injurious to vegetation, with brief notes on their control. For the year 1865 he studied insects, seeds, grains, fibers, silkworms, birds, poultry, and domestic animals; he was, in short, a one-man Department of Agriculture.

In 1865 he attended an entomological convention in Paris and received a gold medal from the Emperor Napoleon III for his original work in entomology. Shortly after his return from Paris he advocated that measures be taken to prevent the importation of injurious pests. "It is well known that several of the insects most destructive to our crops are of European origin, and I would suggest that all foreign seeds and plants imported by this department be subjected to a careful investigation, and if found to be infested by any new or unknown insects, fumigation, or other thoroughly efficacious means of destroying them, should be used before distributing them through the country."

During his years as Entomologist he occupied himself with the Department museum, which was moved to new and larger quarters in 1864. Besides fruit models, it contained mounted birds, insects, plants, botanical specimens, cereal products, fibers, and manufactured articles. It attracted more and more visitors all the time; he had "almost a childish vanity" in showing visitors the museum and explaining the exhibits to them. He sold his collection of fruit models and his other exhibits in 1867 to the Government for $10,000 and gave the Government his collection of 600 mounted birds.

Glover continued to work assiduously on his proposed book, all the while leading an irregular, overworked, and somewhat untidy life. For many years he sent proofs of his copper and stone engravings to prominent entomologists for criticism. T. W. Harris had been so impressed with Glover's earlier work that in 1852 he had tried to get him to illustrate his *Insects Injurious to Vegetation*; Glover apparently was not interested. Glover kept compiling the latest information from various authorities on order after order, such as the Coleoptera, Lepidoptera, Hemiptera, and Neuroptera, and experienced the compiler's difficulty of never being quite up to date. He corresponded with B. D. Walsh, P. R. Uhler, C. V. Riley, F. G. Sanborn, A. R. Grote, and other entomological authorities of his day, receiving from them new material for the Department collections or to be used in illustrating his book. In turn he shipped his correspondents for description new species that he received from collectors in the South and the West.

Baron Osten-Sacken, the noted dipterist, suggested that his book be issued as a popular work, rather than confined to the various orders. Had Glover listened to this advice, it is possible that he might have found a publisher. Such a work, with a selection of his beautiful plates, might have made him world famous, although Dodge thought that "if it had been finished and published in accordance with the author's design, there would be nothing now in entomological literature like it." As it was, he had to print privately limited runs of the plates in various orders, which he distributed gratuitously to his entomological friends and a few others, until in 1887, he published an edition of fifteen copies (author's proofs) under the title *Illustrations of North American Entomology in the Orders of Coleoptera, Orthoptera, Neuroptera, Hymenoptera, Lepidoptera, and Diptera.* This contained a complete set of proof impressions of the 273 copper plates, with 6,179 figures; the copper plates had been engraved and "the proofs colored by him from nature at the time the plates were engraved." The body of the book, other than the plates, was not printed but is a lithographed copy of the handwritten manuscript. Since his penmanship is often minute, the wear and tear on the reader's eyes are considerable.

Dodge (1883) summed up Glover's lack of public success in these words: "His 'work' was his dream, and here for years he accumulated a mass of interesting facts, the publication of which, as discovered, would have made his name as an observer great indeed. Some of these facts have been given to the world in his published reports as United States Entomologist, but the majority were withheld from publication, awaiting the completion of his work—until, from time to time, many of his interesting discoveries were re-discovered and published by the army of careful observers who have come after him and the credit has thereby been lost to him. Perhaps it was his overconscientiousness which kept him from 'rushing into print,' for he often underrated his own judgment in citing the history of insects he had carefully reared and observed, rather preferring to give the experience of another with full credit, than use his own material."

Unfortunately, Glover had very little interest in assembling a collection of insects. He took no care of the insects he collected or figured and even used ordinary pins instead of insect pins to mount them. To him "well-drawn colored pictures were quite as useful" as an insect collection. Needless to say, Riley, Howard, and the other entomologists who succeeded Glover were disappointed in the entomological legacy he left them.

Because of his defective vision during the last ten years of his life,

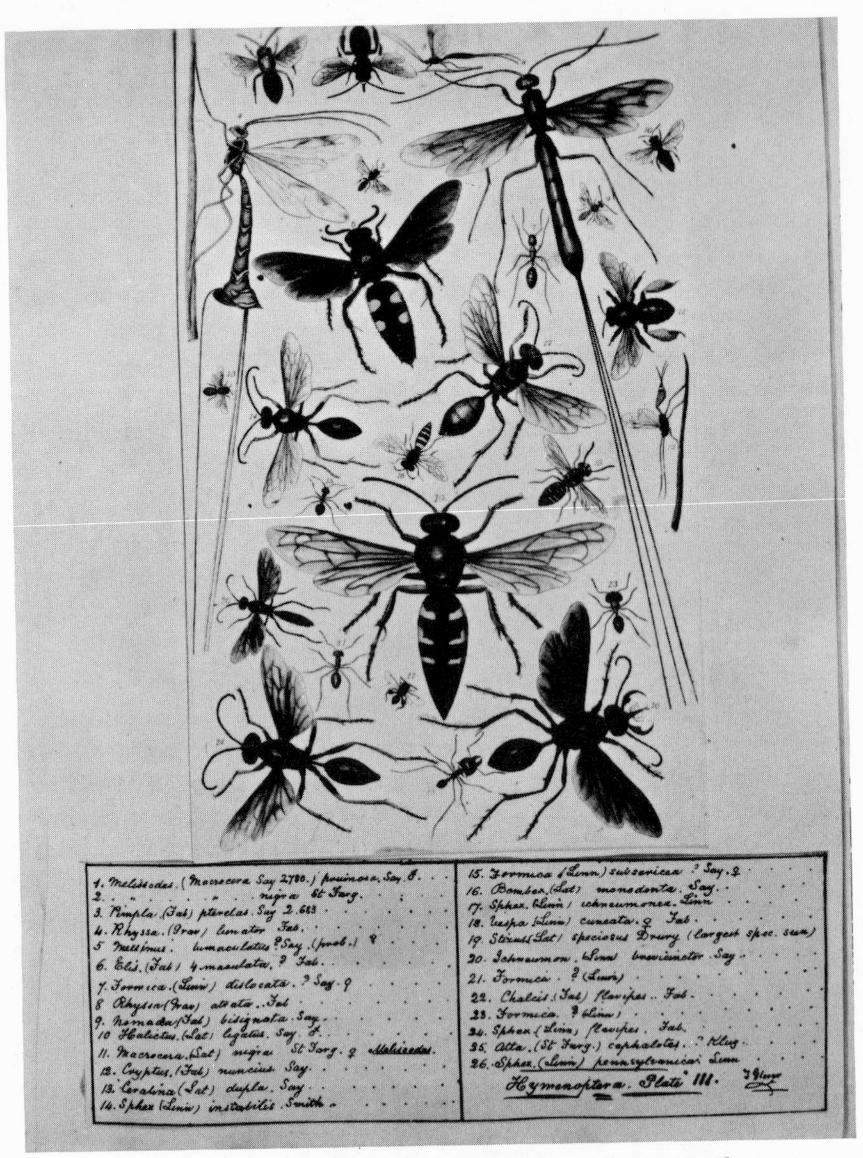

Figure 17. Plate from a collection of Townend Glover's works. Courtesy of the library of the Carnegie Museum

Glover did not like to be on the streets alone, so he stopped attending scientific or social meetings. Then, in the spring of 1878, he became so ill as to be incapable of carrying out his duties as Entomologist. He was succeeded by C. V. Riley, who, when the House Committee of Agriculture purchased the copper plates for Glover's book for $7,500, used his influence to see that Glover got paid.

Blind and feeble, Glover went to live with his adopted daughter in Baltimore. Here his lack of activity weighed heavily on him, and "he felt keenly the neglect of his old friends, some of whom were residing then in Baltimore."

Dodge described some of the personal traits of the man he knew so well. Although serious, he was rarely melancholy or cynical. He never forgot a kindness and rarely forgave an injury. Quick-tempered, his anger was almost uncontrollable once aroused. He was very sensitive to criticism and apparently lacked confidence in some of his abilities, which accounted, in part, for his delay in completing some of his works.

Glover died in Baltimore on September 7, 1883, with his wife and adopted daughter in attendance.

CHARLES VALENTINE RILEY (1843–1895)

Riley was undoubtedly one of the greatest and most controversial figures in the history of American entomology. L. O. Howard (1895, 1930, 1933), who was an assistant to Riley for many years, is one of the

Figure 18. C. V. Riley

most important sources of our information about him. Goode (1896), Packard (1895), Osborn (1937), and others who knew him also wrote about him, so there is no dearth of information as there is about some other prominent entomologists.

It was largely through Riley's efforts that entomology became a prominent part of the United States Department of Agriculture. His organizational ability, his choice of assistants of high competence, and their success against many insects pests, all gave Federal entomology the high reputation it has maintained to this day.

Through the medium of his excellent studies on the life history and control of injurious insects and his excellent illustrations in his nine Missouri reports he, with T. W. Harris, B. D. Walsh, and Asa Fitch, was responsible for making the farmer aware of the importance of insect control in the production of his crops. These studies were considered works of genius by many entomologists of his day.

It was largely through his initiative that the United States Entomological Commission was formed and Federal funds appropriated for the control of such widely distributed pests as the Rocky Mountain locust, the cotton leaf worm, the chinch bug and other pests.

Howard (1930) describes Riley giving a lecture at Cornell in 1871. "I was . . . about fourteen years old, and I was greatly impressed by Professor Riley's picturesque appearance (he was tall, slender, romantic in appearance, with long wavy hair and a luxuriant moustache, looking much more like an Italian artist than like an American economic entomologist). I was not especially impressed by his lectures, since he was not a good speaker, but was charmed by his wonderful ability to illustrate his lectures with blackboard drawings. His crayon sketches were very clever, and if I remember rightly he had acquired the faculty of drawing with both hands at the same time . . ."

And Goode (1896) adds that Riley "was a man of singularly striking appearance and agreeable presence. No one who had once seen him could forget him. Active and graceful, his bearing was such that, though perhaps not more than five feet ten inches in height, he seemed much taller. He never lost the easy, independent carriage which he had acquired during his early life in the West, and there was always something unconventional and picturesque about his costume and appearance. The broad-brimmed sombrero-like hat, dark in winter, light in summer, which he almost always wore, seemed in keeping with his swarthy complexion. He looked like an artist or a musician, and indeed he possessed the artistic temperament in a high degree."

Herbert Osborn, a student at Iowa Agricultural College, says that on first meeting Riley in 1876 or 1877, he "struck me at once as a man of remarkable enthusiasm, with magnetic appeal in his zeal for study and

I was much encouraged by the few words of suggestion that he stopped to give me."

Osborn also describes an incident which reveals how athletic Riley was. "I have seen him pick Miss Murtfeldt (who was a cripple) up under his arm and run up the stairway of a steamer on Lake Minnetonka. I know it was not a light undertaking, for a year later, (and I do not think Miss Murtfeldt had gained much in weight), I attempted to help her into a cab and narrowly escaped dropping her in the gutter."

Riley was born in Chelsea, London, September 18, 1843. He attended school in England, as well as in Dieppe, France, and Bonn, Germany, where he won prizes for sketching butterflies. This early training in Europe explains his great familiarity with foreign languages.

According to Skinner (1895), "The early loss of his father, and the care at school of a younger brother, developed in young Riley a self-reliance and sense of responsibility which gave a practical turn to his views, and convinced him that the classical education he was getting lacked many elements of utility. So at the age of seventeen, he sailed for New York, where after a seven weeks' voyage, he arrived with little means and 'a stranger in a strange land.' " He went to work on a farm in Illinois about fifty miles from Chicago, but left the farm for Chicago after about three years because he was literally working himself to death, as is shown by the following note (Goode, 1896) in a rural publication:

". . . Young Riley was simply too enthusiastic and too bent on excelling in everything. He took no rest. Often he would be up, actually getting breakfast ready to relieve the womenfolk, and milk half a dozen cows before the others were about. When others were resting at noon in the shade, he would be working at his flowers under a July sun. There was not a sick animal of the three hundred on the place that he did not understand and help. He kept a lot of bees, got hold of the best bred colts and some of the best heifers in the county, secured a good quarter section, and spent his Sundays reading, sketching, and studying insects. Three years of this increasing effort under the trying climatic extremes of central Illinois broke the young fellow's health, for it was too great a contrast to his previous life, and with everyone telling him he was wasting his talent he finally concluded to give up the idea of farming. But had his health not failed him, my opinion is that he would be a farmer today, and a successful one too, for he has intense love of rural life."

Riley, even as a youth, was an ardent collector of insects. While on the farm he had sent articles to the *Prairie Farmer,* his first article being on the housefly, his second on the May beetle, and his third, in 1863, which was illustrated, described the egg and larval stages of the Colorado potato beetle.

In Chicago he had a difficult time finding suitable employment; he tried his hand at a number of jobs, including cigar-making, pork-packing, and drawing portraits of fellow boarders. He finally obtained employment with the *Evening Journal* and later with the *Prairie Farmer,* where he was at first a reporter and an illustrator and later editor of the entomology department. After a six months' stint during the Civil War as a private in the 134th Illinois Volunteers in Kentucky and Tennessee, Riley rejoined the *Prairie Farmer* as entomological editor. During the next five years he wrote almost two hundred articles, mostly entomological, for this and other farm journals.

Goode reports that, "Before entering the army he had made the acquaintance of the man whom he joined in establishing the *American Entomologist.* This friend, who was senior editor until his death, was Dr. Benjamin D. Walsh, State Entomologist of Illinois, and it was Walsh to whom Riley always alluded as his master and the man to whom he was most indebted for his early training and inspiration."

In 1868, when he was twenty-five years old, in cooperation with Walsh, Riley started the *American Entomologist.* Howard (1930), referring to this magazine, says, "Of late I have been reading through these two volumes, and I believe that few people have estimated the extraordinary influence that Walsh must have had over Riley in those early days. Remember that Riley had no university training but that he had a genius for illustration and undoubtedly a genius for observation and study. It was an enormous advantage for him to be associated with Walsh, and it is to his credit that he took advantage of this association, and that he adopted, in so far as he could, his ideas and his ways of thinking, and also to a certain extent, his way of writing. In those days he must have been a tremendous worker, and he must have been filled with enthusiasm. . . .

"One does not wonder at the impression Walsh made on Riley. Contrast the education and the temperaments of the two men, and their relative ages. Walsh was a man of the highest education that England could give him; he had a keen mind, and was a master of style. He had a sense of humor denied (according to American tradition) to most Englishmen. It is no wonder that the young, impressionable Riley should have formed one of those great enthusiasms of youth for such a character, and it is perhaps no wonder that the inspiration of this collaboration went a long way to make him the strong entomologist that he undoubtedly was."

In 1868 Riley was appointed State Entomologist of Missouri on the recommendation of Walsh and N. J. Colman of *Colman's Rural World.* Howard (1930) says, "He held this position for nine years and published a series of annual reports which in many ways were far ahead of anything of the sort that had been done in this country. His illustrations were much better than any that had been printed before; the accounts of the different insects were based on careful study; the type was large and the print good, and both the farmers and the entomologists were treated to a kind of document that, taken in all, was unequaled. The previously published reports by Glover, Fitch, and Walsh, with their small type and inferior illustrations, suffered greatly in comparison, not so much in subject matter as in readability and attractiveness." These reports had illustrations of complete life histories, most of them done by Riley.

Meiners (1943) says that "Riley intended his reports primarily for the farmer and every effort was made to keep style and language within the understanding of his readers." Among other things, Riley recommended, in vain, the closing of the Canadian border to imports of potatoes from the United States to prevent the entry of the Colorado potato beetle. He also noted that in Europe the wheat midge was controlled by the parasitic chalcids, and he recommended that these be brought into the United States.

According to Meiners, "Riley had no college training in entomology. This subject did not appear in the curriculum of any of the colleges of his day. One must, therefore, marvel at his thorough knowledge of the subject. The accuracy of his description of insects and their life histories have, in most instances, stood the test of time. He probably studied Kirby and Spence's 'Introduction to Entomology,' Harris' 'Insects Injurious to Vegetation,' and Westwood's 'Introduction to the Classification of Insects,' which gave him the technical knowledge necessary to write on the subject, but one great textbook that he studied and knew better than all others and which was the key to his whole success, was Nature herself."

While State Entomologist of Missouri, Riley lectured once a week on entomology at the University of Missouri and at the newly founded Agricultural and Mechanical College. According to Meiners, "Because of this university connection he soon became known as Professor Riley, a title that remained with him for the rest of his life and the one by which he was best known." From 1870 to 1872 he also lectured on entomology at Kansas State College, which still has some excellent wall charts drawn by him.

While in Missouri, Riley became interested in the outbreak of the grapevine *Phylloxera* that was ravaging the vineyards of France and other wine-making countries in Europe. He recommended the use of American vines as the root stock for the susceptible European vine. This largely solved the problem, and he was awarded honors and a gold medal by the French government.

Riley received an honorary M.A. degree from Kansas State in 1872, an honorary Ph.D. from the University of Missouri in 1873, and served as president of the Academy of Science of St. Louis in 1876 and 1877.

All this time, Riley, with the help of Otto Lugger, Theodore Pergande, and others, was making brilliant contributions to the knowledge of the biology of insects. He studied, among other subjects, the life cycle of the thirteen- and seventeen-year races of the periodical cicadas, the remarkable relationship between the yucca moth and its pollination of the yucca flower, the life histories of blister beetles and their primitive triungulin larvae, and the caprification of the fig.

Like Walsh, he was an ardent supporter of Darwin and published his views on evolution in a number of papers and periodicals. Darwin wrote Riley in 1871 that in his annual reports "There is a vast number of facts and generalizations of value to me, and I am struck with admiration at your power of observation. The discussion on mimetic insects seems to me particularly good and original." Riley also corresponded, as a friend, with Alfred Russell Wallace, Herbert Spencer, Henry Bates, and other eminent men.

In the years 1874–1876 many of the Western and mid-Western states were devastated by plagues of locusts, the so-called Colorado grasshoppers or Rocky Mountain locusts. In some states their destruction of crops was so serious as to cause wholesale starvation among many pioneer families. Riley studied the Rocky Mountain locust, in Kansas, made many original observations, which he published in his annual reports for the State of Missouri from 1874 through 1876, and proposed that a national commission be appointed to study the locust. In view of the urgency of the problem, in 1876 Congress appropriated $18,000 for a commission of three men to investigate the Rocky Mountain locust. These were to be under the guidance of the Director of the Geological Survey of the U.S. Department of the Interior. Riley, then thirty-four, was appointed Chairman of the Commission, A. S. Packard, Jr., secretary, and Cyrus Thomas, treasurer. All three were well-known entomologists. Thomas was a specialist on grasshoppers and State Entomologist of Illinois; Packard was the well-known author of a *Guide to the Study of Insects.*

In addition to the $18,000 appropriated in 1876, another $20,000 was appropriated for 1878 and 1879, or a total of $38,000 in all. The Commission published five volumes and seven bulletins. The first two volumes were on the migratory locusts, the third on a variety of injurious insects, the fourth on the cotton-leaf worm and the bollworm or corn earworm, and the fifth on forest insects. The bulletins covered a diversity of pests, besides those named. In the volumes on locusts, Riley recommended their use as food. He gave some recipes for their preparation, and he was able to vouch for their tastiness because he had sampled them.

One year after the Commission undertook its duties, the locust plague largely disappeared. The public attributed the disappearance, at least in part, to the activities of the Commission. It seems, however, that Nature had co-operated. In any event, Riley's reputation as well as that of the other members of the Commission zoomed, and in 1878 Riley was appointed Entomologist to the United States Department of Agriculture succeeding Townend Glover, who had resigned because of illness.

Riley served only nine months in this position. He resigned in a fit of temper early in 1879, largely because of a feud with his superior, who resented Riley's bypassing him to approach Congress directly for increased appropriations. Professor John Henry Comstock, of Cornell University, who had been employed the summer before by Riley to study the cotton caterpillar in the South, took a two-year leave of absence from Cornell and succeeded Riley.

L. O. Howard, who became an assistant to Riley in 1878, was not too sorry to see Riley replaced by Comstock. Howard (1930) says, "This was a very pleasant change for me, and the atmosphere of the office became distinctly more agreeable. Pergande and I stayed on, but [E. A.] Schwarz went with Riley, who opened offices in his house at the corner of Thirteenth and R Streets, where they continued the work of the United States Entomological Commission."

In 1881, upon the assassination of President Garfield, Riley was reappointed Entomologist of the Department of Agriculture, and Comstock returned to Cornell to continue his distinguished career. Riley brought E. A. Schwarz, B. P. Mann, and Dr. W. S. Barnard with him to the Department. Pergande and Howard were retained, the latter only after some deliberation by Riley. In 1881 a division of entomology was created in the United States Department of Agriculture, which became in 1904 the Bureau of Entomology, in 1934 the Bureau of Entomology and Plant Quarantine, in 1953 the Entomology Research Branch, and in 1957 the Entomology Research Division.

Howard (1930) writes, "Then followed thirteen years of slow but steady growth and accomplishment. The conditions of work in the entomological organization were not of the pleasantest kind, owing to the personal characteristics of the chief. He was a restless, ambitious man, a great schemer, and striving constantly to make his work appear more important. He was ambitious to build up a large organization. Unfortunately, he made many enemies. Some of these were in the Department of Agriculture itself while others were in Congress."

During this time, Riley became the first to recommend the establishment of the Office of Experiment Stations. These Experiment Stations played a phenomenal part in the advance of scientific agriculture in the United States. He also advocated the establishment of a branch of economic ornithology, which later became the Bureau of Biological Survey and finally evolved into the Fish and Wildlife Service.

The Entomologist and his assistants of course were primarily concerned with insects of economic importance, a few of the more notorious being the chinch bug, the Hessian fly, the codling moth, the plum curculio, the hop aphis, and several destructive grasshoppers.

One of Riley's greatest triumphs was his initiation of efforts to collect parasites and predators of the cottony-cushion scale, which was destroying the citrus industry in California. He sent Albert Koebele to Australia, and in 1888 Koebele collected a predator of the scale known as the Vedalia beetle, *Vedalia cardinalis,* now *Rodolia.* These beetles were introduced into California and carefully cultured by D. W. Coquillet. The success of the Vedalia in wiping out the ravages of the cottony-cushion scale was almost miraculous and gave great impetus to the study of biological control for the reduction of injurious pests. However, Howard (1933) notes one unwanted consequence of the Vedalia success in 1888 and 1889. "In fact it is safe to say that progress in the battle against injurious insects on the western coast of the United States was set back for ten years or more on account of the supreme reliance on this method of fighting pests and the consequent abandoning of every other means and every other line of research."

During Riley's regime progress was made in the development of arsenical and pyrethrum insecticides, kerosene emulsions, and the use of hydrocyanic gas (HCN) as a fumigant. Dr. Barnard invented the Cyclone Spray Nozzle, afterwards called the Riley Nozzle.

Howard (1930) notes that Riley deserves great credit for the progress made in economic entomology during his Federal service. However, all was not well in Riley's department. "Intimately associated with all of this work as I was, for he soon made me his principal assistant, I can find myself admiring some of his qualities very greatly; but at the same time there was during the entire period an amount of dissatisfaction and of unrest and of unfair treatment of subordinates that made the period anything but happy. He quarreled with A. J. Cook over the invention of the kerosene emulsion; he quarreled (and quite justly) with the California people over credit for the introduction of the Australian ladybird (see Doutt, 1958); and he quarreled with many other persons and organizations about many different things. It is all past now, but I cannot help thinking that much more would have resulted if he had encouraged independent work on the part of his associates and if he could have exercised more tact in many ways. It must be said that for a number of years before his accidental death he was in very bad health. He had frequent headaches, and was troubled with insomnia. This accounts for much of his restlessness. He could sleep on a long railway journey, and could not sleep in his bed at home. Moreover, he found that he could sleep in a barber's chair better than he could in his own bed. It seems not to have occurred to him to install a barber's chair in his house, but after a sleepless night he would often go to his barber and pay by the hour for a chance to make up lost sleep. All this must have affected his disposition seriously, and must have accounted for some of his rough angles."

Riley was undoubtedly a hard-driving, extremely ambitious man who accomplished much. How much greater a man of such ability would have been if he had given a little credit to his assistants for the part they played in his successes. Howard (1933) writes, "Professionally I hadn't much chance. In those days it was the custom of scientific men to take all the credit for work done by their assistants. This was considered quite ethical, in fact, the proper thing to do. The assistants accepted the situation, not because it was right, but because there was nothing they could do about it. I seldom looked at a scientific paper published during that period without wondering who really wrote it." For example, Otto Lugger, Theodore Pergande, and William Macwitz, as well as others helped Riley in the studies for his nine Missouri reports. They received little or no credit for their efforts.

Riley had always been sensitive to criticism, and the entomological periodicals of his day contain examples of his feuds with various entomologists. Although there were very few who questioned his ability as an entomologist, in other respects he was not held in such great esteem. For example, one entomologist not too long after Riley died wrote, "He had been an actor and seemed to think he was the all in entomology (and he was good—see his 2-vol. Missouri reports). He could not give credit to anyone and when some young upstart stood up and explained his findings, Riley was immediately on his feet and expounded his findings 'I have long suspected that to be the case.' For several generations when you wanted to *fry* anyone down to size you repeated this quote—they got it."

Howard (1933) writes "that he travelled a great deal, leaving me in charge at Washington during his many and frequently long absences. Gradually he allowed me to publish shorter articles under my own signature, and in 1889 he was able to begin the publication of a periodical journal known as *Insect Life*." Six volumes were issued under the joint editorship of Riley and Howard.

If Riley was at times a hard man to work for, he also had his good moments, as for example at meetings of the Entomological Society of

Washington (Howard *et al.,* 1895): "Enough has not been said of Dr. Riley's social qualities. Away from his work he was the most approachable and genial of men, and this side of his character reached its highest plane perhaps at the meetings of this Society. At the close of our stated program it is our custom to spend an hour in general conversation, and here one saw one of the most delightful phases of his character. Official cares were thrown aside and all the geniality of his nature came to the front. It is probable that the picture of Riley which will last longest in the minds of most of us will show him seated at the head of his own hospitable table at some one of the many meetings of the Society held at his house, discussing in his versatile way almost any subject from politics to ethics, his face beaming with animation and good humor." He and Howard were among the founders of the American Association of Economic Entomologists, which in 1953 became part of the largest entomological society in the world, the Entomological Society of America.

Besides his entomological pursuits, Riley had a little time, especially in his later years, to pursue some hobbies. He was a very accomplished horticulturist, and the garden of his home was one of the most beautiful in the city. He relaxed by drawing and painting, and his portraits in chalk and water color revealed his great talent for the fine arts.

By 1893, his strength, nerves, and energy were exhausted by his years of incessant industry, the responsibilities of his office, and the corrosive politics of his time, and he wanted a change. When, that year, a professorship in entomology was open at Oxford University, he made strenuous efforts to get it, and failed. And in 1894, frustrated and disappointed at being bypassed for the position of Assistant Secretary of Agriculture, he again resigned from the Department of Agriculture. Although his salary was relatively small as Chief in Entomology (his title at the time), being only $2,500 per year, he had no financial worries because he had married into a family of means.

After many years of active collecting, Riley had acquired over a hundred and fifteen thousand mounted insect specimens consisting of about twenty thousand species. He also had an alcohol collection of 2,850 vials and 3,000 slides of minute insects in Canada balsam. He gave part of this to the United States National Museum in 1882. He was then made honorary Curator of Insects. In 1885 he was appointed Assistant Curator of the Museum, thus becoming the Museum's first Curator of Insects, whereupon he gave the Museum his entire entomology collection. This was the nucleus for what is now one of the largest collections of insects in the world. Riley's specimens were in excellent condition and well-mounted because of Lugger's and Pergande's expert care.

Riley planned to study during his retirement the insects in the

National Museum and to devote his time to special problems which he had been unable to consider in the past. Moreover, he intended to collect his notes and write a textbook on entomology. But on September 14, 1895, as he was riding rapidly on his bicycle, the wheel struck a granite paving block dropped by a wagon. He catapulted to the pavement and his skull was broken. He never recovered. He was fifty-two when he died, leaving his wife with five children. The record of his accomplishments as well as the man himself will be long remembered by entomologists everywhere.

LELAND OSSIAN HOWARD (1857–1950)

Every student of the history of entomology in the United States is indebted to Howard for his absorbing book, *A History of Applied Entomology (Somewhat Anecdotal)*, and for his informative obituaries of entomologists in a variety of journals. However, J. R. de la Torre-Bueno (1948) gives other reasons why entomologists hold Howard in such high regard.

Howard was a man of wit, a raconteur, a diplomat, an accomplished administrator, and a great economic entomologist. He was largely responsible for making the Bureau of Entomology an important part of the United States Department of Agriculture. He was liberal and just, and a friend to his subordinates. He always encouraged independent work and publication on the part of his staff and did not sign his name to the work of others. He has told the story of entomology and ento-

Figure 19. L. O. Howard

mologists in three books. According to Torre-Bueno, "the real enjoyment of these stories is in listening to him telling them with joy and a dry wit."

Howard was born in Rockford, Illinois, on June 11, 1857. His father, O. G. Howard, was a lawyer, and his mother was the former Lucy Dunham Thurber, both of Delhi, New York. His family, after a few years, moved to Ithaca, New York, from Rockford, Illinois, and it is in Ithaca that he grew up. The beautiful country around Ithaca stimulated his boyhood desire to collect things. He was encouraged in his interests by his parents, who bought him Mary Treat's *The Butterfly Hunters.* When he was ten years old his father gave him T. W. Harris' *Insects Injurious to Vegetation,* with the Sonrel engravings, which young Howard found an exciting and stimulating book. When he was fifteen his father died, and his mother was left with three young children, George, ten, Frank, six, and Leland.

Howard was educated in private schools in Ithaca and entered Cornell University after passing an entrance examination. He started a course in civil engineering because his mother's professorial friends told her there was no future in the study of natural history. His career in civil engineering very shortly foundered on the hidden reefs of differential calculus, so he switched to natural history without telling his mother. "I was J. H. Comstock's first laboratory student, and I spent many hours in his laboratory. I studied botany, geology, and chemistry, and also went in for history—perhaps especially—the French, Italian and German languages.

"I had first met Comstock before entering college. I was out collecting and for the first time had found the very beautiful Huntera butterfly, flying rather abundantly in a field of blossoming buckwheat. A strange young man walked up and said, 'C-c-catching insects?' I replied affirmatively, and he said, 'M-m-my name is Comstock, and I t-t-teach entomology in college. C-c-c-come and see me.' This was the beginning of a life-long friendship. I went to see him, gave him insects for his newly started collection, and read his books, for the first time making the acquaintance of [P.] Lyonnet, [R. A. F. de] Réamur, [J. O.] Westwood, [W.] Kirby and [W.] Spence, [J.] Rennie, John Curtiss, Fitch, Walsh, and a lot of others; for Comstock was already beginning a library in entomology which has since become one of the most important in America." (Howard, 1933.)

Howard graduated from Cornell University in 1877 without having been exposed to applied entomology, because this science was just coming into its own. As a boy, in 1871 or 1872, he had heard Riley deliver two or three lectures on entomology and had been greatly disappointed by its emphasis on agriculture. However, he was not disappointed in Riley's ability as a blackboard artist.

For his thesis, Howard studied the respiratory system of the hellgrammite, the larva of the dobson fly. Apparently he considered the study of entomology and related courses in natural history as a kind of preparation for the study of medicine, and, in fact, did some graduate work in premedical courses. In the meantime, however, Comstock recommended Howard as an assistant to Riley, and Howard (1933) reports: "I must say, however, that the practical importance of entomology did not appeal to me, and I studied insects simply as fascinating forms of life. Of course I was not alone in this, and no one at that time had the faintest idea of what insects were then doing to the human race, and still less of what they were likely to do in the future. In fact, when in the early summer of 1878 Professor Riley began to correspond with me about going to Washington I looked upon this suggested post as simply an agreeable way of earning a living until I could go into something bigger and broader."

Riley was about thirty-five when Howard first began to work for him. To Howard's disappointment, Riley used him as a clerk; Howard even took dictation in a bastard shorthand and wrote letters in longhand, for the typewriter was just beginning to come into general use.

Later, Howard compiled a manual on silk culture from French and Italian texts, which was published under Riley's name. This was standard procedure for many years. It can be said that when Howard eventually became Chief Entomologist this practice was discarded.

When Riley was succeeded by Professor Comstock, Howard found his former professor a welcome change from Riley, as we have said. Comstock brought his recently acquired wife with him. Because of the payments they were making on a house in Ithaca, she worked too, as a clerk in the Department, and they lived very frugally.

Many years later Mrs. Anna Botsford Comstock (1953), who achieved fame in her own right both as a teacher of natural history and as an illustrator, wrote of their days in Washington as follows: "Too often some good play would come to one of the theaters during the last week of the month, when we were all reduced financially to the price of our meal tickets. Under such distressing circumstances, Howard always came to the rescue: He had inherited a watch which he did not use; he would take the watch to a pawn shop, get a six-dollar loan upon it, and we would all go to the play, enjoying it all the more because of the way we achieved the wherewithal. On the first of the month we paid up and the watch was redeemed and laid aside for the next emergency."

Of his work with Comstock, Howard (1933) wrote, "We worked very ardently, and I wrote many papers, some of them based on careful, original investigation. Much to my discomfort, Comstock cheerfully assumed the authorship of these papers, and I entered no protest. I had worked up two of them as a basis for a thesis for a Master's degree at Cornell, but when Professor Comstock asked me to let him publish them in his Annual Report, I consented." In 1883 Howard received his Master's from Cornell for his thesis, "The Morphology of Chalcididae." And it should be noted that after Professor Comstock returned to Cornell he was very careful to credit his colleagues and students when they collaborated with him.

Comstock became ill in Washington, so that it was necessary for him to seek a warmer climate. He chose Florida and there began his famous studies on scale insects. These he shipped to Howard in Washington, who then began to study the parasites of scale insects. In 1880 Howard published his first long paper under his own name, "Report on the Parasites of the Coccidae in the Collection of the Department of Agriculture." About this time also, Howard began to make some of his first extensive field trips, to Virginia to help fight the army worm in timothy grass, and to New Orleans to investigate sugar-cane borers.

With the change in administration in 1881, Comstock was replaced by Riley, and in the political reshuffling that followed his return, Howard expected to be fired. However, Riley kept him on and sent him on three field trips, one after another, to discourage him and to get him out of the office. On one trip he studied the rice insects on some of the sea islands off the coast of Georgia. Howard worked under Riley for the next thirteen years, from 1881 to 1894, eventually, as we have seen, becoming his first assistant. Through his own efforts and those of others, he brought in parasites of the brown-tail moth, the gypsy moth, the European corn borer, and the Japanese beetle and became an expert taxonomist of certain parasitic Hymenoptera, the Ichneumonidae, the Braconidae, and Proctotrupoidea.

Insects were by no means Howard's only interest in life. He belonged to a bicycle club that took outings on tall bicycles. He attended the theater and opera and frequented German beer gardens and music clubs. In fact, he was something of a musician himself, having sung in his college glee club and played the bass viol in an orchestra, and it was in a choral society in Washington that he met a Miss Dora C. Clifton, whom he married in 1886.

Howard at this time was studying medicine at night, but, as he says, marriage and the malaria he had picked up proved too much for him, and he eventually dropped his medical studies though not his interest in medicine.

For some reason or other Riley was never entirely satisfied with Howard, and Howard learned later that Riley had offered his job to S. W. Williston, the well-known dipterist and, later, notable paleontologist. Williston declined, apparently being well aware of Riley's temperament. It is clear that Riley and Howard were not entirely compatible; for example, E. A. Schwarz (1929) in a letter of September 20, 1894, to H. G. Hubbard states, "Prof. Riley is said to have returned from Europe but has not yet made his appearance at the Department. Upon hearing of his return, Mr. Howard has escaped into the mountains of New York and will be absent until October 1st."

When, in this same year, Riley resigned again, Howard (1933) decided that "Here apparently was my chance. But the authorities, knowing that I had been what is termed a 'loyal assistant,' feared that I might be a difficult person, and so they sent out letters to James Fletcher of Canada, S. H. Scudder of Boston, and to A. J. Cook of Michigan, asking for suggestions as to the most competent man to succeed to the office." They recommended Howard. He was made Chief of the Division in June 1894.

Riley and Howard were among the founders of the Entomological Society of Washington in 1884. What the Society lacked in numbers it made up in the quality of its membership. Howard (1909) wrote of it:

> In those early days entomology and beer went together. There were good reasons for this. Marx, Schwarz, Heidemann, Pergande, Lugger, Schoenborn, Ulke, were all Teutons, as their names indicate: John B. Smith's real name is Johann Schmidt. At that period the German university idea dominated scientific America. There were thousands of young Americans in the German universities each year. The American university students adopted in part the customs of the German university students.
>
> The after meetings of the Entomological Society were interesting. The conversation was good; the refreshments were unlimited in quantity but limited in kind; you could have light beer or dark beer, and that was about the extent of the variation. It was my custom to order two cases of beer, each of 24 bottles, for an average attendance of 7 or 8, and I always made the arrangement with the grocer to return those bottles which were not empty, as well as the empty ones, but it soon became a standing joke between us that it was unnecessary to make any provision concerning the unempty bottles. I am not sure that this custom, which no longer holds, was a good one. I am not sure that it was a very bad one. So far as I know, it never seriously affected the health of any of the members, but on the whole perhaps it was unfortunate and I am inclined to believe that the present method is best. I should dislike to see some of the younger members of the Society drink as much beer as some of us did at their ages, and while I would not vote the prohibition ticket as Banks does, I believe that Banks was about right when the Society met at his house for the first time and he gave us hot lemonade and cold lemonade and some very excellent raisin cake. It is true that a few glasses of beer will make a stupid remark sound witty, but there was no necessity for such stimulus to the imagination in the old days, because all of the remarks were witty.

Howard (1933) reports that in 1881 Dr. A. F. A. King published an essay in which he theorized that malaria was a mosquito-borne disease. He came to Riley's office to discuss this. "To my eternal shame, I confess that we both pooh-poohed the idea, and as a matter of fact, no one took any stock in it until Ross's triumphant demonstration in Calcutta seventeen years later."

Even before Ross's discovery, Howard had been interested in mosquitoes. As a boy he had found that kerosene applied to the surface of a bucket of rainwater killed the mosquito wrigglers. In 1892 studies conducted by him in the Catskills reaffirmed this finding. He determined the life histories of a number of mosquitoes, including that of *Anopheles quadrimaculatus* Say before it was known to be a carrier of malaria. Dr. Walter Reed, while undertaking the famous studies which established the role of *Aedes aegypti* as a carrier of yellow-fever, constantly corresponded with Howard about mosquitoes.

In 1901 Howard produced a book on mosquitoes entitled, *Mosquitoes, How They Live, How They Are Classified, and How They May Be Destroyed.* With Dr. H. G. Dyar and F. Knab he was coauthor of a four-volume work, *The Mosquitoes of North and Central America and the West Indies,* two volumes of which appeared in 1912, and two in 1917. According to Howard (1933), "Dr. Dyar was fortunately a man of some wealth, and he financed several expeditions to regions that had previously been unknown to us. . . . Knab, bright fellow, talented investigator and clever artist that he was, died in 1918 from a disease contracted, curiously enough, from the bite of an insect in Brazil years before."

Howard found some hitherto unexpected uses for mosquitoes:

> This reminds me of an interesting thing that occurred while I was still in Texas. I arrived at San Antonio late one night and went to my old hotel, the Menger, only to find that there was a big convention in town and that all the rooms were occupied. The room clerk, an old acquaintance, said that he could give me a cot in a room with three other men. I accepted and went up with my luggage. As I was preparing for bed (there being no one else in the room at the time), I heard footsteps in the hall, the door opened and in came the other three men. I pulled out my hand lens, knelt on my cot and began to examine a spot on the wall.
>
> "What are you doing?" said one of my roommates.
>
> "I have just found a yellow fever mosquito," I said.
>
> "My God!" said one of them, and they collected their luggage and left, leaving the whole room to me.

Howard was interested in other insect vectors of disease than mosquitoes. He greatly publicized the disease-carrying propensities of the housefly and wrote a book which was published in 1911 and entitled *The Housefly—Disease Carrier.*

During his stewardship of the Entomological Service the public became more and more conscious of the depredations of insects, especially the San Jose scale, the gypsy moth on forest trees, and the cotton boll weevil. As insects began to assume a more prominent role in the public eye, Congress and the states appropriated more money for their control, and the entomologist and the science of entomology became increasingly prominent.

While all this was taking place, Howard made numerous trips to Europe, first to find parasites of the gypsy moth and later to attend meetings and for a variety of reasons. With his gift for languages and natural amiability, he soon became well acquainted with many of the entomologists in Europe.

As a result of his contributions to medical entomology, Howard was awarded a degree by George Washington University, of which he was especially proud, an M.D. *honoris causa.* Later he received honorary doctorates from Georgetown, Pittsburgh, California, Toronto, and Rutgers. In 1894, he was elected President of the American Association of Economic Entomologists; in August 1898 he was elected permanent secretary of the American Association for the Advancement of Science, an office which he held for twenty-two years. This permitted him to meet scientists from all over the United States and from abroad, and it also augmented his relatively small income.

He was an active member of the Cosmos Club of Washington, which had many distinguished members, and Howard often regarded this club as his "second home." Here he was admired for his wit as well as for his ability with the billiard cue.

On October 1, 1927, after almost fifty years of service, Howard retired as Chief of the Bureau of Entomology. The retirement dinner in his honor was so well attended that there was no room for the women. Not to be outdone, the ladies "gave him another dinner at which he was the only man." He was able to stay on for another four years as principal entomologist. He writes (1933), "The whole period of my official career was full of interest, and I led a thoroughly happy life. I was hardly conscious that I was working, since I was doing the things that I most wanted to do. I saw a great service built up, and as head of this service for more than thirty-three years many honors came to me."

Mrs. Comstock (1953) wrote of the days Dr. Comstock and she were in Washington with Howard: "Those who have had the pleasure of knowing Leland Howard will understand how much his society meant to us in those days. He was one of the most interesting men in Washington, which is saying much, and he had the power of winning admiration and affection to a remarkable degree. His wit has been the prized alleviation of after dinner speeches for three-score years."

For the next nineteen years after Howard's retirement he continued to live in Washington; then he moved to France, and later returned to spend his last years in Bronxville, New York, where he suffered from poor eyesight and an ailing back. He died on May 1, 1950, when he was almost ninety-three years of age.

Besides his books on mosquitoes and flies, Dr. Howard was author of *The Insect Book* (1901), the previously mentioned *A History of Applied Entomology* (1930), *The Insect Menace* (1931), and *Fighting the Insects: The Story of an Entomologist* (1933).

Howard took great pains to encourage young entomologists. In this respect J. J. Davis (1958) writes that it was a regular procedure for Howard to write letters to them whenever they published on a worthwhile project.

We can conclude our remarks about this notable entomologist with Torre-Bueno's (1948) note: "Personally, he is rather short with quite a bald head and a charming crooked smile. There were other entomologists of great attainments during his active service, but none had so powerful an impact on world-wide study of harmful insects, not alone on this country but likewise in Europe, perhaps to a greater degree than here."

Osborn (1937), Walton and Bishopp (1937), Gahan *et al.* (1950), Anon. (1950), Clark (1952), and Bishopp (1957) are some of the other sources of these biographical notes on Howard.

CHARLES LESTER MARLATT (1863–1954)

It was fortunate for entomology, particularly economic entomology, that so forceful and competent an individual as C. L. Marlatt succeeded Howard as Chief of the Bureau of Entomology. It is true that Townend Glover, F. M. Webster, and probably others urged that steps be taken to prevent the introduction of injurious pests into the United States, but it was largely due to the intelligent and persistent efforts of Marlatt that the Plant Quarantine Act of 1912 was finally passed.

An entomologist who knew Marlatt well and who asked me not to use his name, wrote that

> My own personal relations with Dr. Marlatt were most friendly and I frequently sought his advice and dealt with him extensively on quarantine matters when he was Chief of the Federal Horticultural Board which he and Karl Kellerman dominated.
>
> Marlatt was an excellent administrator, probably the best entomology as a unit ever had.

Figure 20. C. L. Marlatt

He was apparently wealthy, had a house on 16th Street well designed for entertaining, in a then fashionable neighborhood; drove an electric brougham; traveled extensively; spent part of each summer in the north.

He played golf and had several favorites from among the entomologists with whom he usually played once a week.

He entertained the Washington Entomological Society occasionally in the early days at his home, but I don't now recall any entertainment after he became Chief.

He was decisive and brusk and even might have been considered arrogant. Just the opposite of Howard and much like Lee Strong. He was not gregarious but quite selective in his associates. He was not friendly but not unfriendly and had a wry sense of humor.

He was tall and spare not thin; he wore tailored clothes, fine shoes and hats and was most precise in his appearance, never sporty or sloppy in the least. For instance his trousers had side pockets that opened at the top like sailors' trousers and like all old time trousers; the colors dark brown and subdued banker mixtures—carefully creased hat in winter and a fine Panama in summer . . .

Actually my relations were primarily business relations so many of my impressions possibly would have undergone change had I known him socially. I think few entomologists knew him socially.

In an undated letter to Professor Herbert Osborn (1937), Dr. Marlatt describes his juvenile efforts in the field of entomology:

My earliest recollection, probably dating back upwards of sixty years, was of the time at Manhattan (Kansas) when Professor Riley, at that time entomologist of the State of Missouri, used to come to the college annually to give the course of lectures on insects harmful to horticulture and agriculture. These lectures were open to the public, and were well attended and frequently illustrated by the Professor by rapid offhand sketches, greatly adding to the interest. On one of these occasions, my brother and I and a neighbor boy were taken by the Professor on a collecting trip which was my induction into entomology—but hardly that because my mother was very much interested in entomology and botany and had, before my time, made a considerable collection of the insects of the region and was still very much interested in the subject which accounts for my being taken to the lectures.

I have only a very slight recollection of this trip, except that Professor Riley came under a large tree by the creek and began examining the leaves and became very much excited to find them coverd with galls. I know now that this tree was hackberry and the galls represented the small leaf galls which I myself studied later. Although his audience was a small one, in size and numbers, he immediately appealed to the "gallery" as was his habit and stamped around making various ejaculations of wonderment as to why he did not recognize the cause of this injury offhand, but finally, the probable determination apparently came to him and he expressed it rather volubly, indicating that he should have known it at once!

This, then, was the beginning of entomological instruction or education in connection with the Kansas State Agricultural College at Manhattan. Very possibly, there was some work done there prior to that time and my mother, who was a professor in the Bluemont College, which later became the nucleus for the State College, I think must undoubtedly have given some instruction in the field of natural history.

Undoubtedly Professor Popenoe was the first man to really start definite entomological work at the college and experiment station in any important way. His coming to the institution was about the same time as the beginning of my college experience, and, as you know, I later became assistant in entomology and horticulture. Popenoe was a mighty good collector, and had a wonderful knowledge of local insects, keeping himself in training so that he could identify all the common species collected by his students at sight, and thus giving the impression that his knowledge of insects was unlimited. The annual meetings of the State Academy of Sciences brought all science workers together, a considerable number of whom were entomologists, and it was at these meetings, which were attended by Popenoe and myself, among others, that I met Snow and several of his students, including Kellogg.

Marlatt received his B.S. from Kansas State in 1884, and an M.S. in 1886; most of his entomological studies were under the direction of E. A. Popenoe. In 1887 Marlatt was made Associate Entomologist at Kansas State College under Popenoe and illustrated a paper for him. The skill Marlatt showed in delineating insect figures attracted the eye of C. V. Riley, who brought him to Washington in 1889 as Assistant Entomologist and artist. Fortunately for Marlatt, Miss Lily Sullivan was hired as an entomological artist about the same time, and he was thus able to apply his talents to research.

In 1894 Marlatt became First Assistant Entomologist and Assistant Chief of the Division of Entomology, a position he held until 1925, when he became Associate Chief. According to Cory *et al.* (1955), "When Dr. Howard retired in 1927 Dr. Marlatt became Chief of the Bureau, adding the responsibility to his already heavy load as Chief of the Plant Quarantine and Control Administration. After serving approximately two years in this dual capacity, at his request he was relieved of the responsibility of the management of plant quarantine enforcement that he might devote full time to the administration of the Bureau of Entomology."

Marlatt devoted his marked abilities to a great variety of problems, publishing papers on insect oviposition, morphology, life history, hibernation, and control, and on a new species of sawflies. He also wrote popular and technical publications on insecticides and their application, as well as on cattle, wheat, shade and forest trees, and household pests. The periodical cicadas were of special interest to him; he wrote twenty papers on them.

During his long career, he became an expert on sawflies and scale insects. He described many species of sawflies that attacked food plants. With the inroads of the San Jose scale, he became very much interested in the Coccidae. In 1900 he assumed charge of the Government's collection of Coccidae and established new methods for organizing and keeping collections of scale insects. For this reason many collectors sent him type or cotype material for safekeeping. His work in coccid taxonomy was a definite advance in this field.

He wrote a number of papers on the destructive San Jose scale, which was introduced into the United States in the 1870's. In 1901–1902, largely at his own expense, he traveled in Japan and China in search of the native home of this insect pest. Years later, he assembled the notes on the trip made by him and his wife and wrote a very interesting book, *An Entomologist's Quest: The Story of the San Jose Scale* (1953).

When he was in Japan, the chirping crickets in their tiny cages both attracted and distracted him, and he wrote in this book:

> At this season of the year one's attention is attracted everywhere in Japan by the shrill singing of crickets, kept in tiny bamboo cages as we do song-birds. Passing through one of the streets at night, an immense din reached my ears which I at first was unable to explain. I later discovered that it came from the stand of a vendor of song crickets (Suzumuchi) where hundreds of insects of various species were singing noisily together, thus advertising the vendor's wares and business. The best singers represent two distinct species, one, the

common form, is something like our tree-cricket (*Oecanthus*), but having smoky-colored wings and white antennae. Its song is very shrill and is made with the wings held at right angles to the body and elevated perpendicularly, giving the appearance of a large sail or fan. The note is intermittent, rising slowly to a very shrill, piercing sound, and then stopping suddenly. The larger singing insect is a yellow-coated katydid, about two inches in length. The singing of the last insect is continuous for a considerable interval, and sounds like the rattle of an old sewing machine, and is very loud. I purchased two cages, one with the larger songster and the other containing three or four of the *Oecanthus*. The din made by these insects in my room was so great that I was compelled to put the cages out in the yard for the night to enable me to sleep, and at the end of my stay in Tokyo, I was very glad to give them to a more appreciative Japanese friend. Nearly every house in Japan at this season of the year will have a cage of these singing insects, and in going about the streets in the evening, one's ears are assailed by their song at every step. The crickets are caught in the open, but seem to be half-domesticated and take readily to confinement. They are fed on cucumbers and other green vegetables and on fruits, and will live for a month or six weeks, if not longer.

After a long careful search he determined the native home of the San Jose scale. "In China, the proof of its being a native of Northeastern Asia was convincing; in that it occurred on native fruit in regions where no importations of foreign plants had been made, and that in those regions the Scale was largely kept in control by natural enemies. The importation of plants from this region, as now known, brought the San Jose Scale to the United States." He suggested that the scale be called the Chinese scale, but the name, San Jose scale, was too well established.

As a result of this journey, Marlatt and other investigators introduced the predacious ladybird beetle, *Chilocorus similis* Rossi into the United States, where it serves as an important predator of the San Jose scale in those areas in which it has become established.

"The spread of the San Jose Scale into the Mississippi Valley and eastwards to the Atlantic Coast was a great stimulus to protective quarantine action between States, though too late to be of much value." Legislation had been introduced as early as 1897 to protect the growers from the invasion of imported pests, but it was years before it passed because no one followed up the progress of this legislation through Congress. On January 29, 1909, Marlatt, with the approval of Dr. L. O. Howard and the Secretary of Agriculture, submitted a bill to Congressman C. F. Scott. Although the bill passed the House without any great opposition, a lobby of Eastern nurserymen delayed its passage through the Senate for three years. (Marlatt, 1953.)

Later, a new bill written by Marlatt was started through Congress, but was badly defeated. "The few Congressmen who voted for the measure," wrote Marlatt, "were mostly from the Western States, which had been supporters of the Bill from the beginning. The opposition in the House, on the other hand, was very strong, and couched in what might be described as threatening or abusive language directed at the author of the Bill!

"The reaction of my Chief of Bureau, Dr. Howard, was: 'Now that you have had your fingers burned, drop the effort.' My reply was: 'A new Bill will be introduced tomorrow!'

"During the next year I made this promise good, by personal contacts with leaders of the opposition, with whom I had not previously been acquainted. I was able, through full explanations, to change the opposition of these leaders into strong support. The most useful new friend of the Bill was Mr. [J. R.] Mann of Chicago, the Floor Leader of the House." The Bill passed the House, and then the Senate, and was signed on August 20, 1912, by President Taft.

Marlatt was appointed head of the Federal Horticultural Board which administered the Plant Quarantine Act and served in this capacity from 1912 to 1929. During the four years that passage of the act was delayed, the following major pests invaded the country: the European corn borer, the Japanese and related Asiatic beetles, the Oriental fruit moth, and the citrus canker—all monuments to shortsightedness of selfish individuals.

Two eradication programs were successfully completed under his leadership, the Mediterranean fruit fly in Florida and the date scale in California. The eradication of the former species which first appeared in Florida in 1929, was almost at the "sunset" of his exciting and useful career. He accepted this challenge, determined that nothing short of eradication should be the goal. Displaying his usual ability to select the right man to do the job in the field, he appointed Dr. Wilmon Newell, Plant Commissioner of the State Plant Board of Florida, who was in complete accord with his eradication idea. These two made an unconquerable team and within 18 months from the date of its discovery the much feared pest was indeed eradicated from 1,022 properties in Florida, at a cost of seven and one-half million dollars. Had this program been led by men with less foresight and determination than was shown by Drs. Marlatt and Newell, this pest might have remained with us as a permanent unwelcome guest. It is men of this caliber that make us proud of our profession.

As an administrator and organizer he had no superior in the fields of entomology and plant quarantine. He was a firm believer in choosing of personnel with care, but after discussing the work to be performed in a general way with the appointee, he then gave the worker wide latitude in the exercise of judgement. This promoted loyalty and intense interest in the work. Further, he had the ability and the personality to bring the needs of entomological and plant quarantine work to his superiors and members of Congress. (Cory *et al.,* 1955.)

Dr. Marlatt received an honorary degree from his alma mater in 1922 and many other honors during his career. He was president of The Entomological Society of Washington in 1896–1897 and president of the American Association of Economic Entomologists in 1899. He died on March 3, 1954, at the age of ninety-one, leaving a wife and four daughters.

FRANCIS MARION WEBSTER (1849–1916)

Professor S. A. Forbes (1916) said of Francis Marion Webster, "His record as an entomologist is probably unparalleled in this country as an example of unusual success and usefulness against heavy initial handicaps."

In attempting to enlarge upon the information given in the brief obituaries for this entomologist, I wrote to one of his former associates, Professor J. J. Davis. Professor Davis replied that he was "on the staff of F. M. Webster from 1910 to 1919 [1916 is the correct date]. He was a kindly man. He was an ardent Mason and Doctor Howard used to make fun at him for marching in a parade of Templars with his uniform. He owned and operated a large farm in northern Illinois and was a self-made entomologist. Naturally he was a strong advocate of farm practices for insect control. He dealt with cereal and forage crop pests and repeatedly told me he had no trouble with such common pests as cutworms, white grubs, Hessian fly, etc. because he practiced good farm practices. This impressed me greatly, and when I became interested in

Figure 21. F. M. Webster

truck garden insects and later orchard problems, I found garden practices and orchard practices were important in pest control . . . Webster was a better entomologist than most now-a-day persons who claim to be entomologists."

Walton (1916) describes Webster as follows, "Personally, Professor Webster was genial in manner, frugal and abstemious in habit and extremely simple in tastes; of exceeding honesty; in speech most temperate and he had acquired a literary style that was at once direct, lucid and forceful. He was also a most practical man, possessing a broad knowledge of agricultural methods and was therefore enabled to see his scientific problems from the viewpoint of the farmer."

Walton says that Webster "evinced a tremendous interest in his work and was able through sheer force of character to transmit this quality to his entire staff of investigators, each one of whom was made to feel that his superior took a lively and intensely human interest not only in his work but also in him personally." The younger men who worked with him remembered him for "his kindly interest, generous viewpoint and sound advice. He evinced absolutely no trace of that petty jealousy regarding credits in the publication or results which mars the character of some otherwise truly big men of science. On the contrary, he was ever ready to sacrifice both time and labor in assisting his men in their efforts."

Webster was born in Lebanon, New Hampshire, on August 2, 1849. Forbes (1916) says his parents moved to DeKalb County in northern Illinois when he was four years old, and he passed his boyhood on a farm. "The death of his father when he was fifteen years old left him largely to his own resources, and he had little formal education. Marrying at twenty-one years of age, he supported himself by manual labor in the town of Sandwich for a few years, after which he bought a farm in his own county, and lived there for the eight years following." In the fall of 1881 he sought employment in the field of entomology. He wrote, "There are but two ways of becoming a naturalist, one to cheat yourself out of sleep and Sundays, which is the way I have been doing for ten years, and the other, getting scientific employment, as I wish to do now."

Even before he worked as an entomologist, he had collected beetles and had written articles on insects for the *Prairie Farmer* and other farm publications as well as for the *Bulletin of the Brooklyn Entomological Society* and other biological publications. In October 1881, when he was thirty-two, he was hired as an assistant in the Illinois State Laboratory of Natural History at Normal. He took with him his wife and children and his collection of 2,500 named species as well as the experience of a keen observer and a practical agriculturist.

From 1884 to 1892 he was a special field agent to the United States

Department of Agriculture and did some of his best research during that period under C. V. Riley. He studied grain pests intensively. From 1886 to 1890 he investigated the plagues of buffalo gnats in the valley of the lower Mississippi River and sought ways to suppress them. In 1888 he went with Koebele to Australia to find natural enemies of the cottony-cushion scale or citrus-fluted scale. It was on this trip that Koebele discovered the Vedalia beetle.

During his career, Webster was connected with the Agricultural Experiment Stations of Illinois, Indiana, and Ohio, and the first *Purdue Experiment Station Bulletin* was Webster's bulletin on the Hessian fly. In 1903 there appeared one of his major publications, *USDA Bulletin 42*, "Some Insects Attacking the Stems of Growing Wheat, Rye, Barley, and Oats, with Methods of Prevention and Suppression."

Webster joined the U.S. Bureau of Entomology in 1904 and in 1906 became head of the section of Cereal and Forage Crop Insects in the Bureau of Entomology; eventually he directed eighteen field stations, his headquarters for seven years being Purdue University in Lafayette, Indiana. Webster's ability as an entomologist was widely recognized and in 1897 he was chosen president of the American Association of Economic Entomologists. On January 2, 1916, he suddenly died of pneumonia, just a few days after he was chosen president of the Entomological Society of America.

Howard (1930) notes that he was a prolific writer with more than six hundred titles in publications of all kinds. "He discovered parthenogenesis, dimorphism, and alternation of generations in the genus Isosoma.

"Webster was instrumental in the calling of the first convention of a national horticultural quarantine law (1897). It was the beginning of the agitation which resulted in the passage of the Federal Horticultural Law in August, 1912, fifteen years after that first convention."

HENRY GUERNSEY HUBBARD (1850–1899)

In the short span of his life, H. G. Hubbard achieved a good measure of fame for a number of entomological accomplishments, but none greater than for his publication, *Insects Affecting the Orange*, which appeared in 1885. "It is doubtful whether the Department of Agriculture had ever published quite such an admirable report," wrote Howard (1930). "It was fully illustrated, and covered more than 200 pages. Looking it over today, one marvels. The writer knew his subject so well; he was so keen an observer, and so broad a thinker, and yet at the same

Figure 22. H. G. Hubbard

time he was so practical. It is a monumental publication, and today stands out among the publications of the Department. Hubbard knew his insects and he knew his crop. He knew the parasites of his insects, and he knew plant diseases. Moreover, he knew enough of chemistry and enough of machinery so that he was able to point out exactly what the orange grower could do and should do. Nothing at all equal to it has been published in any country."

Hubbard was born in Detroit, Michigan, on May 6, 1850. According to Schwarz *et al.* (1901), "His father, Bela P. Hubbard, a prominent citizen of Detroit, and a man of strong scientific tendencies, was one of the founders of the American Association for the Advancement of Science and was deeply interested in botany, forestry, arboriculture, and archaeology." It is not surprising that with such a background the son should be interested in natural history.

Hubbard attended private schools in Cambridge, Massachusetts, and studied under private tutors in Europe. In 1869 he entered Harvard, where he came under the influence of such outstanding scientists as Louis Agassiz, Asa Gray, and N. S. Shaler, the geologist. His interest in insects was encouraged by H. A. Hagen, who was on Harvard's staff, as well as by C. R. Osten Sacken, who was doing some voluntary work on the Diptera. While at Harvard, Hubbard met G. R. Crotch, the English coleopterist and lepidopterist, who died of consumption in 1874 at the age of thirty-two and E. A. Schwarz, the German coleopterist, who became his closest lifelong friend.

"During the winter of 1873–74," wrote J. B. Smith (1899),

Messrs. Hubbard and Schwarz systematically investigated the hibernating quarters of the Coleoptera near Cambridge, turning up species theretofore undreamed of, in utterly unheard of numbers. Mr. Hubbard's share of this became the property of the Cambridge Museum, and is known as the "Winter Collection." Hubbard graduated in 1873 and was a graduate student under Hagen in 1873 and 1874. Shortly after Agassiz died, Hubbard and Schwarz went to Detroit and started the famous collection which eventually went to the U.S. National Museum. They collected in and around Detroit in the summer and fall of 1874, and in Florida in 1875 and had a tremendous collection ready for the annual meeting of the American Association for the Advancement of Science in August, 1875. This meeting was held in Detroit under the presidency of Dr. John Lawrence LeConte, the noted coleopterist. Many noted entomologists attended the meeting and Messrs. LeConte, Scudder, Grote, Lintner, Osten Sacken, and Riley, and others were the guests of Mr. Bela Hubbard, the father of H. G. Hubbard.

The little outbuilding in the Hubbard grounds, containing the collection of insects, immediately became a centre of interest, the material there stored being unparalleled for wealth of specimens and with so many new forms that Dr. LeConte declared that it made it necessary to re-write part of his *Classification*. Here was formed that personal friendship with Dr. LeConte which lasted to the death of the latter, and no better correspondents did Dr. LeConte ever have than the Messrs. Hubbard and Schwarz.

In 1876 Hubbard and Schwarz made a series of expeditions to the region of Lake Superior. Hubbard later collected termites for Hagen in Jamaica. In 1879 he was appointed Naturalist to the Geological Survey in Kentucky, where he studied the fauna of the Mammoth and other caves. When two of his brothers drowned in a sailing accident in Florida Hubbard went to Crescent City to manage the property of one of them.

In 1880, after Riley saw the great progress Comstock had made in the study of scale insects and other citrus pests, he appointed Hubbard a Special Agent of the U.S. Entomological Commission. In 1881 Hubbard began an intensive study of citrus pests, and this resulted in his well-known bulletin, *Insects Affecting the Orange*. Among his great accomplishments was the perfection of a kerosene soap emulsion as a practical means of combating scale and other citrus pests. In time he became a great authority on subtropical horticulture. Hubbard and Schwarz together collected throughout Florida and in Hot Springs, Arkansas, in and around Yellowstone Park, and in the Pacific Northwest. In 1894, shortly before Riley resigned as Chief of the Division of Entomology, he and Hubbard went together to the British West Indies. In 1895 they studied the fauna of the Florida Land Tortoise or "Gopher." After the "great freeze" of 1896, which killed orange trees worth thousands of dollars, Hubbard investigated the ambrosia beetles that infested the dead and dying trees.

Although Hubbard had been connected with the U.S.D.A. for many years, it was mostly in a nominal position because of his ill health. He had long been afflicted with tuberculosis. In 1896, for the sake of his health, he went to the Lake Superior region and then to Arizona. Schwarz joined him in 1897 and 1898, and there they collected insects, particularly the fauna of the giant Saguaro cactus.

Howard *et al.* (1928), writing of the Hubbard-Schwarz partnership, says that shortly before Hubbard's death, "he and Schwarz went together into the dry climate of Arizona," where "they constantly carried on their entomological observations, and we like to think of the two together on their short excursions, preparing their own meals in the lower mountains; Hubbard so weak that he could only sit and study the things around him while Schwarz did the camp work. It was on such an excursion that Hubbard found the first queen of *Termes* in this country. He was that sort of a man and Schwarz was that sort of a man. Put either of them anywhere, under the most unfavorable circumstances, and they would see things that no one else ever saw, and their knowledge was such as to appreciate the importance or nonimportance of what they saw. Had Hubbard stayed with Schwarz down there, his life might have been prolonged, but he unwisely went to Detroit for a Christmas with his family and the end came soon." Hubbard died on January 19, 1899, leaving a wife and four children.

He was held in very high regard by his fellow entomologists as a collector, investigator, and a person of fine character. He was unselfish, kind, amiable, and unpretentious. Not only did he know insects, but he was very well acquainted with botany, horticulture, birds, and various animals. He was both a field entomologist and an applied entomologist of great ability.

After Hubbard's death, Schwarz continued to befriend Hubbard's children.

ANDREW DELMAR HOPKINS (1857–1948)

A. D. Hopkins was the first forest entomologist in the United States. He developed the principles of forest entomology, organized research and methods of control of forest insects, and studied their classification and biology in order to control them.

"Near the close of a hot Saturday afternoon late in June 1909," wrote S. A. Rohwer (1950),

a large attractive man completed, with the satisfaction of an expert, a game of "cow-boy" pool and joined the group of observers in the Cosmos Club. As he

Figure 23. A. D. Hopkins

approached his resemblance to the nationally known silver-tongued orator, William Jennings Bryan, was commented on by more than one of my associates: This was my introduction to Andrew Delmar Hopkins. It was the beginning of a friendship—sometimes strained by differences in opinion—that endured and improved through the years.

Dr. Hopkins was a keen observer, original in thinking and ideas, forceful, energetic, generous and tolerant. He was positive and somewhat aggressive in matters on which he had fixed opinions, but quick to recognize and accept such qualities in others. He had little patience for those who agreed to avoid an argument, but he expected those who did not agree to defend their views with facts. Where differences of opinion or interpretation existed he fairly pointed out his views and observations to those who disagreed with him. He was patient and thorough in his studies and expected the same from those who were associated with him. The keenness of his observations and his almost uncanny intuition, however, often caused others to consider him impatient—if not unreasonable. He appreciated, but did not worship, the value of academic training. Not infrequently he expressed the thought of the well known phrase "college is a place where pebbles are polished and diamonds are dimmed."

Although Hopkins had no formal university training and his only doctorate was an honorary one conferred on him by West Virginia University in 1893, he was an imaginative and practical entomologist. Dr. L. O. Howard says he was entomologist of the West Virginia Agricultural Station in Morgantown from 1890 to 1902 and was Professor of Economic Entomology in West Virginia University from 1896 to 1902. He became one of the two presidents of the American Association of Economic Entomologists in 1902, the year when the annual meeting was changed from midsummer to winter. He joined the United States Bureau of Entomology the same year and was placed in charge of the Section of Forest Entomology when it was established in 1904. He resigned as chief of the Section in 1923 and continued his work in the Bureau, doing pioneer research on bioclimatics, a subject that greatly interested him.

While head of the Section of Forest Entomology he wrote a monograph of those important enemies of pine trees, the pine beetles of the genus *Dendroctonus,* establishing important principles and practical control methods from their biology.

Of two important principles named after him, one is the Hopkins Host Selection principle, which states that the females of many different insects select for oviposition the same species of tree on which they and their ancestors were reared throughout many generations. The other is the Hopkins Bioclimatic law, which shows that biological events occur in the spring four days later for each degree of latitude northward, five degrees longitude eastward, or for every 400 feet above ground. Therefore, if the stage of insect or plant development is known for one locality, the stage of development for the same species in other areas can be estimated.

Dr. Hopkins' biographers give, unfortunately, only very brief glimpses of him. E. A. Schwarz, in a letter to Dr. L. O. Howard, writes of a collecting trip in Williams, Arizona, in which he remarks, "Our entomological collections are fast increasing. The insects affecting Coniferous trees are of course taking the front rank in interest and I wish Dr. Hopkins with his herculean strength and big axe were here to help us."

Dr. W. Dwight Pierce, in a personal communication to me, wrote, "A. D. Hopkins was across the hall from me for several years and I was very interested in his work on the Scolytid beetles, which I left alone because they were his field. He introduced anatomical studies in his work, of which I highly approved. He laid the foundations for weevil anatomy."

Some notable entomologists worked under Hopkins—for example, the taxonomists A. G. Böving, Carl Heinrich, W. S. Fisher, C. T. Greene, L. H. Weld, F. C. Champlain, and H. E. Kirk, and the forest insect specialists, J. M. Miller, F. P. Keen, and J. C. Evenden.

According to Rohwer, when Hopkins ceased his administrative work on forest insects in 1923, he "remained associated with the Bureau of Entomology for several years and worked on bioclimatics at his farm near Kanawha Station, West Virginia, with the assistance of the late M. A. Murray. Later his relation to the Department of Agriculture was transferred to the Office of Experiment Stations and the Forest Service. During this period he completed what he considered his most important publication, *Bioclimatics, a Science of Life and Climatic Relations* (Misc. Pub. No. 280, U.S. Department of Agriculture, January 1938).

Hopkins died at his home in Parkersburg, West Virginia, on September 22, 1948.

Besides the persons I have cited, Brooks (1925), Snyder and Miller (1949), and Stemple (1966) have prepared biographical and bibliographical material on Hopkins.

FRANK HURLBUT CHITTENDEN (1858–1929)

F. H. Chittenden was an economic entomologist who established his reputation through his studies of vegetable and stored-product pests. Perez Simmons tells us that Chittenden's work was the "starting point of stored-products entomology in this country."

Mrs. Doris H. Blake (1951) has written a revealing account of Chittenden, for whom she worked in the U.S. Bureau of Entomology for many years, and most of the following is from her portrait of him.

He was born in Cleveland, Ohio, on November 3, 1858. He grew up in the small town of Elyria, Ohio. His father died when he was quite young, and his mother supported her two children, Frank and his older sister, by teaching school.

Frank studied entomology at Cornell under W. J. Barnard while John Henry Comstock was with the U.S.D.A. in Washington, D.C. Apparently at some period during his school days he studied under Comstock because it is said that Comstock "offended him by requesting that he cease

Figure 24. F. H. Chittenden

whistling while he worked in the laboratory. However, Mrs. Comstock came to him privately and said 'I like your whistling, Frank, and I hope you won't stop.' " Always the individualist, Chittenden felt Barnard did not have a great deal to offer him and he did pretty much as he pleased. As a result he did not receive a degree, but a licentiate instead.

Upon leaving Cornell he was associated for a short time with the Brooklyn Museum. Here he was one of the founders of the Brooklyn Entomological Society and an editor of *Entomologica Americana.* In 1891 he joined the Bureau of Entomology in the U.S.D.A., undoubtedly influenced to make this move by L. O. Howard, one of his former classmates. He brought to Washington his mother, who lived to be ninety, and his sister. He never married.

In 1904 through the efforts of William Jacob Holland he was given an honorary degree of Doctor of Science by Western University of Pennsylvania (now the University of Pittsburgh) and thereafter was called "Doctor" by his staff.

Mrs. Blake describes her first impressions when she entered his office:

> On the wall sides of the room stood glass-doored cases full of books and Schmitt boxes. On every desk, table, and even chair were piled high in the greatest confusion heaps of books, boxes, and papers, and everywhere too were glass jars covered by cotton cloth in which were live insects. At the desk next to the window sat a huge man. He arose to greet me, over six feet in height, erect, deep-chested, small hipped. He had a rosy-cheeked florid complexion, curly gray hair, and dark hazel, almost brown eyes. He wore a shiny dark navy suit and an old-fashioned wing collar with a made-up tie. It was characteristic of him that he never had learned to tie a four-in-hand or even a bow knot for his shoe lacings. He smiled widely showing magnificent strong teeth and greeted me in a big hearty voice. But when he shook hands, his grip was curiously shy. His hands were small and the fingers delicate and fine for so large a man. . . . As one of the handful of government entomologists that had been in the Bureau almost since its beginning his knowledge not only of the injurious insects of his particular field but of all economic species was prodigious. He had written innumerable Farmers' Bulletins and longer treatises on life histories of economic pests. His studies have been the basis of much later work. Even the beautiful illustrations made under his direction are still reproduced in various bulletins and textbooks. His little green bound book on *Insects Injurious to Vegetables* was one of the first of its kind in this country.

Chittenden also did much of the editorial work on the U.S.D.A.'s periodical *Insect Life* and in 1917 he was placed in charge of the Bureau's truck-crop investigations. He accumulated over the years a very large collection of insects. He was especially interested in the slender-snouted weevil, *Curculio,* which breeds largely in acorns, and he received acorns from all parts of the country to enable him to rear and study the emerging insects.

Howard (1930) observes that "he had a broad knowledge of insects and filled an important niche in the service. Later he became a man of prominence and built up a large branch of the service. His many articles and reprints showed great care and minute knowledge. It is probable that no one had a greater knowledge of the insects affecting truck crops in the United States than did Chittenden."

Chittenden was noted for being difficult to get along with. A number of sources indicate that he was on strained terms with many members of his staff as well as with such eminent entomologists as T. Pergande, Nathan Banks, and E. A. Schwarz. Mrs. Blake writes that although Schwarz and Chittenden had nothing to do with one another, Chittenden "was eager to hear every detail of my interviews with Schwarz and would sit rapt as I related what had been said." Mrs. Blake adds, I soon learned that behind his superficial roughness and caustic tongue, there was real kindness. Maddening as his irritability was, I knew that it was only skin deep. After a trying morning he would come around to my desk with an apple or bag of peanuts, and, in this unspoken apology, I would sense his contrition."

Chittenden was still a member of the Bureau of Entomology when he died on September 15, 1929, in Washington, D.C., at the age of seventy.

WALTER DAVID HUNTER (1875–1925)

It is fortunate that Dr. L. O. Howard wrote two obituaries of Walter David Hunter, for if there is any other printed source of information about this entomologist it has escaped me.

Hunter was born in Lincoln, Nebraska, on December 14, 1875. Howard (1925) writes that "Hunter's father, who was a lawyer and ranked high in his profession, died, a young man, in April 1880, when our former president [of the Entomological Society of Washington] was four years old. He entered the preparatory school of the University of Nebraska at the age of fourteen, and graduated from the University with the degree of A.B. in 1895, before his twentieth birthday.

"There were four children in the family (Walter being the second), and all seem to have been born naturalists. Their mother writes me that Hunter's field work was begun long before he entered the University. She states: 'There was not a fence corner within eight miles that the children did not know what birds, plants and insects could be found there. Eight miles was about the limit of our old white horse. Later on they extended their knowledge on foot many miles more.'

Figure 25. W. D. Hunter

"In the University, he soon began work under Prof. Lawrence Bruner, first on ornithology and taxidermy, but he was soon led by this teacher's enthusiasm into a close study of insects. He seems to have been the most capable and promising of Bruner's students, since he stayed with him after graduation and became an instructor, continuing his work all the time and receiving the degree of Master of Arts in 1897."

Professor Bruner, an outstanding specialist on migratory locusts for the U.S.D.A., left Lincoln, Nebraska, in 1897 to work in Argentina on the locust problem there. Hunter was left in charge of the grasshopper survey; this happened again in 1898. After Bruner returned to the University of Nebraska for the second time, Hunter became his assistant, but shortly thereafter, because of a lack of funds in the University, it was necessary to let Hunter go. He took a job in entomology for $500 a year at the Iowa State College of Agriculture, at Ames, Iowa.

Hunter's family did not think entomology offered the young man a secure future and attempted to steer him into a business career, or law or teaching, but Hunter could not be dissuaded or discouraged.

Because of his fine work on migratory locusts in 1897 and 1898, Hunter was appointed a Special Field Agent in 1901 of the Division of Entomology of the U.S.D.A. to study the boll weevil in Texas. He set up his headquarters in Victoria, Texas, and in a short time was accepted as a Texan. While in Victoria, he married Mary P. Smith. In 1905, when the boll weevil began to invade the northern part of Texas, he moved his headquarters to Dallas. In 1909 the main boll weevil laboratory was established in Tallulah, Louisiana. There some outstanding entomologists were associated with him, such as W. E. Hinds, A. W. Morrill, J. C.

Crawford, W. A. Hooker, W. W. Yothers, A. C. Morgan, W. D. Pierce, C. E. Sanborn, F. C. Bishopp, F. C. Pratt, J. D. Mitchell, R. A. Cushman, Wilmon Newell, C. E. Hood, E. S. Tucker, T. E. Holloway, G. T. Smith, G. N. Wolcott, B. R. Coad, and others.

Howard (1925) writes, "It is unnecessary here to go further into the details of his cotton boll weevil work. It was magnificent. It was monumental. Probably never before had any species of insect been studied as intensively. Year after year the work was carried on, and there was early developed a series of recommendations which, had they been followed by the planters, would have retarded very greatly the progress of the weevil and would have saved the country hundreds of millions of dollars. Through this work Hunter became very well known to the people of the South, and especially to the people of Texas, and he gained a high place in their esteem.

"Realizing that every effort should be devoted to the boll weevil alone, Hunter was instructed to discourage general collecting by his force, but, as he once said, 'What are you to do? With a lot of enthusiastic entomologists coming suddenly into a region with a fauna and flora absolutely novel to them, you can't keep them from collecting.' He was perfectly right. . . . They had never seen many of the insects that were flying about them, that came to their lamps at night, that crawled over their working tables, that even got into their food at dinner. To them it was like picking up rare jewels. . . ."

Hunter and his associates found a number of practical control measures. He recommended the planting of early maturing cotton, but most of the hidebound farmers would have none of it. Nevertheless the entomologists found methods to control such insects as rice, tobacco, and sugar cane pests.

"His work on the pink bollworm of cotton and his success in bringing about the extermination of this insect in Louisiana and its near extermination in Texas is well known, but very few of us in the North began to realize the difficulties of this accomplishment. Facing pronounced opposition on the part of many people, his tact and his firmness and his resourcefulness eventually brought about the desired results, but I fear, at the expense of his health."

Howard (1925B) continues, "It is difficult to analyze the way by which he impressed and attracted many different sorts of people. He had a distinctly calm and judicial manner, but was not slow in his decisions. He never openly antagonized a man except in an important or critical situation. He was a man of great tact, and was what we commonly call a good mixer. These qualities in part accounted for his success with the southern people and for his success as an administrator, which was very great. He had the respect and affection of his many assistants to a marked degree. As a public speaker he was forceful if not eloquent. He commanded instant attention and held his audiences."

F. S. Packett (1925), describes Hunter in connection with his pink bollworm studies: "A recognized student with a remarkable memory, always in possession of all available information concerning any subject under discussion, he was a leader in any assembly that he attended and by force of character and ability to constantly see the main issue, unclouded by details, impressed others with his views and conclusions."

Later in his career, he and some of his associates began to study the pests of livestock and the medical aspects of entomology. Thus Texas fever, and Rocky Mountain spotted fever and ticks, as well as flies and mosquitoes came under their scrutiny.

He was elected president of the American Association of Economic Entomologists in 1912 and received an honorary LL.D. in 1916 from Tulane University.

Howard says that he died suddenly in El Paso, Texas, on October 14, 1925.

4 Early Entomologists of Canada

A simple bard of Nature I
Whose vernal Muse delights to chant
The objects of the earth and sky,
The things that walk, the things that fly
And those that can't.

Anon.

Canada is separated from us by geography but many of the insects are the same. This is one reason that American and Canadian entomologists share problems, interests, and organizations. Here are a few Canadian entomologists whose efforts have contributed greatly to our knowledge of insects. Later Canadian entomologists are considered elsewhere in this work.

THE ABBÉ LÉON PROVANCHER (1820–1892)

The Abbé Provancher was one of Canada's pioneer entomologists, botanists, and horticulturists. He was also an ornithologist and a specialist on molluscs, worms, and other animals. He was born at Bécancour in the Province of Quebec on March 10, 1820, and was educated for the Catholic priesthood. For a time he was Curé of Portneuf, and Maheux (1922) reports that in 1847 "he devoted himself to the service of some hundreds of Irish immigrants, stricken down with an epidemic of typhus. His heroism upon this occasion gives an idea of his unselfish character. Though nervous and rather irascible, he concealed under a coarse appearance the heart of a true friend, always frank and generous."

Because of ill health he was forced to abandon some of his activities, and was transferred to Cap Rouge, near Quebec, where he found time to study the natural sciences. At first he was chiefly interested in botany; he wrote on botany and horticulture and conducted experiments with fruit trees. In 1860 he was transferred back to Portneuf and wrote on the cultivation of fruit trees as well as on their insect pests. His botani-

Figure 26. The Abbé Léon Provancher

cal work *Flore Canadienne* appeared in 1862, and in 1869 he began the publication of *Naturaliste Canadien,* completing the twentieth volume in 1891.

Provancher commenced his *Petite Faune Entomologique du Canada* in 1874. It was completed in four volumes totaling 2,506 pages. The volumes are entitled: *Coléoptères and Additions; Orthoptères, Nevroptères, Hyménoptères; Hémiptères;* and *Additions and Corrections to Hyménoptères.*

The Abbé described hundreds of species of Canadian insects, including 923 species of Hymenoptera alone, as well as many Hemiptera. Because of his isolation from entomological libraries and collections he made many errors, which, says Essig (1926) "are small enough, compared to the difficulties he encountered."

He also wrote on travels to Jerusalem and the West Indies, on education, agriculture, and many other subjects. Many students became interested in natural history through his efforts. He was honored by a number of Canadian and American societies and received an honorary doctorate from Laval University.

He died at Cap Rouge on March 23, 1892. Besides Maheux (1922), Essig (1931) and Harrington (1909) prepared biographical notes on him.

WILLIAM SAUNDERS (1835–1914)

Like Léon Provancher, Saunders also pioneered in Canadian entomology and horticulture and was the very first in the field of Canadian economic entomology. Howard (1930) says that he helped originate the

Figure 27. William Saunders

experimental farm system of Canada. He also helped to establish the Central Experimental Farm near Ottawa and was its first director. Osborn (1937) notes that it was because of Saunders' "strong personality and active leadership" that the original Entomological Society of Canada, later the Entomological Society of Ontario, was established.

Goding (1894) described the entomologist as being five foot ten, "with a symmetrical figure," weighing about 175 pounds. "His hair is dark brown, his eyes blue. He is one of the most approachable of men, with a look of kindness ever beaming from his genial countenance, yet with a quiet dignity which forbids familiarity."

Saunders moved to Canada from Devonshire, England, with his parents when he was twelve years old. Although he had little formal education, he obtained some technical training in chemistry and became a druggist in London, Ontario. According to Bethune (1914), "His agreeable manners, thorough honesty and untiring industry brought him a fair measure of success. His love of nature led him to the collection of wild plants and insects which could be found in abundance in the neighbourhood, and he became an ardent student of Botany and Entomology. Finding many medicinal plants readily obtainable, he began the preparation of fluid extracts, which were so pure and reliable that they soon became widely and favourably known among the medical profession and led by degrees to the establishment of an extensive and lucrative business, both wholesale and retail." Later the wholesale business was managed by W. E. Saunders, the eldest of his six sons.

Saunders was active in this business for twenty-five years, but at the same time maintained his agricultural and horticultural interests on his farm near the city. In the meantime he had become a professor of Materia Medica at the University of Western Ontario in London and was president and founder of the Ontario College of Pharmacy. He was also an active member of the American Pharmaceutical Society, becoming president in 1877–1878, and a Fellow of the American Association for the Advancement of Science. His frequent attendance at A.A.A.S. meetings, says Bethune, "caused him to have a widely extended friendship with notable men of all kinds by whom he was highly esteemed and respected."

Bethune states: "The writer's acquaintance with Dr. Saunders began more than fifty years ago, when we were both young men, and soon ripened into a warm friendship, which has continued unbroken until now during all these years. In those early days, when the study of Entomology was so difficult owing to the scarcity of books on the subject, we were in constant correspondence, helping each other in every way we could, and spending each summer some days together, comparing notes, studying specimens and making collecting expeditions. Many happy hours we spent together in early morning tramps to the ponds and woods about London, and in the evening when his day's business was over, in examining the captures we had made. At that time there were few in Canada who took the least interest in the objects which to us afforded the keenest pleasure, but as time went on we found here and there a congenial spirit, and were led on in 1862 to attempt the organization of an Entomological Society [of Canada]. This was successfully accomplished during the following spring, and last year [1913] the completion of half a century's work and progress was celebrated."

In 1868 Saunders and Bethune founded the *Canadian Entomologist* and were the "sole contributors" to the first two numbers. Bethune was the editor for the first five years; then Saunders succeeded him and managed the magazine until he moved to Ottawa in 1886. Three years earlier, Saunders' book, *Insects Injurious to Fruits,* was published in Philadelphia. The second edition appeared in 1892. This book augmented his already solid reputation as an entomologist.

W. E. Saunders (1939), has the following to say about his father: "Looking back to my earliest entomological recollections, I am struck with the number of noted entomologists that came to London to visit my father. Among the earliest of these was Augustus R. Grote, a rather dapper man of perhaps forty-five years, who visited us on several occasions. Another noted visitor was C. V. Riley who had an official position at Washington. Riley was a spare man of much good humor, and I well remember hearing him tell of being out in the Western States at the time of the locust plague, and his glee as he told us of having eaten them and they were 'not at all bad when fried.' "

W. E. Saunders describes his father as the "moving spirit of the Society in London." As a young man Saunders had accompanied his father on collecting trips, and he says his father beat the foliage and collected the specimens in an umbrella. His father was especially interested in Lepidoptera; he raised caterpillars and wrote careful descriptions of them.

In 1886, after visiting experiment stations in the U.S. at the request of the Canadian Government, Saunders was appointed Director of the Experimental Farms of the Dominion in Ottawa. "In this new sphere of labour" says Bethune, "he applied himself with his wonted vigour, and in the course of a few years was mainly instrumental in bringing these establishments into thorough working order and into a high standard of excellence."

He distinguished himself in many fields. He made important studies on kerosene sprays, arsenicals, and other insecticides. His knowledge of fruit insects and their control was second to none. He was an authority on forestry, insectivorous birds, and beekeeping. He developed hardy varieties of raspberries, gooseberries, grapes, and improved varieties of wheat. He was a highly regarded scholar and teacher of pharmacy. He was a successful businessman and an inspiring governmental administrator. Naturally, a man of his ability was the recipient of many national and international honors. He died on September 13, 1914, after a prolonged illness, in London, Ontario.

CHARLES JAMES STEWART BETHUNE (1838–1932)

C. J. S. Bethune, who with William Saunders founded the original Entomological Society of Canada in 1863, later called the Entomological Society of Ontario, and the *Canadian Entomologist,* was one of Canada's foremost entomologists. For much of his ninety-four years he was the prime mover in the development of Canadian entomology.

He was born on a farm in West Flamboro Township, Upper Canada (now Ontario), on August 12, 1838. He came from a family of ministers; his father, the Rt. Rev. A. N. Bethune, was the second Bishop of Toronto; his grandfather, the Rev. John Bethune, "came from Skye to North Carolina in 1774 and ministered to a Loyalist regiment during the revolutionary War. After coming to Canada with the Loyalists the Rev. John Bethune opened the first Presbyterian Church in Montreal." (Anon., 1932.) C. J. S. "graduated from Trinity College in 1859, at the age of 21 with first class honours in classics and mathematics. He received his M.A. degree in 1861 and the degree of D.C.L. in 1883. . . .

Figure 28. Rev. C. J. S. Bethune

After spending nine years in the Anglican priesthood he became head-master of Trinity College School, Fort Hope, in 1870. He remained in this position until 1899. During these years he built the school, rebuilt it after it was destroyed by fire and made it one of the great schools of Canada."

Dr. Bethune was President of the Entomological Society of Ontario in 1871–1876, 1890–1893, and again in the Jubilee year of 1913. The first issue of the *Canadian Entomologist* appeared on August 1, 1868, through the joint efforts of Saunders and Bethune, with the latter as editor. He edited the *Canadian Entomologist* and the *Annual Reports* of the Society for almost thirty years.

Dearness (1939), reminiscing about the Entomological Society of Ontario, speaks of Dr. Bethune as follows:

"At the earliest meeting which I can remember, the Rev. Dr. Chas. J. S. Bethune, F.L.S., F.R.S.C., was the leading systematist. He seemed to know the scientific name of every insect that was submitted to him. His enthusiasm brought him to every meeting all the way from Port Hope, where he was for a time principal of a boy's college. One of his boys that he brought with him to one meeting was William Osler who became the celebrated Sir William Osler of Oxford University. When trying to interest the first year medical students in biology, I used to tell them how Sir William laid the foundation in part of his wide training in the study of insects."

Dr. Bethune's influence on entomology in Canada increased when he occupied the Chair of Entomology and Zoology at the Ontario Agricultural College in Guelph in 1906. He remained head until his retirement in October 1920. Many of Canada's outstanding entomologists were his pupils.

Spencer (1964) studied under Bethune, and his recollections are recorded below:

I took some course work with Dr. Bethune and was closely connected with him when he suffered from a cataract and was operated on first one eye and then the other. For a year during his blindness I was his amanuensis and attended to all his mail and odd departmental duties. He had a prodigious memory for references to literature, entomological history and scientific names. He always spoke in a soft, gentle voice and students in a big classroom seldom heard what he was saying. One may gauge the character of the man from an incident that happened when I was curator of the society's collections and was working over a cabinet of tropical butterflies. At that time I was impressed with the opportunities of earning a living as an economic entomologist and drawers full of exotic butterflies did not appeal to me. I remarked to him, "Dr. Bethune, why do you suppose these butterflies were created? They occur in the tropics where nobody pays any attention to them and they don't seem to be of any use to mankind." He replied, "Spencer, I suppose they were created because the Lord has an eye for the beautiful."

For such a scholarly, kindly and gentle old man, Dr. Bethune had three rather surprising pet aversions: one was the Salvation Army which once apparently refused to help someone that he sent to them; the second was Cardinal Merry del Val for reasons which I did not comprehend and the third was almost any man from northern United States whom the good doctor described in staccato tones as "a Yankee." Now Dr. Bethune had some very close American friends and it was surprising to hear him speak of Northerners in this way. It is still more remarkable when we recollect that the Entomological Society which he was so instrumental in founding, keenly recognized the material aid and encouragement received from Americans and very early in its history made them Honorary Members of the Entomological Society of Ontario as shown on the following list with the date of their election [Americans listed were E. T. Cresson, W. H. Edwards, Townend Glover, A. R. Grote, G. H. Horn, A. S. Packard, C. V. Riley, S. H. Scudder, B. D. Walsh, Asa Fitch, and others]. It was not until years afterwards that I discovered a possible reason for this dislike; he was the grandson of John Bethune who was born on the island of Skye, Scotland, educated in King's College, Aberdeen, immigrated to South Carolina and was Chaplain to the 84th regiment of Royal Highland Emigrants during the War of Independence; his sympathies with and for the Southern Colonies, later to become the Confederate States, were apparently inherited by his grandson Charles—hence his scorn of "Yankees"—spoken of, however, without an adjective!

The last time I saw Charles Bethune was in the garden of his house in Toronto when he was in his 91st year and getting a bit feeble, but his mind was still active and he discussed at some length an article he just read in the current number of *Science*. He was a great smoker and when he stopped smoking a pipe, he gave me his terra-cotta tobacco jar . . .

Needless to say, Dr. Bethune received many honors and was president of the Entomological Society of America in 1913.

He died on April 18, 1932, his mind keen until the end.

JAMES FLETCHER (1852–1908)

Fletcher became one of Canada's greatest economic entomologists and applied botanists. Through his ability to provide the help the farmer needed and his talent in communicating with the farmer in language he understood, he made the role of the applied scientist essential to Canadian agriculture. It was largely through his influence that governmental officials voted ever larger sums for entomological and agricultural research in Canada.

Fletcher was instrumental in the formation of the American Association of Economic Entomologists and served as its president in 1891. L. O. Howard (1930) reports: "The writer considered Fletcher as one of his warmest personal friends. He always attended the meetings of the American Association of Economic Entomologists. In fact the two of us during a never-to-be-forgotten summer in Washington drafted the original constitution of this organization which was effected in fact at Toronto in 1889. Fletcher's visit to Washington during that particular summer was a great joy to himself and to the men here who met him for the

Figure 29. James Fletcher

first time. He had never been so far south before, and every insect and every flower and every tree and almost every person he met interested him enormously. He would stop colored boys on the street and hold long talks with them. He would spend an hour looking over the bark of a shade-tree. It was almost impossible to get him home to dinner. His enthusiasm was infectious."

Fletcher was born on March 28, 1852, at Ashe, in the County of Kent, England. He attended King's School in Rochester, England. In 1874 he moved to Canada and became a clerk in the Montreal Bank of British North America. After two years he left this position to become an assistant in the Library of Parliament in Ottawa.

His good friend, W. H. Harrington (1909), describes him as a handsome young man "endowed with unusual physical and mental vigor, and his strong vitality and genial nature made him a great favorite with his companions . . ." He was captain of the Ottawa Snow Shoe Club and was fond of tobagganing and skating. During the summer he took many camping trips and collected insect and botanical specimens. He was also "one of the organizers of the Ottawa Football Club and his sturdy form in black and red stripes, was a pillar of strength in the scrimmages . . ."

When Fletcher assumed his position in the Library of Parliament, he was able to continue his studies in botany and entomology, two fields in which he already was proficient. Arthur Gibson (1909) states that before Fletcher went to Canada he knew the butterflies and other insects of his native land. In time he became an authority on Canadian butterflies. He joined the Entomological Society of Canada in 1877 and was an active and prominent member in the organization. His marriage two years later produced two daughters.

Fletcher's knowledge of the fauna and flora of Canada resulted from his great capacity for work. While a clerk for the Library of Parliament he spent every spare moment, weekends and evenings, in furthering his knowledge of natural history. He published his first paper on insects, *An Outline Sketch of the Canadian Buprestidae,* in 1878 and in 1879–1880, his *Flora Ottawaensis.*

Fletcher soon received recognition for his knowledge of insects and plants and in 1884 was appointed Dominion Entomologist to the Department of Agriculture. Now that he was a professional scientist, he began to devote himself to the problems of the farmer. He studied the life histories of many insect pests, their parasites, and their control, and began to issue his well-known annual reports.

He constantly labored before the Committees of Agriculture of the House of Commons for the establishment of permanent experiment stations equipped with a staff suitable to do the job. Harrington (1909) says that it was "therefore a great triumph and cause of joy for him, when [on July 1, 1887] the Experimental Farms were established, and he received the position of Entomologist and Botanist which he so successfully occupied for more than twenty-one years, making for himself a world-wide reputation as a leader in such work."

Upon his appointment, he was transferred from the Library of Parliament to the staff of the Farms. Gibson (1909) says, "He was thus enabled to devote himself entirely to natural history and his work became the great pleasure of his life." For the next twenty-one years he was not only concerned with economic entomology, but also with the applied aspects of botany. He studied the grasses and clovers; the smuts, rusts, and other parasitic fungi; and especially the weeds that afflicted the farmer.

Harrington (1909) writes of his work as Dominion Entomologist and Botanist: "Thorough and painstaking in his investigations, though hampered always by inadequate quarters and insufficient assistance, he had also the ability to present the results in an attractive and simple manner. His position required him yearly to make extensive journeys throughout the Dominion and to address audiences of very varied aims and capacities. He also frequently lectured before learned societies, and delivered addresses to schools and organizations of divers kinds, and having been present on many such occasions I can testify that he invariably charmed his hearers by the simple, yet graphic presentation of his subject combined with his fine voice, his pleasing presence and his genial manner."

Gibson (1909) mentions Fletcher's original studies on the life histories and control of injurious insects such as the Mediterranean flour moth, the hop-vine borer, the pea moth, the peach-bark borer, the wheat jointworm, the San Jose scale, the Rocky Mountain locust, the Hessian fly, the variegated cutworm, and many others.

His work on plants, weeds, and plant diseases equaled his work on insects. He prepared first-class bulletins on farm weeds, wheat smuts, and fungous diseases affecting plants, and produced many other publications. In addition to all his other duties, from 1887 until 1895 he was in charge of the Arboretum and Botanic Garden at the Central Experimental Farm in Ottawa.

Whenever he made an excursion into the field, nature students were eager to accompany him. Harrington (1909) recalls that the "briefest outing with him was invariably interesting, as his knowledge was so extensive and his faculty of observation so trained that there was ever something upon which new light could be given, or which could furnish material for future study. His intimate knowledge of large sections of the Dominion, and his extended acquaintance with scientists and other prominent persons, combined with his remarkable memory and unfailing brightness and geniality, made him a most charming and enterprising companion, either at home or abroad."

Not the least of Fletcher's many kindly aspects was the attention and encouragement he gave young people interested in natural history. Shutt (1909) relates: "My bedroom window commanded one in his office, and night after night for weeks I would retire—and that at no very early hour—leaving his light burning. He was naming botanical and entomological specimens for amateur collectors all over the country, scores of whom probably he was thus encouraging in their studies by his kindly help. He must have been blessed with a strong vitality and much strength, for by sunrise next morning, if the season were summer, he would be out gardening—a work, or rather a pastime for him, of which he was an ardent lover. He took the greatest pride in his garden and nothing gave him more pleasure than the presenting of its products to his friends."

According to Harrington (1909), Fletcher "was not content to be merely a church-goer, but as a lay-reader he took the service whenever necessary in several of the suburban and rural churches." For many years he was superintendent of the Sunday school at Ottawa East and took great interest in the children, who were very fond of him.

Fletcher made friends for Canada wherever he went. Osborn (1937) remarks that he was "of fine physique, of abounding spirits and humor, the life of any party, and could recount personal experiences or tell stories in Canadian French patois that would hold a crowd of listeners for any length of time." He was the recipient of many honors, including an honorary LL.D. in 1896 from Queen's University for his contributions to agricultural science.

After suffering for three years from a malignant tumor, Fletcher died in the Royal Victoria Hospital in Montreal on November 8, 1908.

CHARLES GORDON HEWITT (1885–1920)

Although Hewitt was with the Entomological Branch of the Canada Department of Agriculture for only eleven years before his death, his influence on entomology and conservation in Canada was greater than might be expected from the few years allotted to him.

He was born near Macclesfield, England, on February 23, 1885. His parents were Thomas Henry and Rachel Frost Hewitt. He received his early education at King Edward VI Grammar School in Macclesfield. He then attended Manchester University, where he received a B.Sc. in 1902, an M.Sc. in 1903, and a D.Sc. in 1909. According to Gibson and

Figure 30. C. G. Hewitt

Swaine (1920), "He obtained first-class honours in Zoology at Manchester University, and was university prize man and scholar. In 1902 he was appointed by his alma mater Assistant Lecturer in Zoology, and in 1904–9 occupied the position of Lecturer in Economic Zoology. In 1909 he left England for Canada, having received the appointment of Dominion Entomologist. In 1916, his title was changed to that of Dominion Entomologist and Consulting Zoologist."

Spencer (1964) says that when Hewitt first came to Canada "he created an unfavourable impression. He was reserved, stiff in his attitude and seemed somewhat conceited and diffident—such a marked contrast to his predecessor, the amiable and genial Fletcher. But opinions soon changed for Hewitt showed himself a sound thinker and a great organizer with a remarkable ability for picking the right men for responsible positions."

As soon as Hewitt arrived in Canada he became aware of the importance of quarantine legislation to protect Canada from imported insects, pests, and diseases, and through his efforts Parliament passed the Destructive Insect and Pest Act in May 1910.

"Within a year of his appointment and under his leadership," writes Spencer (1964) "the important Destructive Insect and Pest Act was passed in 1910, an Act which to this day can be modified by regulations by order-in-council to suit any conditions which may arise to prevent the introduction and spread of noxious insects, plant diseases and other agricultural pests in Canada. This Act required funds for maintenance and putting in force, so Hewitt was able to establish twelve small laboratories in all provinces except Prince Edward Island with trained men in charge; he stepped up the importance of Forest Entomology and put J. M. Swaine in charge; he separated entomology from the Experimental Farms Branch and made it a distinct Entomological Branch which he organized into four distinct divisions—Field Crop and Garden Insects, Forest Insects, Foreign Pests Suppression with strategically placed quarantine inspection stations, and Systematic Entomology. These advances at Federal level stimulated parallel developments in the Province which soon appointed Provincial Entomologists . . ."

Hewitt also made arrangements for the collection of parasites of the gypsy and brown-tail moths in Massachusetts and their liberation in eastern Canada. The problem of insects as vectors of disease was of great interest to him, and he devoted a great deal of study to house flies, mosquitoes, ticks, and other disease-carriers. In 1914 his book, *The House-fly, Its Structure, Habits, Development, Relation to Disease and Control,* was published by the Cambridge University Press in England. Among other insect studies, he prepared circulars on the honeybee and the large larch fly, and worked with ants, stable flies, the spruce budworm, and thrips, as well as house sparrows and other birds. Besides his scientific interests, Hewitt was devoted to literature, art, music, and problems in social welfare.

One of Hewitt's greatest achievements was in the field of conservation and in "furthering the treaty between Canada and the United States for the protection of migratory birds." His book, *The Conservation of the Wild Life of Canada,* appeared posthumously in 1921, largely through the efforts of his wife, who was the daughter of Surgeon General Sir Frederick Borden, of Conning, Nova Scotia.

Hewitt's contributions were recognized by his Canadian and American colleagues. In 1913 he was elected president of the Entomological Society of Ontario and in 1915 president of the American Association of Economic Entomologists.

The career of this talented man was cut off at its height by his death from pneumonia on February 29, 1920.

5 Notable Teachers in Entomology

Butterflies flit to and fro
The herdboys are returning home.
Oh, blind old mister teacher
Why don't you let me go?
Chinese folk song

The field of entomology has been blessed with many fine teachers. Some of them have been especially successful in imparting both their knowledge and enthusiasm to their students. We can recognize these teachers by the many competent students they produced and by the strong entomology departments they left behind.

HERMANN AUGUST HAGEN (1817–1893)

The German entomologist, Hermann August Hagen, had a profound influence on American entomology, for he was the teacher of several notable American entomologists, among them John Henry Comstock, A. J. Cook, Herbert Osborn, H. G. Hubbard, and C. W. Woodworth.

Hagen was born in Königsberg, East Prussia (since 1945 Kaliningrad, U.S.S.R.) on May 30, 1817. His father, Carl Heinrich Hagen, was a Royal Councillor and a professor at Albert University. After graduating from a gymnasium, Hermann began the study of medicine at the University of Königsberg in 1836. There he was profoundly influenced by several German teachers, including his zoology professor, M. H. Rathke. In 1839 Hagen accompanied Rathke to Norway, Sweden, Denmark, and various parts of Germany, where they studied the principal entomological libraries and collections. These teachers directed and encouraged his interest in insects, and, according to Henshaw (1894), it is believed that he became attracted to dragonflies because the first specimen he happened to catch proved to be an undescribed insect of that order. In 1839 he

Figure 31. H. A. Hagen

published his first entomological paper, "List of the Dragonflies of East Prussia." In 1840 he received his medical degree from the University of Königsberg, his thesis being on the dragonflies of Europe. He pursued further medical studies in Berlin, Vienna, and Paris, returning to Königsberg in 1843. Henshaw writes, "At the surgical hospital, where for several years he was first assistant, he performed a large part of the operations; among the needy his services, always in demand, were given with that steady tenderness so characteristic of the sympathetic side of his nature."

Despite his heavy load as a physician, Hagen continued to publish papers on dragonflies, collaborating with Baron Edmond de Selys-Longchamps on several. Hagen's *Monographie des Termites* was published from 1855 to 1860.

Baron Osten Sacken made Hagen's acquaintance in the spring of 1856, when the Baron was on his way to North America. The Baron encouraged Hagen to work on the Neuroptera of North America and sent him all the material he had, including collections made on expeditions to the West. As a result Hagen wrote his *Synopsis of the Neuroptera of North America,* which appeared in print in 1861. The manuscript, written in Latin was translated into English by P. R. Uhler, a well-known hemipterist. Osten Sacken helped to correct the proof sheets.

Hagen's great *Bibliotheca Entomologica* appeared in 1862 and 1863. This two-volume work attempted to list all the publications in entomology up to 1862. It was completed only after meticulous and intensive study in the libraries of Germany, France, Belgium, Holland, and England. Hermann Loew (1903), the great German dipterist wrote to

Osten Sacken, "When I see Hagen trudging along with his fifty pounds of manuscript, from town to town, and from library to library, I think to myself, *this is not the kind of work I would have undertaken.*"

Besides dragonflies and termites, Dr. Hagen was interested in Neuroptera in the broad sense, including ant lions, lacewings, psocids, and embiids, as well as in other orders. He also wrote on insects found in amber and other European fossil insects.

Notwithstanding his medical practice and his heavy program of entomological studies, he found time to be a member of the school board and vice president of the City Council of Königsberg.

When P. R. Uhler left the Museum of Comparative Zoology at Harvard to become assistant librarian at the Peabody Institute in Baltimore, Baron Osten Sacken helped persuade Louis Agassiz to select Hagen as Uhler's successor. In 1867 Hagen left Königsberg to take charge of the entomological department of the Museum, and in 1870 he began his university career as Professor of Entomology at Harvard. Through his untiring efforts, the Museum collection became a model for other entomological museums in the United States. When Hagen first came to this country such entomologists as E. A. Schwarz, H. G. Hubbard, who graduated from Harvard in the class of 1873, G. R. Crotch, the Englishman, and Osten Sacken were all working in Cambridge.

Uhler received many letters from Hagen, some in German, some in English—of sorts. The Museum of Comparative Zoology in Harvard contains some of these letters, one of which, dated March 19, 1870, from Cambridge, Mass., follows:

My dear Uhler,

Sometimes it seems to me to be far more separated as in the time, when I was so many thousand miles from here. The mechanical arrangement of a collection is a heavy operation and not at all fitted for some significant labor. But as I have engaged myself for this work, only useful for later entomologists, I will see to fulfill my pledge. Surely I have learned more as I thought before in handling families and genera so far distant from my daily study, and I have worked so much as possible. In this time the Lepidoptera and nearly the half of the Coleoptera are in some manner arranged, or better ready for somebody wishing to undertake a significant examination. Further the whole collection is placed safe in the new boxes, and assured against the attacks of Dermestids. This work was annoying and long, because it was necessary to label in the same time the insects. etc. etc.

Yours very truly . . .

Samuel Henshaw (1894) says that "His lectures, given at rare intervals to advanced students, contained much genuine and exact knowledge, and his many acts of kindness and words of wise counsel will not soon be forgotten by those who enjoyed the facilities of the department under his charge."

Henshaw describes Hagen as "a man of marked character, simple and sympathetic, and if at times somewhat hot and hasty in temper and impatient of opposition, he had also the warmest of hearts and most generous of dispositions. His unostentatious hospitality was enjoyed by many entomologists, who found his life in Cambridge quiet, contented and happy."

One of Dr. Hagen's earliest and most special students was John Henry Comstock. Mrs. Comstock (1953) writes that although Hagen

> gave no lectures in Harvard at that time [1872], he received this student of entomology with enthusiasm and gave him a series of memorable lectures which proved a lasting source of inspiration. When Henry became Professor Comstock, he related this experience:
>
> It was a very warm summer. Dr. Hagen would come into his room, where I had usually preceded him and was at work, take off his coat and vest and hang them on a hook behind the door; then he would light his pipe, which had a long flexible stem, sit down at the table, place the bowl of the pipe on the floor at his side and after a few puffs to make sure that the tobacco was burning well, would say, "Now you kom and I vill you tell some dings vat I know." He used a sheet of paper and a pencil in lieu of a blackboard; I sat facing him at the small table and took notes. These introductory rites took place every morning with almost no variation, and preceded lectures which dealt very largely with the morphology of insects. These lectures were superbly clear and well organized and my notes on them were of very great use many years later, when I gave lectures on this subject in Cornell University. . . .
>
> Henry felt justified in borrowing money for this summer's instruction, something he had never done before. He had asked Squire Simpson of Scriba, the father of his early roommate at Cornell, for a loan of fifty dollars, and it had been granted. Fifty dollars was not a great sum with which to meet expenses at Harvard for a summer, but he had a free room in a temporary building called Zoological Hall, and lived as cheaply as possible. When he found his money dwindling, he told Dr. Hagen that he could not stay very much longer but did not tell him why. The Doctor divined the cause and asked him to take his dinners with Mrs. Hagen and himself, which Henry gladly did, and thus came under the influence of a gracious lady and a charming home.

Howard (1930) says of Hagen: "He was a tall, stout man, whose English was very bad. It was most difficult to understand him. He visited Washington in the early eighties; and that was the first time I saw him. He would not speak to Schwarz. I never learned why. There must have been some misunderstanding between them before Schwarz left Cambridge with Hubbard."

In 1887 Howard and his wife visited the Museum of Comparative Zoology so that he could do some work on chalcids. Howard (1930) writes that Hagen

> was a very courteous man, and insisted on showing Mrs. Howard (then a young bride and very pretty) around the Museum while I worked. Mrs. Howard had brought her sewing, and would have been quite content to be left alone; but Hagen insisted, and Mrs. Howard, without the slightest knowledge of any of the things he showed her and totally unable to understand him, found herself in a predicament which needed all of her tact. As she told it to me afterwards, he presently stopped and said, "You do not understand me." "Oh, perfectly," she replied; and was very much taken aback when he remarked, "Vell, vot did I say?" . . .
>
> During this visit Henry Edwards arrived one day with a very large box of Australian insects. He had recently returned from Australia, where he had been acting for some years (he was an actor by profession) and incidentally collecting all sorts of insects. I was immensely impressed by Edwards, tall, handsome, cordial man that he was. As Hagen exclaimed over this or that extraordinary Neuropter, Edwards would say, "You may have it." *"Ach! meiner theurer freund!"* Hagen would say, and would forcibly embrace him and kiss him on both cheeks. I don't believe Edwards had ever been in Germany, and I think this method of expressing delight was as novel to him as it was to me, and naturally made him very uncomfortable, but he did not show it.
>
> Hagen was very generous, placed everything at my disposal, trusted me perfectly, and helped me all he could. His opinion on many entomological questions was for years the last word in America . . .

Professor Herbert Osborn (1937), who spent one winter studying with Dr. Hagen, gives the following picture of him:

"He was then well along in years but in full vigor and I was at once impressed with his striking personality and the wealth of information at his command. He was doubtless at that time the most learned entomologist in America if not in the world . . .

"As I worked in the same room with the Doctor I had an excellent opportunity to hear his emphatic exclamation of approval or disgust concerning his own work, or that of his assistant, Mrs. Batchelder . . . His 'Ach yes, no! very well' seemed equally appropriate on many occasions and 'Ach very good' was an occasional welcome even if in explosive form."

He believed in a student's making a very thorough study of a single species and constantly provided the student with references to the literature to help him solve his problems. Dr. Hagen wanted his students to study entomology "for the love of science" and not for the purpose of making a living out of their research efforts.

According to Osborn (1937), "He had a very methodical daily program in the laboratory. He came in usually at a very definite hour and after putting away his hat and coat would seat himself at his table, carefully adjust his wig, comb down the hairs to an exact line on his broad forehead, light up his pipe which was very frequently refilled and relighted during the day. (He took great satisfaction apparently in being the only man in the Museum exempted from the rigid rule of 'no smoking' in the Museum.) His time was carefully divided between reading the technical papers that came to his table, writing out bibliographic slips of reference, attending to the disposition of museum material and his monographic work. He gave a series of lectures during my stay devoted mostly to the color pattern of insects, later published as one of his contributions. As my work progressed he seemed much interested in some of the points determined and I figured that I must have 'struck oil' when he invited me to take Sunday dinner at his home."

Dr. Hagen had a genuine interest in his occasional students. He wrote a letter of introduction to Uhler for Herbert Osborn, who wished to see Uhler's collection of Hemiptera, in which he was interested. (January 28, 1882, MCZ collection.)

We can see from the following letter to Uhler of March 6, 1882, that the interests of the entomological library of Harvard were of paramount importance to Hagen:

> I come again asking your assistance. I have to say that I believe it to be proper that Harvard should possess a full copy of Th. W. Harris papers. So I have myself and *entirely* out of my own pocket brought together all except the 16 numbers of the list. Even I have 16 papers not before known from Harris. I have advertised all here in the daily papers and in the advertising catalogues printed monthly for all booksellers in the U.S. I may say that some numbers cost one to $2 each and more.
>
> If you can help me, I am ready to pay for each number $1 or more if needed, for the termites I will go as high as $5.
>
> Half of the papers wanted are published in N. England, some in Boston, but it was till now entirely impossible to get them.
>
> I hope you will not mind that I am obtruding on your time, as it is for the benefit of American students, that I try to bring together a complete American library.
>
> I am very much obliged to you for your kindness to Mr. Osborn. I am glad to see he is a very good and earnest worker.

P. P. Calvert (1893) was another entomologist who profited by his association with Dr. Hagen: "Of Dr. Hagen, personally, the writer can say little, but that is very pleasant. I met him but once, for parts of three days in July, 1890, at the Museum in Cambridge. Although unwell and obliged to rest frequently, he showed me his collections of Odonata, giving me types and other specimens. He was very hopeful of recovery, and talked of the work he would like to do next year. Referring to [J. O.] Westwood, [H. C. C.] Burmeister, [F.] Poey, [J.] Gundlach and de Selys, all then alive and all older than himself, he laughingly said, 'I am the baby at seventy-four.' I may, perhaps, be permitted to add one other personality. In February, 1890, he sent me his unpublished notes on *Leucorhinia,* giving me permission to publish them, and when I wrote him for a title, he wrote 'Synopsis of Leucorhinia' with my name as author, although the work was all his own. None but a generous man would have done so."

Dr. Hagen's opinions were not necessarily always correct. For example, he thought a yeast fungus when sprinkled or fed to insects would result in the dissemination of disease among insects. He insisted that the Hessian fly had not come into New York with the straw bedding of the Hessian soldiers, although A. S. Packard and others had strong evidence that it had. And the following excerpt from a letter dated January 15, 1890, to Professor C. H. Fernald about the gypsy moth, *Porthetria dispar* (Linn.), shows he did not put much stock in the biological control of this pest:

"I myself have seen O. dispar many times in large numbers on property belonging to my father (in Germany), I have helped kill them, for the next year, very few were left or found. Mr. C. F. Freyer one of the most prominent Lepiderologist[s] in 1879 tells a similar story. A species as nearly related as O. moracha, is of course very dangerous; but a species attacking so many things is not dangerous. To import for O. dispar enemies from Europe, would make Massachusetts a laughing stock in the opinion of all scientific men in Europe."

Anyone who studies Hagen's work at the Museum of Comparative Zoology must be impressed with his remarkable ability as an entomological artist and the detail in his sketches, many of which illustrated his entomological studies.

In 1882, Hagen helped make a survey of injurious insects along the route of the Northern Pacific Railroad through California, Oregon, Washington, and Montana. He was greatly impressed with the beauty of much of the country he traveled through.

Calvert (1893) notes that Hagen, worked unceasingly, from 1868 to 1890 to arrange the collection of insects in the Museum, "a very large part of which, tedious as it is, was performed by his own hands." He also had his problems with collectors, as indicated in a letter he wrote on January 24, 1885, to Uhler, wherein he complains that T. L. Casey was keeping types from the collection J. L. LeConte lent him for study. In 1875 Hagen gave his own collection to the Museum.

In September 1890, after several years of ill health, Dr. Hagen was stricken with paralysis. "His wife's devotion, strength, and cheerfulness have been simply magnificent. For more than three years she never once spent six hours, winter or summer, away from the house. She had to lift him, move him in his bed, dress him, and never flagged or faltered. She seems surprisingly fresh and healthy-minded after all she has gone through, and is certainly an admirable woman." (Osten Sacken, 1903.) On November 9, 1893, he died, and his wife returned to Königsberg.

JOHN HENRY COMSTOCK (1849–1931) and ANNA BOTSFORD COMSTOCK (1854–1930)

There are a number of noted husband-and-wife teams in entomology but probably none as famous as John Henry and Anna Botsford Comstock. Comstock was the first entomologist at Cornell. He was a dedicated teacher and also the author of widely-used books on insects and spiders. He was very ably assisted by his artist wife, who achieved fame through her ability as a wood engraver and a teacher of nature study. They had no children of their own, and were celebrated for their kindness to the children of others, the students at Cornell.

The following account of the Comstocks is largely drawn from *The Comstocks of Cornell,* an autobiography by Mrs. Comstock, printed more than twenty years after her death and edited by Professor G. W. Herrick and Ruby Green Smith.

Ebenezer Comstock and Susan Allen Comstock, John Henry's parents, were of New England and New York State stock. Upon their marriage they migrated to what is now Janesville, Wisconsin, and there purchased a farm, where, on February 24, 1849, John Henry was born.

His father, an adventurous young man, was smitten by the gold bug that same year. Believing that the quickest way to pay off his heavily mortgaged farm would be to find gold, he joined a train of covered wagons heading for the golden West. The scourge of cholera ended this dream quickly; he died on the River Platte, and only many years later did Mrs. Comstock find out what had happened to her husband.

Shortly after Ebenezer's departure Mrs. Comstock lost their farm by foreclosure, and the next few years she worked as a housekeeper to support herself and her young son. After spending some time in Wisconsin and Ohio, she moved to New York when John Henry was four years old. She became ill, and John was placed in an orphan asylum for about a year, after which he stayed with several relatives, including an uncle, a minister, who "whipped the devil" out of him. John, who was a nervous, sensitive child, stammered, and his uncle attempted to curb this affliction with beatings—needless to say, he failed.

This quick-tempered, active, high-strung boy led a lonely life until, when he was eleven years old, he had the good fortune to be taken in by a kind-hearted and understanding couple, Rebecca and Captain Lewis

Figure 32(A).

Figure 32(B). (A) and (B). J. H. Comstock. Courtesy of E. I. McDaniel

Turner of Oswego, New York. The latter was a master of sailing vessels on the Great Lakes. Their three grown sons were all sailors, and when the whole family was together during the winter, "Hurricane Hall" was an exciting and aptly named place.

"Ma Becky" and "Pa Lewis" treated young Henry as if he were their own. For his work on the Turner farm he was given board, clothing, and three months of winter schooling, but best of all, the affection he needed so badly in view of his longing for his mother, who, in the meantime, to support herself, had become a nurse and could only rarely come to see her son. In an environment ideally suited to nurture a love for nature, Henry cared for everything outdoors except snakes and walking sticks.

In 1863, when he was fourteen, Henry tried to enlist in the Union forces, but because of his youth and small stature he was turned down.

The following year his mother left for California with a patient, in the vain hope of finding out something about her husband, who by then had been gone for fifteen years. Living among sailors as he did, Henry naturally decided upon a career as a sailor. Ma Becky, in her tactful way, encouraged and taught him to cook, because she knew the life of a steward was not nearly as strenuous as that of a common sailor on deck. Thus it was that Henry became a cook on vessels sailing the Great Lakes.

"At the end of his first season of sailing," says Mrs. Comstock, "Henry bought his clothing for the coming year and started home with twenty dollars. But at Cape Vincent someone gave him a counterfeit ten dollar bill, and it was a sorrowful boy who arrived at Hurricane Hall with just half of his savings with which to buy his books and to pay other expenses."

A kindly teacher encouraged Henry to study algebra and aim for a higher education. Henry proved to be an eager and apt student.

The following summer he was once again on a schooner and once more a steward. He wrote to his mother: "I have a healthy family to cook for and they have good appetites. I have baked up five barrels of flour since I came aboard the 'Thornton' this spring. Today I have been baking bread, huckleberry pies, green apple pies, and gingerbread, in rough weather." He earned enough to buy his own clothes and pay for his schooling during the winter at an academy, and the following winter, after a summer of sailoring, he attended a seminary.

His mother, having finally learned of her husband's death, married a mine-owner and hotel-keeper in Forbestown, California, and in the winter of 1868 they invited Henry to come to California. Henry was very much tempted on account of his health, but when he found that there was no state university in California and that the curriculum at the College of Santa Clara in San Jose was largely classical and theological, he decided to stick to the East.

Henry, who had dodged the pitfalls of a sailor's life because of a truly religious nature, was warned by a clergyman to avoid newly-founded Cornell as the den of the devil himself. This immediately aroused Henry's interest, so he wrote for the catalogue of Cornell University and found the courses in science tempting indeed.

By means of a book called Wood's *Botany* Henry had become familiar with the flowers of the Great Lakes region. He had no special interest in insects until on July 1, 1870, in a bookstore in Buffalo, a clerk led him to the science shelf.

> There was no botany of flowerless plants there, but there was another book that caught Henry's attention—Harris' *Insects Injurious to Vegetation.* This admirable book on entomology was superbly illustrated with many engravings by Henry Marsh, which made the volume a classic in the history of that art, and with eight colored steel plates by J. H. Richard. The text is a model of lucidity. Henry was breathless over the discovery of this book, for he had not known that there was such a science as entomology. The price of the book was ten dollars. That was too much for a sailor lad, saving money to go through college, so he rueful̲y went back to his ship to get supper for the men. But he could think of nothing else save that wonderful book. That night he could scarcely sleep from thinking of it. The next morning, as soon as breakfast was out of the way, he drew ten dollars of his pay from the captain and hurried to the bookshop with a terrible fear that someone else had bought the book meanwhile. But it was there and it became his own and "walking on air," he went back to the schooner, which was being loaded with coal. Since coal dust pervaded everything, he covered the book with paper and handled it with care, so that no smudge could deface it. He began to study it with an intensity of interest that absorbed every moment of his spare time. He learned the orders of insects with the book propped up before him while he was washing dishes. From that time on, he desired most of all to become an entomologist. But he thought he could not make a livelihood in that way! He could earn his living by being a doctor or a teacher, but the study of insects would be his chief interest in science. (*The Comstocks of Cornell.*)

Comstock entered Cornell University in the fall of 1869, one year after the opening of the University, little dreaming that he would spend sixty years there instead of four. He had been there only five weeks when he came down with malaria, which he had contracted while working around the Great Lakes the summer before. He did not re-enter Cornell until the fall, after another summer of sailing. He worked his way through school at every possible job, from that of unskilled laborer in the construction of one of Cornell's halls, to that of laboratory assistant to a professor.

One reason Cornell had attracted Henry was the listing in its catalogue of a course in entomology. However, there still was no professor of entomology in 1872, so thirteen students, aware of Henry's knowledge of insects, petitioned the faculty to permit him to present a course of lectures in economic entomology. Most of the students were further advanced than Henry in their schooling, but none of them had studied entomology.

In the summer of 1872 Henry studied entomology at the Museum of Comparative Zoology of Harvard under Dr. H. A. Hagen, from whom he profited greatly, particularly in the field of insect morphology.

Needham (1946) describes the first day of instruction:

> I remember that Comstock told me that when he and Hagen sat down at opposite sides of the little table over which came most of the instruction of that marvelous summer, Hagen began by saying:
>
> "Do you shbeak Cherman?"
>
> "No sir."
>
> "Do you shbeak French?"
>
> "No sir. Sorry."
>
> "Do you shbeak Latin?"
>
> "No sir."
>
> "Vell den," (with a sigh of resignation), "I guess ve vill half to shbeak English. Come now I vill tell you some tings vot I know about entomolochy."
>
> So the work proceeded, with Hagen lecturing and making rapid sketches on loose sheets of paper that lay before him on that little table, and Comstock listening and taking notes.

When he returned to Cornell in 1872, he continued his spare time jobs and gradually rose from janitor to chimes-master.

Henry's success as a teacher of entomology in his sophomore year was so encouraging that he decided to become a teacher of the sciences rather than a doctor. In 1873 the Board of Trustees of Cornell appointed him Instructor in Entomology. He wrote to his mother: "This will be a fine position for me. It will not interfere with my studies, so I shall be able to keep on with my classes as before. My work will consist of a course of lectures in the spring term, private instruction in my laboratory, and care of a collection of insects. I shall spend the summer vacation in the field making collections and studying the habits of insects. My salary will be for the first year five hundred dollars, which is enough to support me nicely here." He also taught invertebrate zoology for a short time in the absence of the regular professor.

When Ma Becky died in 1873, Henry mourned her as if she were his own mother, and thereafter visited Captain Turner whenever he was able.

During that year, Henry lectured twice a week at Cornell and had special students in his laboratory, which was a renovated chimes-master's study. The laboratory was without a microscope until John Stanton Gould, an eminent lecturer in agriculture, gave his own microscope to the laboratory.

Soon Henry's small insect cabinet was crowded with specimens, and President Andrew D. White of Cornell, who apparently was quite interested in Henry's efforts, out of his own pocket purchased some new cases for the collection.

In 1874 Henry received his B.S. from Cornell, the only degree he ever got; he never had the time to earn an advanced degree from another institution.

A syllabus of his lectures, first published in 1876, was a prototype of his textbook, *An Introduction to Entomology*. That same year he was made Assistant Professor of Entomology at a salary of $1,000 a year. L. O. Howard, who specialized in entomology in his senior year, was, as previously indicated, one of his students.

In 1875 Henry became interested in one of his freshman students, Anna Botsford. During the Christmas of 1876 he visited her parents. Anna's parents liked the ambitious, industrious, and very restless young professor, whom they called "Harry." Anna wrote, "I remember clearly Harry's personal characteristics in those days. He was very active and moved with a rapidity that gave the observer a breathless sensation. He was unconsciously restless; even when he sat reading, he moved constantly, much to the detriment of his trousers, as I was to discover later. . . . My father, who loved to tease me, asked *sotto voce,* 'Anna, does he ever sit down and sit still?' "

Anna Botsford was born in a log cabin in Cattaraugus County, New York, on September 1, 1854. She was raised on a farm and was intimately acquainted with all its chores. She attended private schools and at the age of fourteen served as a substitute teacher. She later attended a seminary, but resisted being force-fed religion. She writes: "It was not until I went to Cornell, where no one questioned my beliefs, that I became tolerant. I had met intolerance with intolerance. Cornell taught me again the lesson taught me by my parents, to respect the spiritual experiences and religious beliefs of others."

In 1875, the year after she entered Cornell she became a student in Comstock's class. Comstock asked permission to eat at her table, and it was not very long before they were collecting insects together.

In 1878, C. V. Riley invited Comstock to go South for the summer to study the cotton worm, *Alabama argillacea,* which was ruining many of the cotton planters. Comstock made his headquarters at Selma, Alabama. When, later, yellow fever became rampant through much of

Figure 33. A. B. Comstock

the South, Selma was quarantined, much to the dismay of his betrothed, Anna Botsford. Despite a touch of malaria, he was married on October 7 of the same year. After a brief honeymoon, Henry escorted his bride to the little house he had built commanding a view of Cayuga Lake, on what is now a part of the Cornell campus.

Mrs. Comstock wrote: "While Mr. Comstock's work as a lecturer and teacher was quite apart from my duties, we worked together then as later. He used to help me wash the dishes and I used to go to the laboratory and help him in all possible ways. I enjoyed putting the laboratory in order. I wrote his business letters, at his dictation, and made diagrams to illustrate his lectures in invertebrate zoology, using holland curtain cloth and oil paints. I also made stencil outlines of his lectures. Thus we worked hard during the days; evenings, we had company or went to receptions or lectures at a rate that seemed, later, veritable social intoxication."

Mrs. Comstock continues, "During our first year together, Harry and I began an enterprise that we carried out as consistently as possible for forty years, the entertaining of Professor Comstock's students in our home. We invited his laboratory class of nine to tea; I was the cook, and the menu was scalloped oysters, chocolate cake, lemon layer cake, pickles, jelly, sliced oranges, tea and coffee." Mrs. Comstock does not tell us is that she usually prepared for twice as many as she invited because the students often brought their friends.

When, in April 1879, Comstock became Chief Entomologist in the

U.S. Department of Agriculture, his staff consisted of L. O. Howard as Assistant Entomologist, Theodore Pergande, who had been Riley's valued assistant, George Marx, the staff artist, who studied spiders, and a copyist. Mrs. Comstock helped as a volunteer in the office before she became a paid assistant, and she and Howard learned how to type on one of the two typewriters in the Department.

Comstock's first job was to complete his studies on the cotton worm and to write a report on it. Accordingly, he appointed William Trelease of Cornell to go to his former headquarters in Selma and continue investigations on this pest. Because of the difficulties Riley had with Commissioner LeDuc of the Department of Agriculture, Riley would in no way co-operate with Comstock regarding the use of data and specimens on cotton insects obtained when Riley was Entomologist. This eventually resulted in a coolness between the two.

Howard says that when Comstock came to Washington he was very nervous and in ill health. It was feared that he had tuberculosis, for he was coughing up blood. Because of this and his interest in citrus insects, Comstock went to Florida during the winter of 1879–1880. There he began his studies of scale insects, and he began to send collections of citrus and other pests back to Washington. A man who had been with him in Florida for several days told Mrs. Comstock that he was "the most economical man I ever saw traveling at government expense." At the end of February, Comstock returned to Washington, where, in 1880, he published his *Report on Cotton Insects.* It was well received by cotton farmers, members of Congress, and entomologists.

In July of the same year, the Comstocks left for California to continue the studies of scale insects infesting citrus trees. For the first time since he was thirteen, Comstock saw his mother in Marysville, California, near Sacramento, and also met his stepfather. In San Jose, Comstock found a new scale that was ravaging the orchards. This he named the "pernicious" scale, *Aspidiotus perniciosus,* but unfortunately for the city of San Jose, it became known as the San Jose scale. Years later C. L. Marlatt showed that the original home of this scale was China. According to Comstock, *Aspidiotus perniciosus* was the most injurious orchard pest he had ever seen. Not only did the scales suck the living sap, but they also poisoned the cambium, eventually killing the tree.

In the fall of 1880 the Comstocks returned to Washington, and he began seriously to study the taxonomy of scale insects. Anna Comstock made drawings of the pygidia (rear ends) of the female scale insects; Comstock's system of classification evolved with the help of these drawings.

When, after the assassination of President Garfield in 1881, Comstock was replaced by C. V. Riley as Entomologist, he returned to Cornell. Though somewhat disappointed that he could not realize his ambitious plans for entomology in the Federal Government, he was happy to go back to the Cornell campus with all its opportunities. Moreover, he was given a Federal appropriation of $1,500 to complete his work on scale insects. His intensive work on this subject while preparing a textbook for his classes, with all his other activities, made it necessary for him to work nights in the laboratory. Using his skill in carpentry, he also built his own insect cabinets, which were the forerunners of the famed Cornell insect cabinets.

His *Report of the Entomologist of the United States Department of Agriculture for the Year 1880,* which came out in 1881, contained the first simple classification of scale insects based on the posterior segments of the body as drawn by Mrs. Comstock as well as L. O. Howard's article on the parasites of scale insects. This work helped establish the reputations of Comstock and Howard as entomologists and of Mrs. Comstock as an illustrator of insects. With the adoption in 1882 of some of Comstock's insect outlines by other universities, his influence as a teacher was given further impetus.

Later that year he and Mrs. Comstock attended a meeting of the American Association for the Advancement of Science in Montreal, Canada. "Mr. Comstock read a paper before the Agricultural section and C. V. Riley, who was present, attacked him quite rudely. My husband kept his temper and answered in a few quiet words that made controversy impossible. Later, several of the entomologists congratulated him on his manner of dealing with the matter." At the next Association meeting attended by Comstock, Riley was in a most conciliatory mood.

In the meantime Mrs. Comstock continued her education at Cornell with the purpose of earning a degree. Also, though she had no previous background in the art of engraving, she purchased wood-engraving tools and a booklet of directions so that she could make engravings for her husband's proposed *An Introduction to Entomology.* Her efforts in wood engraving in time brought her international fame. In fact she was one of the few women ever to be elected to the American Society of Wood Engravers. She received her degree from Cornell in 1885, her thesis being *The Fine Anatomy of the Interior of the Larva of Corydalus cornutus.* This is the "hellgrammite," the aquatic larva of the grotesque dobsonfly.

In the summer of 1885 Comstock held the first of his renowned summer courses. Students collected and studied insects in the field two mornings each week and then mounted and studied them in the laboratory.

Mrs. Comstock, in the meantime, continued her studies in wood engraving, and finally went for a short period to Cooper Union in New

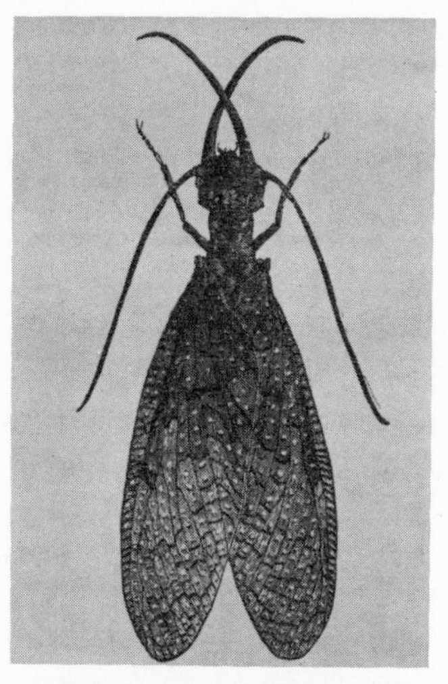

Figure 34. Woodcut of the male dobson fly by A. B. Comstock. Adopted from John Henry Comstock, *An Introduction to Entomology*. Copyright 1933, 1936, 1940, by Comstock Publishing Co., Inc. Used by permission of Cornell University Press

York City to work with John P. Davis, a well-known teacher of wood engraving.

In the spring of 1886 Comstock began to give instruction in beekeeping; he maintained five hives of bees for this purpose.

Whereas President White of Cornell had been quite sympathetic to the science courses in the University, the attitude of his successor, President C. K. Adams, was so discouraging that for a time Professor Comstock thought about giving up entomology and perhaps taking up medicine. He had even begun to send for catalogues from medical colleges. At this critical point, the trustees of Cornell gave him a substantial increase in salary, and he returned to work on his entomology textbook with renewed enthusiasm.

When, in 1887, the Hatch Act establishing experiment stations in land-grant colleges was passed, Comstock acquired enough funds to start an "insectary." He coined the word himself and defined it as a greenhouse where plants could be grown to feed insects in order to facilitate the study of their habits. The funds also provided for a microscope and two student assistants.

The first part of *An Introduction to Entomology* was published in 1888, in a 234-page paper-bound edition containing many wood engravings executed by Mrs. Comstock. By 1940 it was in its ninth edition.

Comstock now put himself on a strict regimen: he began to retire at

8 P.M. and rise at 4 A.M. By this means he was able to devote himself to his writing with a fresh mind and without interruption before classes began.

C. V. Riley came to inspect the insectary in 1890 and to observe Comstock's methods of studying living insects. Riley remarked to Mrs. Comstock, "It is better for Comstock that he came back here when he did, for this is one of the best entomological positions in the country and a much happier one than the Washington position."

In 1891 Dr. David Starr Jordan, a former student in Comstock's first entomology course at Cornell and president of Stanford University, asked Comstock to take the chair of entomology at Stanford at a salary well above what he was getting at Cornell. Comstock declined because he did not feel equal to the task of establishing a course in entomology at another college. Nevertheless, it was arranged that he lecture on entomology at Stanford during the winters of 1891–1893.

During their third winter there the Comstocks were joined by Vernon L. Kellogg, who had done graduate work at Cornell and who later attained fame as entomologist, biologist, author, and administrator. Upon Comstock's recommendation, Kellogg was made head of Stanford's Department of Entomology.

At the end of 1894, the Comstocks published their *Manual for the Study of Insects,* which contained 701 pages and 797 figures and cost only $3.25. This publishing venture was so successful that it led them to set up the Comstock Publishing Company.

In 1897 Appleton published a more rudimentary book by the Comstocks called *Insect Life,* which was written especially for the elementary teacher. The following year Mrs. Comstock, who had acquired some experience in teaching nature study in the local schools, was appointed Assistant Professor of Nature Study in the Cornell University Extension Division.

When, in 1894, Anna's father became too old to manage his dairy farm properly, Comstock bought it. Under Comstock's management the farm became a show place for more than twenty years. Mrs. Comstock remarks that her husband wrote to Dr. David Starr Jordan in 1896: "Anybody can be a professor, if he studies hard enough along one line, but it takes a man of wide knowledge of all the sciences, to be a farmer."

Comstock had been interested in spiders for many years and had devoted much of his spare time, such as it was, to observing them and their webs in the field. In 1903 he went South to do more field work there in several states, taking photographs of the webs on plates and collecting notes.

That same year the Comstocks began work on another book, *How to Know the Butterflies,* which appeared in print in 1904. Nevertheless

Comstock continued his study of spiders, the results of which finally appeared in *The Spider Book,* published in 1912. In July 1905 he wrote: "I worked on spiders in the early morning (4:45 to 7); lectured at 8; went into the field for a reconnaissance from 10 to 12; and took the class into the field from 2 to 5 P.M. It would have been a delightful day's work if it had not been so long."

In 1909 Mrs. Comstock began writing her *Handbook of Nature Study,* designed as a textbook in natural history. Although Professor Liberty Hyde Bailey and Comstock himself advised her against the project, she persevered in her efforts. It was published in 1911, and by 1939 it had appeared in twenty-four editions and in eight languages.

After living for thirty-three years in the Fall Creek Cottage on the Cornell campus, the Comstocks sold it to Cornell University in 1910 for $5,000 (it had cost them $3,000) and purchased a new house in Ithaca.

On December 6, 1913, Comstock, then almost sixty-five, resigned from his position as Professor of Entomology and General Invertebrate Zoology so that he could reserve his energy for more study and more writing. His former students thereupon presented him with $2,500 for the establishment of a memorial library of entomology. He turned the money over to Cornell University, which set up the John Henry Comstock Memorial Library of Entomology.

A year later Mrs. Comstock published her *Pet Book* about the behavior and care of common pets. Not until July 1919 did Mrs. Comstock receive appropriate recognition for her efforts in making the observation of nature a part of the school curriculum: she became a full professor at Cornell in the Department of Nature Study, which she had developed. In 1921, at the age of sixty-seven, Mrs. Comstock gave her last lecture at the University.

Finally, in 1920, *An Introduction to Entomology,* the summation of a lifetime of studying and teaching entomology, appeared in print. Well received by both teachers and students, it is to this day a prime source-book in entomology, having gone into nine editions.

Even when Comstock was nearly seventy, as his wife reports, he "was working steadily eight or ten hours each day, trying to unravel the tangled skein of the evolutionary relations of insects as shown by their wing venation. As soon as he worked one thread free, at one turn of the reel it would tangle again. Never had he been more vigorous mentally than now."

Comstock's active career came to an end on August 6, 1926, when he suffered a brain hemorrhage. Mrs. Comstock wrote: "There are no words to describe his bravery and patience and cheerfulness after this calamity which, for us, ended life. All that came after was merely existence."

In the meantime Mrs. Comstock developed a serious heart condition, and she was seriously ill through 1929 and 1930. Despite her own struggle, she was her husband's helpmeet and companion almost until her death on August 24, 1930. Comstock himself lingered on until March 20, 1931. The ashes of both lie beneath some old oaks, overlooking the Lake Cayuga they both loved.

Needham (1931) writes of Comstock: "In his day he taught entomology to more than 5,000 students. Practically all of these at some time or other entered the hospitable Comstock home. All the entomological world knows how John Henry and Anna Botsford Comstock worked together for more than half a century; how they supplemented and aided each other; how common were their interests; and how mutual was their labor. All know also, how generous was their hospitality. Many savants from foreign shores were their guests. Many struggling students found under their roof-tree a second home."

Comstock was short and slight in build; Mrs. Comstock was a large woman with a somewhat overpowering personality and a protective attitude toward her husband. Because of his stutter, he usually kept in the background; when he was asked to tell a story, he often let his wife tell it.

Besides the halls and libraries dedicated to them, as well as other honors from Cornell, they were honored in the names of some of their students, such as L. O. Howard, F. H. Chittenden, M. V. Slingerland, V. L. Kellogg, W. A. Riley, A. D. MacGillivray, J. G. Needham, R. N. Chapman, R. H. Pettit, E. P. Felt, and A. L. Quaintance.

Generations of students have been nurtured on the books written by the Comstocks. Through these students and teachers the Comstocks' warmth, kindness, enthusiasm, and love of nature have been transplanted to all parts of the world.

Besides the autobiography by Anna Botsford Comstock (1953), other worthwhile references on the Comstocks are Howard (1930, 1933), Herrick (1931), and Needham (1931, 1946).

ALBERT JOHN COOK (1842–1916)

Professor Cook taught one of the first formal courses in entomology in the United States and left a lasting imprint on the development of entomology in Michigan and California.

E. O. Essig (1931), one of Cook's students, has provided much of the following information about him.

He was born in Owasso, Michigan, on August 30, 1842. He received an education in agriculture and biology in Michigan Agricultural Col-

Figure 35. A. J. Cook. Courtesy of E. I. McDaniel

lege and graduated in 1862, receiving an M.S. two years later and an honorary D.Sc. in 1905. He continued his graduate training in 1867–1868 at Harvard under Louis Agassiz and H. A. Hagen. In 1867 Cook became an instructor in mathematics at his alma mater, and taught a half-year course in entomology. In 1869 he became professor of zoology and entomology. He was an enthusiastic teacher and organized an excellent insect collection with the help of his students and collectors in other parts of the United States. Essig states that he was "a leader in the Michigan Horticultural Society and active in the Farmers Institute, both important organizations in Michigan at the time, and carried on a voluminous correspondence on farm problems in the state." Cook produced a kerosene-soap emulsion in 1877, which C. V. Riley claimed as his invention, but L. O. Howard notes that both of these workers may have been anticipated by a Mr. Henry Bird of Newark, New Jersey, who apparently made the first such emulsion in 1875.

Cook was responsible for introducing Paris green for control of the codling moth in 1880. Professor Eugenia I. McDaniel of Michigan State University wrote me that she was told Cook had "stood on a street corner eating a recently sprayed apple (sprayed with Paris green) and told his companions if he was alive at 3 P.M. the next day it would be safe to use the sprayed fruits (their idea for checking tolerance was a little different from ours)." Cook also introduced crude carbolic acid emulsions and carbon disulfide as insecticides.

In 1876 there appeared the first edition of his *Manual of the Apiary;* this was so popular that, by 1910, it had gone into nineteen editions. He also wrote shorter works such as *Silo and Silage, Maple Sugar and the Sugar Bush, Birds of Michigan,* and many bulletins and articles.

He was one of the original founders of the American Association of Economic Entomologists, and in fact acted as chairman of the organizational meeting in August 1889 during a meeting of the American Association for the Advancement of Science in Toronto, Canada.

In 1894 he left Michigan to become professor of biology at Pomona College in Southern California. Here, with the assistance of Charles Fuller Baker, a prominent entomologist and teacher in his own right, he helped organize a strong department of biology and entomology. It was to C. F. Baker and A. J. Cook that Essig dedicated his book, *Insects of Western North America.*

Cook soon became recognized as an entomological and horticultural authority in California. He was active in farmers' clubs and pomological organizations throughout the state and helped the farmers to obtain telephone lines and insurance, to organize co-operative exchanges, and to receive scientific advice on orchard problems. He also secured aid in 1907 from the U.S. Bureau of Entomology to investigate HCN in citrus fumigation.

Essig states that, "As a teacher of both botany and zoology he was a remarkable success, and in cooperation with C. F. Baker from 1908–1911, did some very splendid work training students in entomology. He was also responsible for securing the financial backing necessary for the *Journal of Entomology,* the *Journal of Economic Botany,* and the *Report of the Laguna Beach Marine Laboratory.*"

"I stopped at Pomona once in the late nineties," observed L. O. Howard (1930), "and found that Professor Cook had taken off his beard and had become young and active and most enthusiastic. There was a spirit of vigor about the institution that was very pleasing. Part of this was doubtless due to Baker, but Professor Cook's personality seemed to have been completely changed."

In 1911, after eighteen years at Pomona College, Cook's reputation throughout California prompted Governor Hiram Johnson to appoint him State Commissioner of Horticulture. Cook took such outstanding men with him as E. O. Essig in entomology and H. S. Fawcett of Florida in plant pathology. On his departure from Pomona the College presented him with the following tribute:

"The loss to Pomona College will be felt most keenly by all of the students who have known Prof. Cook personally. His genial nature, his great heart, his tremendous and infectious enthusiasm, his keen interest in the personal welfare of every student under him—these things have made him greatly beloved to all. His interest in his students has never, through all the years, been a perfunctory one, but always a living, active interest . . . an interest that not only helped them to find their life work . . . whatever it might be, but ever afterward supported and encouraged them to great efforts. In a quiet way, unknown to the public, he has even financially assisted deserving students to complete their work, and for this he has been repaid in some things beyond the value of money—loyalty and love."

As Commissioner, according to Essig, "He at once reorganized the quarantine, insectary, and executive departments, founded the Monthly Bulletin, and did everything in his power to make the office useful to the farmers of the state. In 1912 he exerted great influence and pressure to secure the passage of the national quarantine law which is of great importance to California and other agricultural states." His duties involved talking on horticultural and entomological subjects to innumerable audiences and publishing hundreds of articles in the California farm press.

Essig writes that Cook's first three years in office were "stormy," but his sincerity and efficiency gradually won over his opponents and he was reappointed to office by Governor Johnson in 1916. However, he became ill this same year. He went East for treatment and died in his old home in Owasso, Michigan, on September 29, 1916.

CHARLES HENRY FERNALD (1838–1921) and HENRY TORSEY FERNALD (1866–1952)

One day during the 1890's when L. O. Howard (1910) was riding along a Massachusetts road with Professor C. H. Fernald, the latter said to him, "I wonder whether after I shall have gone people will think of me as a systematic entomologist or rather as an economic entomologist." Howard says he replied instantly, "You forget probably the biggest work you have done and the best work, and that is as a teacher." This statement applies not only to Charles Henry Fernald, but also to his son, Dr. Henry Torsey Fernald.

Charles Henry was born on March 16, 1838, at Fernald's Point, Mount Desert Island, Maine, the son of Eben and Sophronia Fernald. Their farm with bountiful plant and animal life and the nearby sea with its teeming marine life provided an environment designed to arouse Charles's interest in nature. His father was a shipowner, and Charles aspired to becoming a ship captain. Braun (1921) says that he spent his summers from the age of fifteen to twenty-one at sea and the winters studying and teaching. At the age of twenty-one he entered Maine Wesleyan Seminary to prepare himself for a sea captaincy. Here his interests broadened with the varied courses he took. Three years of study pre-

Figure 36. C. H. Fernald

pared him for entrance to Bowdoin College, but upon the outbreak of the Civil War he joined the Union Navy. He served in the Navy from 1862 to 1865, where he rose to the rank of ensign. Just before setting out to sea, he married Miss Maria Elizabeth Smith, of whom we will have more to say later. While in the Navy he was able to take the courses required by Bowdoin for the B.A. degree, which he received. After the war he became a principal of Litchfield Academy at Litchfield Corners, Maine. Later, while also principal of Houlton Academy in the same state, he began the study of geology, botany, zoology, and entomology on his own. In 1871 he obtained his M.A. from Bowdoin. In those early years he depended a great deal upon his correspondence with various authorities for enlightenment about his interests. According to Braun, "In 1871, he was made a professor of natural history in the Maine State College (now the University of Maine), and was called upon to teach botany, physical geography, human anatomy and physiology, comparative anatomy, veterinary science and zoology, with special attention to entomology, geology and mineralogy." At various times he also taught algebra, trigonometry, agriculture, and horticulture. He was truly a one-man university. "During this time (1871–1886)," says Braun, "insects came more and more to occupy his attention, as is shown by the titles of his papers. While connected with the Maine State College, he wrote *Grasses of Maine, Butterflies of Maine* (1884), *Sphingidae of New England* (1886), and *A Synonymical Catalogue of the Described Tortricidae of North America* (1882). The conclusions in the last-named paper were based on his studies and comparisons of specimens in the museums of this country and Europe. Numerous other shorter entomological papers appeared during this time, chiefly notes and descriptions of new species in the Pyralidae and Tortricidae. At the close of his service here, the Main State College conferred upon him the degree of Doctor of Philosophy. While he was a professor at Maine State College, he once attended a summer school on Penikese Island in Buzzards Bay under the leadership of the great zoologist, Louis Agassiz."

Early in his career Fernald had become interested in economic entomology, and in 1875 his paper, "Destructive Insects—Their Habits and Means of Preventing Their Depredations" appeared in the *Third Annual Report of the Maine State Pomological Society.*

According to L. O. Howard (1910), when in 1872 Fernald began to teach a regular course in entomology at Maine State College, Dr. H. A. Hagen at Harvard was the only other teacher of the subject, and he "had only an occasional student." Apparently Howard had forgotten that A. J. Cook started to give courses in entomology at Michigan Agricultural College in 1867.

Howard (1933), recalling his days under Riley at the Department of Agriculture, draws an amusing picture of Fernald: "We were constantly rearing little moths, and the American who then knew the most about these forms was Professor C. H. Fernald, of Amherst, Massachusetts. During one of his summer vacations (I think it was in 1880) Fernald and his wife came to Washington and spent part of July and August in the Department of Agriculture, working in the same room with me, Comstock and his wife being absent in California, where Comstock had gone to round out his knowledge of the scale-insects of the United States. The Fernalds were delightful people, Professor Fernald having a somewhat boyish humor that filled me with joy. At that time beggars and selling agents of all kinds had the run of the government offices, and I shall never forget the dear old Professor's method of getting rid of them. When an unmistakeable beggar entered the room, Fernald would immediately jump to his feet, rush over to him, grasp him by the hand, and say, 'How do you do? I am so glad to see you. How did you leave the folks at home?' Almost invariably this so embarrassed the beggar that he turned and left."

In 1886 the trustees of Massachusetts Agricultural College at Amherst (now the University of Massachusetts) invited Fernald to become Professor of Zoology and Lecturer in Veterinary Science. Fernald demurred because he felt his knowledge of veterinary science was inadequate. Finally, it was agreed that he teach this subject until a qualified individual could be found, and in August of the same year he and his family moved to Amherst, where he served as Professor of Zoology until his retirement in 1910.

According to W. E. Hinds (1933), "As a pioneer teacher in a field of vast and growing importance, he will be remembered always among entomologists. As a teacher his general theory appeared to be both comprehensive and intensive: 'Know something about everything, and everything about something.' His teaching was scholarly, sound, and interesting. His own enthusiasm for careful first-hand study of any insect species in the field and in its natural environment and relationships was contagious. He emphasized also the importance of the power of self-directed, original investigation."

C. A. Peters, Professor Emeritus in Chemistry, University of Massachusetts, has written me that, "I first saw Prof. Fernald at a Farmers Institute in Worcester about 1890. My father took me. I was 15 years old. A kindly soft spoken gentleman who evidently knew what he was talking about. I wondered how one man could know so much."

J. Robert Parker, one of Fernald's students, has written me that "My contacts with him were in a 1909 graduate course on the history and philosophy of entomology. There were only a few of us in the class and sessions were informal. Professor Fernald encouraged us to give our own opinions and to disagree with his—in other words he taught us to think. He was always friendly and very much interested in our personal problems. We all admired him."

When Federal funds, under the Hatch Act, were appropriated, Fernald became the Entomologist in the "Hatch Experiment Station." When in 1895 his colleague R. S. Lull was appointed assistant in the Department of Zoology, Fernald was able to devote himself largely to entomology. In 1899, with the cooperation of and advice from many leading entomologists in the United States and Europe, he organized at the Massachusetts Agricultural College a graduate course in entomology suitable for the M.S. and Ph.D. degrees, even though the College as a whole had at that time less than two hundred students.

Leopold Trouvelot, a French astronomer at Harvard University, brought into this country egg-masses of the gypsy moth for some studies connected with the pébrine disease of silk worms, which was then ruining the silk industry in France. In 1869 these escaped from his house into some waste land near the town of Medford. The results were described by Howard (1930):

He notified the scientific public, but nothing was seen of the gipsy moth, which remained, however, gradually increasing, on this waste land until 1889 when a tremendous plague of caterpillars almost overwhelmed the little town. The numbers were so enormous that the trees were completely stripped of their leaves, the crawling caterpillars covered the sidewalks, the trunks of the shade trees, the fences and the sides of the houses, entering the houses and getting into the food and into the beds. They were killed in countless numbers by the inhabitants, who swept them up into piles, poured kerosene over them and set them on fire. Thousands upon thousands were crushed under the feet of pedestrians, and a pungent and filthy stench arose from their decaying bodies. The numbers were so great that in the still, summer nights the sound of their feeding could plainly be heard, while the pattering of their excremental pellets on the ground sounded like rain. Valuable fruit and shade trees were killed in numbers by their work, and the value of real estate was very considerably reduced. So great was the nuisance that it was impossible, for example, to hang clothes upon the garden clothesline, as they would become covered with the caterpillars and stained with their excrement. Persons walking along the streets would become covered with caterpillars spinning down from the trees. To read the testimony of the older inhabitants of the town which was collected and published by a committee, reminds one vividly of one of the plagues of Egypt as described in the Bible.

At first the residents thought the caterpillar was a native species, but Mrs. C. H. Fernald identified it as the gypsy moth of Europe.

In 1890 a Gypsy Moth Commission was organized by the State of Massachusetts. The technical entomological work was under the direction of Fernald and the field work under the capable E. H. Forbush at the State Board of Agriculture. Through the excellent co-operation between the state legislature and its competent directors, the gypsy moth was almost eradicated by 1901. Unfortunately, politics entered the picture at this crucial point, appropriations were stopped, and the gypsy moth once again became rampant. Now, both state and Federal appropriations are needed to keep the gypsy moth from spreading from the East into other parts of the nation.

Braun (1921), in discussing the economic aspects of Professor Fernald's work, states that "His greatest work in economic entomology centered around the fight against the Gypsy Moth in Massachusetts. As scientific director of this work, and as entomologist of the Hatch Experiment Station, he was active in the successful controlling of this pest during the ten-year period from 1889 to 1899. He brought to this work executive ability combined with a broad conception of the problems and difficulties involved. Accounts of this work are embodied in the annual *Gypsy Moth Reports* and the more complete volume on *The Gypsy Moth* (1896), prepared in connection with E. H. Forbush, field director of the work." Incidentally, chemists employed on the gypsy moth were responsible for the discovery of lead arsenate, an arsenical that does not burn the foliage, as did some of the arsenicals previously used.

Because of the impetus in gypsy moth work, economic entomologists were in demand, and this in turn resulted in the development of entomology in Massachusetts and other state colleges. Professor Fernald and later Dr. Fernald took advantage of these opportunities and made the State college at Amherst an outstanding institution for the training of entomologists. Through the efforts of the Fernalds, many of their students were well trained for governmental service as economic entomologists, and many important workers in state and Federal entomology were graduates of Massachusetts Agricultural College.

During the late 1890s Fernald was also concerned with another European pest, the brown-tail moth; fortunately it proved to be a lesser problem than the gypsy moth and was largely controlled by imported parasites.

By 1899 Fernald had become so burdened with teaching the senior course in entomology, with the development and teaching of the newly-organized graduate courses, with his work in the Experiment Station, and his supervision of the entomological phases of the gypsy moth program, that it was necessary to obtain an experienced and capable individual to help him with these duties.

The man chosen for the job by the president of the Massachusetts Agricultural College was Fernald's son, Dr. H. T. Fernald. Dr. Fernald had taught zoology and entomology successfully for nine years at Pennsylvania State College (now University) and at the time was also State Zoologist of Pennsylvania, working mainly on the control of insect pests and aiding the farmers of that state in control methods.

To induce Dr. Fernald to leave the position he was then holding, that of full professor, the trustees of the Massachusetts Agricultural College established a Department of Entomology and offered him a full professorship. Dr. Fernald accepted and went to Amherst September 1, 1899. Thus Dr. Fernald was Professor of Entomology while his father, whose main interests were entomology, was Professor of Zoology. Peters tells us "entomology blossomed in the sun, and zoology lay dormant in the shade" because of the special interests of the Fernalds.

Mrs. Maria E. Fernald, Professor Fernald's wife and Dr. Fernald's mother, was a collector in her own right. She was born at Monmouth, Maine, May 24, 1839, and graduated in the first class of the Maine Wesleyan Seminary and Female College. She married Fernald on August 24, 1862, and assisted in his studies of moths. She became quite expert with the Family Tortricidae, and at one time helped identify the Tineid moths in the U.S.D.A. collection in Washington while Comstock was the Entomologist. Dr. Fernald notes that his mother "began her interest in entomology when my father began his studies in that subject. She used to go out in the fields and woods with him at Orono, collecting, but a little later restricted her efforts to collecting by lamp in the window at night and by 'sugaring' when moonlight prevented using the light. In these ways she captured a large number of moths, many of them rare, and a few new species, some of them described by [H. K.] Morrison and [A. R.] Grote."

Professor Peters tells us that Fernald described his wife's skill by saying she could "put her hand out the window, take the moth, put it in a cage & not disturb a single scale." For a time Mrs. Fernald collected moths for Dr. W. J. Holland of the Carnegie Museum, Pittsburgh. Possibly her greatest claim to fame is Bulletin 88 of the Hatch Experiment Station, *A Catalogue of the Coccidae of the World,* which was very useful to Professor T. D. A. Cockerell and other workers on scale insects. For further notes on Mrs. Fernald see E. P. Felt (1920).

According to Felt the Home of Professor and Mrs. Fernald on Hallock Street in Amherst, was "a delightful center for both social and scientific activities."

With the arrival of Dr. Fernald, Professor Fernald was able to devote more and more of his time to the development of graduate courses in entomology, and from 1908 to 1910, the year he retired, he was director of the Graduate School. Although he had taught entomology exclusively for the previous fifteen years, he had never held the title of Professor of Entomology.

Professor Fernald naturally followed with great interest the planning and erection of a new building which was to house the entomology department. Although it was named in his honor, he never taught a course in it. After he retired, he worked for several years on a study of Pyralid moths, but was unable to complete it because of his failing eyesight.

Mrs. Fernald died on October 6, 1919; her husband followed on February 22, 1921. It was W. E. Hinds, the first Ph.D. in entomology at the Massachusetts Agricultural College, who said the "graduates and the work which they have done form the most real and lasting memorial to Charles Henry Fernald."

Henry Torsey Fernald was born on April 17, 1866, at Litchfield, Maine, while his father was the principal of the Litchfield Academy. C. P. Alexander (1953) says that his "boyhood and early manhood were spent in Orono, Maine. Many of his summers were spent on Mount Desert Island, at the birthplace of his father on Fernald Point, near Southwest Harbor."

Because both of his parents were active collectors, Henry soon developed an interest in insects. He received his B.S. degree from Maine State College in 1885 with a key from Phi Beta Kappa and his M.S. in 1888. Since his parents wanted him to receive the best possible graduate instruction he entered Johns Hopkins and earned his Ph.D. in 1890.

That same year he married Minna R. Simon, who shared his interests for more than sixty years. They were the parents of a son and two daughters, none of whom took up the study of insects professionally, but who became prominent in the arts and economics.

From 1890 to 1899 Fernald was Professor of Zoology at Pennsylvania State College as well as State Entomologist, and State Zoologist for the last year of his stay, after which he became Professor of Entomology in the newly established Department of Entomology at the Massachusetts Agricultural College. He was Professor of Entomology and Head of the Department until 1930, when he retired because of an asthmatic condition. He moved to Orlando, and later to Winter Haven, Florida.

Alexander (1953) reports that Fernald received many entomological honors; he was president of the American Association of Economic Entomologists in 1914 and a charter member of the Entomological Society of America. His fields of investigation and research "included life-history studies of various insects, investigations of insecticides including foliage burning by arsenicals, and in the field of insect taxonomy, important studies on the Sphecid wasps, upon which group he published various papers, particularly on the species of the New World. In 1913 he went to Europe where he visited many of the leading museums and studied the various types of Sphecidae described by earlier workers. Near the close of his active career he presented his important collection of these wasps to the United States National Museum."

His well-known text, *Applied Entomology,* published in 1929, went through five editions, the last two being coauthored by Dr. Harold H.

Figure 37. H. T. Fernald

Shepard, who recalls how as a student, after one of Mrs. Fernald's fine suppers, the three of them would spend the evening poring over their stamp collections.

The *Fernald Club Yearbook* for 1952 paints a picture of him as seen by his students. In it, J. R. Parker writes: "I am also thankful that I could study under Dr. Fernald and his illustrious father, Professor C. H. Fernald. I was one of the few who had the privilege of learning from both of these great teachers. In my day we always spoke of the elder Fernald as Professor and his son as Doctor. They were a splendid team. Doctor Fernald was the precise teacher of fundamentals in both taxonomic and economic entomology; Professor Fernald gave us an insight into the history and philosophy of entomology.

"Doctor Fernald took a keen personal interest in his students and made every effort to place them in positions. The large number who became professional entomologists reflect his success in this respect. His interest did not cease when his students took jobs and were on their own. He followed their professional careers and family lives and was always available for friendly and helpful counsel by correspondence or personal interviews. I believe this continuous interest was the quality of his personality which endeared him to so many of us."

H. N. Worthley writes in the same yearbook: "One year Dr. Crampton decided to brush up and sit in Dr. Fernald's graduate lectures on tropisms. In discussing the olfactory sense one day, Dr. Fernald recounted an experience while trout fishing. He was smoking a cigar (his friends differed about their contents). As the current carried his line downstream a slight breath of air carried a thin line of cigar smoke down with it. Suddenly a butterfly fluttered across the stream and through the line of smoke. Immediately it turned, and fluttered back through the smoke—back and forth repeatedly. As Dr. Fernald paused to let the appreciation of this observation sink in, Dr. Crampton put a mild question. 'Dr. Fernald,' said he, 'could it by chance have been a cabbage butterfly?' "

Dr. Louis Pyenson remembers Fernald as the slightly elderly professor, with gold-rimmed spectacles and a white goatee, demonstrating how a praying mantis caught a grasshopper. He "took an eraser and placed it on the edge of his lectern. Then with his hands held up in front of his head, he stealthily approached the eraser. When within striking distance, he reached out quickly and grabbed the eraser."

Even after retiring to Florida, Fernald was interested in the progress of his "boys" and corresponded with many of them. He had the remarkable ability of making "outstanding entomologists of all students" and the students in turn never forgot how much they owed to him. And entomology, as a whole, has profited by the enthusiasm and dedication both Fernalds instilled in several generations of entomologists.

Dr. Fernald died on July 14, 1952, and was buried beneath the palms and live oaks in Palm Cemetery, Winter Park, Florida.

FRANCIS HUNTINGTON SNOW (1840–1908)

Vernon Kellogg once said of Snow, "He made first-class men out of the best of us, and something at least worth while out of the worst of us."

Figure 38. F. H. Snow. From C. K. Hyder, *Snow of Kansas*. Courtesy of the University of Kansas Press

This is not surprising, considering Snow's competence as a teacher, minister, entomologist, ornithologist, botanist, geologist, paleontologist, mineralogist, and chancellor of the University of Kansas. For forty-two years he was one of the men largely responsible for guiding this small pioneer college, at first largely a preparatory school, toward its present status as a great university. Anyone interested in Snow should read C. K. Hyder's masterful biography, for it is from Hyder (1953) that much of the following has been gleaned.

Snow was born in Fitchburg, Massachusetts, on June 29, 1840. His mother died when he was eleven, and he was raised by his stepmother, Margaret Pollack Snow, who treated him as though he was her own. His father, a successful businessman, a stubborn idealist, and an independent thinker, was antitobacco, a prohibitionist, an advocate of women's rights, and an ardent abolitionist.

"Frank's early journals," says Hyder, "show him engaged in an amazing variety of useful chores. His father's few acres of land included a large orchard and garden. Though the tract was known locally as 'The Garden of Eden,' rocks and weeds abounded. If Frank was not paring apples or squashes, sorting apples in barrels, killing and picking chickens, oiling the harness, or performing some other task connected with household, yard, garden, or field—on one or two occasions we find him sewing patchwork for a quilt!—he was likely to be sawing wood."

Although Frank was a normal, healthy boy, he became slightly deaf after an attack of scarlet fever. He began his schooling in Troy, New York, and attended a free high school in Fitchburg from 1854 to 1858, during the exciting days before the Civil War, when he became very much interested in politics. The fiery antislavery views of his father definitely influenced his thinking.

Frank entered Williams College in Massachusetts in 1858, when Mark Hopkins was its president. The curriculum emphasized classical languages and mathematics, with lectures on the sciences and humanities. During the years he attended Williams he began to show his bent for natural history through his interest in flowers, minerals, and geological formations and his love for hiking. He was quite active in the college Lyceum of Natural History, where seminars were held every two weeks on assigned topics. Although Frank received the highest grades the college had given any student, he was decidedly not a bookworm, but a born student leader constantly involved in the college's social and athletic life. In 1862 he was the natural choice for valedictorian of an outstanding class. Frank was then twenty-two, about five and a half feet tall, and weighed all of 120 pounds.

After graduation, he took a job at the high school in Fitchburg, but his career as a school teacher was quickly derailed when some youngsters took advantage of his deafness to "cut up." Then for a short time he assisted his father in his office. But this did not satisfy this very ambitious, industrious, and religiously-inclined youth. Shortly thereafter he attended Andover Theological Seminary and, being an avowed pacifist, joined the U.S. Christian Commission and assisted chaplains and aided the wounded in hospitals, often working on the battlefield. Frank was usually in the middle of things and his journal described the following scene at Lee's surrender:

> Within 15 minutes after Gen. Lee had left the house the most enthusiastic cheering was commenced among the rebel troops and was thought at first to be from the same cause as that of our own soldiers, but it was afterwards ascertained that it was the sight of Genl Lee which inspired his men with such enthusiasm; but a great many of them shed tears when he told them of

his surrender to Genl Grant. I chanced to be standing within a rod of Genl Grant when he dismounted from his horse upon his return from the consummation of his first year's work as Lt. Gen. of the U.S. armies. He seemed almost as cool and unmoved as when I saw him ride to the front at Petersville [*sic*] one day last week. While his staff officers were looking about for a drink of a more *spiritual* nature, he satisfied his own thirst by drinking from a tin cup wh[ich] he filled with water from a pail wh[ich] stood near by. Then sitting down on Gen. Gibbon's camp chair he cut off a twig from a little bush at his feet and began to whittle. In a few seconds he turned to Gen. Gibbon & drily remarked "I think we will begin to go home tomorrow." Gen. Gibbon got enthusiastic at once, and shouted to his comrades of Grant's staff, "Hurra, boys, hurra! We're going home." In striking contrast was the immovable Grant, though even *he* gave evidence of satisfaction which could not be concealed. Night closes upon over a hundred thousand happy Union soldiers who can now lie down between their blankets for the first time since the opening of the war with the surety that there will be no fighting on the morrow.

The year 1866 was a momentous one for Snow. He was ordained a minister, he fell in love with Jane ("Jeanie") Appleton Aiken of Andover, and the embryonic University of Kansas at Lawrence offered him the professorship of Mathematics and Natural Science. He was the third professor appointed to the faculty of the newly-founded college. His salary was $1,600 a year. With great reluctance he left his "Jeanie" in Andover. They were not married until the summer of 1868. They had six children, besides one who died in infancy; all six seemed to have their father's penchant for collecting and natural history.

Lawrence then was essentially pioneer country, though the Indians were friendly. In the spring it was full of wild flowers and also very muddy. The University, one of the first coeducational institutions, started really as a preparatory school, all the students being of high school age.

One of his female students was particularly impressed "by the cheerful and kindly manner of one who was so boyish-looking that a stranger . . . approach[ed] him, supposing him to be a student . . . to ask to have the faculty, particularly Professor Snow, pointed out. She noticed how Snow would patiently repeat, with no trace of annoyance, his clear explanation of mathematical principles. Later she observed his contagious enthusiasm in the teaching of science. In the absence of Professor Robinson, Snow could teach Latin and Greek as competently as mathematics."

A fellow professor asked a question and then answered it: "What was the secret of Professor Snow's success as a teacher? . . . He was friendly, he was kind, he was patient, he was honest, he was lucid and forceful in speech, he was enthusiastic; and above all and under all was his manliness and fearlessness. These characteristics won for him the respect, the admiration, and the affection of hundreds of students who came under his tutelage. In after years we loved to recall, not so much that we had studied botany and zoology at the University, as that we had studied them under Professor Snow."

Hyder (1953) summarizes his attributes by noting that he was exacting, outspoken, with at times "sharp-spoken" convictions. He was spontaneous, with a boyish sense of humor and an unusually sunny disposition. "He was a lover of beauty, whether found in nature, art, or human character." He was a man with an unusually flexible mind and every inch an educator and scientist. Like his Indian pony, Snow "was always at a gallop," bursting with radiant energy. In addition to his professional work, he preached in many nearby communities, conducted Sunday school classes, and began to probe the mysteries of the plant, animal, and mineral kingdoms in Lawrence and vicinity. His courses consisted of botany, geology, zoology, meteorology, physiology, and comparative anatomy, besides a special course in natural history.

At first he was the only naturalist in Kansas with an interest in birds. His *Catalogue of the Birds of Kansas* first appeared in 1872. His paper, *The Relation of Birds to Horticulture,* which showed that quail were avid eaters of grasshoppers, was published in 1876.

Snow was interested in almost everything, including meteors. Knowing that a meteor shower was due on the night of November 12 or 13, 1876, he gave a lecture on the subject. He writes that when the shower of meteors began to fall he "sent two of the boys to ring the Methodist bell, that all the people might see the splendid sight. I have become quite a lion in this community and have received many congratulations upon my success in getting up a shower of meteors to illustrate my lecture." After this Merlin-like success, the city of Lawrence contributed a small sum of money to help the university build an observatory.

Snow began serving in 1873 as a meteorologist for the State Board of Agriculture and continued his weather observations for thirty-five years. These are among the oldest continuous records in the United States.

No one living in Kansas at that time could long remain oblivious of the voracious grasshoppers. Snow writes that one day in 1866 when he was traveling in Kansas, his train sometimes could not move because the crushed bodies of grasshoppers prevented traction of the wheels on the rails. The grasshoppers were so thick in 1874 and 1875 that Snow could recount catching 190 grasshoppers with a single sweep of the net. One old woman who tried to protect her cabbages by putting them under a tub later found a tubful of hoppers. One solution C. V. Riley proposed for the overabundant grasshopper population was to incorporate the creatures in certain recipes; however "a dessert compounded of baked locusts and honey" never gained the public's favor.

Besides his catalogue of birds, Snow prepared catalogues of plants and lists of butterflies and of the Coleoptera of Kansas. In his paper, *The Fishes of the Kansas River as Observed at Lawrence* (1875), Snow mentions seeing a blue catfish weighing 175 pounds and reports the story of a 250-pounder caught in 1856. "The fish had been landed with the aid of a steamboat towline and a yoke of oxen."

In the meantime Snow and his students had accumulated a collection of birds and insects that drew visitors to the University like a magnet. By collecting and exchanging herbarium specimens, insects, minerals, birds, fossils, and the like, he gradually assembled a nucleus for an outstanding University museum.

Snow's reputation as a field naturalist was based on twenty-six summer field trips he took from 1876 to 1907 through Kansas, Colorado, New Mexico, Texas, and Arizona. Many of his students accompanied him on these trips and learned more and forgot less than in any class thereafter. Both boys and girls "roughed it" on those expeditions, and one young lady wrote, "While his interest in collecting . . . never abated, we learned of his tender care for his invalid wife, his solicitude for the welfare of his young son, his cheerfulness in all circumstances, his winsomeness, his love for music, and for all the best things in life. We learned to know him as a man, and found him as fine and interesting as we had known him as an instructor."

On his second expedition to Wallace County, Kansas, he captured several hundred specimens of the tiger beetle, *Amblychila cylindriformis* Say, which was extremely rare from the time Thomas Say first collected one in 1823 in the Rocky Mountains. "Snow and his companions discovered that its favorite haunt at night was the holes of animals like the kangaroo rat. After Snow had sold (as high as $25 each) a considerable number of specimens not needed in the University collection, the price went down; whereupon he desisted from further sales till the price went up again, having something of a corner on the market. The exchange value of the beetles enabled Snow to make additions to the beetle collection of the University; and the two students who had accompanied him and who had been promised part of the proceeds from the beetles they captured earned a substantial part of their college expenses for a year."

While he was collecting in New Mexico in 1881, the Apaches went on the war path and threatened the campers. He and his collectors hastened to get out of the area. He made his way out with his little boy Will, his Winchester rifle, and thirteen boxes of insects. Their teamster was murdered by the Apaches.

"Much lore about the Snow expeditions has grown up," wrote Hyder. "Snow usually warned new recruits that the coming trip would be a test of their mettle. He expected everybody to do his share of the necessary work. One job he usually reserved for himself: he was boyishly but justifiably proud of his ability to make good muffins, or gems, as they were sometimes called."

He collected insects incessantly, often around the campfire, into all hours of the night. "The pleasant gatherings around the bonfire at night, the songs, the lively conversations, sometimes the informal talks or stunts, the friendly comradeship lingered in many students' memories."

On these trips Snow climbed mountains and fished whenever he got the chance, besides collecting insects, birds, plants, fossils, minerals, and meteorites. Despite his ceaseless activity he always came back refreshed.

In 1881 Williams College conferred on him an honorary Ph.D., and in 1885 the legislature named the Snow Hall of Natural History in his honor. In 1889 Snow's title was "Professor of Natural History and Director of the Natural History Museum." However, in 1890 he was persuaded to accept the chancellorship of the University. With great reluctance he gave up his career as a teacher and scientist.

Snow was not a taxonomist, and many of the insect specimens collected by him and his students found their way into the hands of specialists. Undoubtedly, their efforts were in large part responsible for the identification and classification of much of the insect population of the West. For a number of years he was the editor of *Psyche,* published by the Cambridge Entomological Club.

Snow devoted much of his energy to the applied aspects of entomology in Kansas. Since he was well known as an economic entomologist, the Kansas farmers believed he was the logical man to be chancellor of the University. "In 1882 he had been appointed entomologist to the State Board of Agriculture. He had answered hundreds of inquiries, helping farmers to cope with the insects which damaged their fruit or laid waste to their crops. Besides writing many articles on troublesome insects, Snow enccuraged Vernon Kellogg, who in 1890 took over the teaching of entomology, to continue furnishing such information. Kellogg's *Common Injurious Insects of Kansas* acknowledges the author's extensive obligation to Snow." (Hyder)

The problem of the biological control of chinch bugs through fungus infections was one of Snow's pet projects. Healthy chinch bugs were infected in the laboratory and distributed through many states. Although, at times, this method of biological control appeared to be quite effective, this was the case only when environmental factors for propagation of the disease were especially favorable.

For many years he studied the writings of Darwin and in time accepted his teachings. This took great courage because of the clergy's hostility to Darwin and his ideas. Snow lectured for many years on evolution, and his views were accepted because of his deep religious faith.

As chancellor he strengthened the faculty of both science and the humanities. Samuel W. Williston gave up his position at Yale to accept the chair of geology and paleontology in the University of Kansas. In 1901 Snow obtained an appropriation from the legislature for a natural history museum.

He occasionally shucked the responsibilities of his office by participating in sports such as croquet, tennis, baseball between faculty and seniors, and speeding on his bicycle. "One day," Hyder relates, "he collided with a cross-eyed man whose eyes he had been watching. 'Why don't you look where you are going?' the man angrily gasped. 'Why don't you go where you're looking?' came Snow's swift rejoinder."

In 1889 Professor Snow suffered a calamity from which he never completely recovered. His son, Will, a reporter for the *San Francisco Chronicle,* accidentally drowned while on a tugboat, and Snow suffered a nervous collapse. He spent the years of 1900 and 1901 on leave, recuperating at Stanford University in the company of Vernon Kellogg and the David Starr Jordans. In 1901 he resigned the chancellorship of the University of Kansas and became Professor of Organic Evolution, Systematic Entomology, and Meteorology there. He still continued his teaching and his summer collecting expeditions.

Once again, in 1907, his nervous illness recurred, and in the summer of 1908 he and a student went to the lake region of Wisconsin. Here, on September 20, he suddenly died. He was buried in Oak Hill Cemetery, Lawrence.

About forty species of insects bear Snow's name and he and his students collected some quarter of a million specimens, comprising 21,000 species.

In addition to Hyder other important references to F. H. Snow are Williston (1908), Green (1908), and Kellogg (1909).

HERBERT OSBORN (1856–1954)

Although Osborn wrote three books on the history of entomology, the reader will find that he will have to go to other sources to learn much about him. I am indebted to the Knulls (1954) and to Dr. E. P. Breakey, a former student of Osborn's, for much of the following.

Herbert Osborn was born March 19, 1856, in Lafayette Township, Wisconsin; both of his parents were natives of Massachusetts. His father, Charles Paine Osborn, went to Wisconsin in 1840, where he married

Figure 39(A).

Figure 39(B). (A) and (B). Herbert Osborn. (A) Courtesy of J. J. Davis

Harriet Newell Marsh. Herbert was the third child in a family of four. When he was seven, the family moved to Fairfax, Iowa, a prairie land full of wildlife and a source of joy to a boy with an eye for living things. When he was fifteen he won a prize of fifty cents at a county fair with a case of stuffed birds.

Osborn (1937) says that his "experience as a boy in an amateur printing office with typesetting and composition; learning carpentry and practical horticulture with my father; some drawing and painting lessons in early youth; and serving as clerk in a drug store where I became acquainted with many drugs and chemicals; as well as experience in general farming" proved very helpful to him later in his career.

Illness and the "Depression of 1873" cut short his attendance at Grinnell College, and it was 1876 before he entered Iowa Agricultural College, now Iowa State College. There he studied under Professor E. C. Bessey, an enthusiastic biologist with a good knowledge of botany and some acquaintance with entomology. According to Osborn (1937), "Possibly the reason that he encouraged me particularly to devote myself to zoology and entomology was that he desired to be relieved from the teaching of that branch of study. Possibly he did not think me a good enough prospect as a botanist to keep me in the field." Bessey was the first to use London purple as an insecticide and gave it this name to distinguish it from that other arsenical, Paris green. Osborn received his B.S. in 1879. The day after graduation he was made an "assistant" in the College. The following year he won his M.S. and three years later married Alice I. Sayles, by whom he had two sons and three daughters. He was made professor at Iowa Agricultural College in 1885.

Osborn (1937) describes his background and training in entomology in these words: "In my own case there was little to guide me in the method of teaching entomology except the methods I had learned in connection with my studies under Professor Bessey in zoology and botany. I had, while still an undergraduate, taken charge of laboratory work in zoology, which consisted at first (1876) in the use of [D. S.] Jordan's *Manual of Vertebrates* in the identification of birds, mammals, reptiles, and fishes, and since no keys were available for insects I prepared a *Key to the Common Genera of Insects* that was used in my laboratory course in 1880 and for some years following. Also I gave lectures on the common insects of special economic importance in the region and these lectures were expanded from year to year using the reports of Riley, [William] LeBaron, and [Cyrus] Thomas with material adapted from [A. S.] Packard's *Guide to the Study of Insects,* and from personal studies of insects of the locality. During the winter of 1881–'82 I was with Hagen at the Museum of Comparative Zoology at Cambridge and I attended the lectures he gave which, however, were very few, and consisted of a series of talks on color and color pattern in insects, later published as one of his papers on insect coloration." Dr. Hagen thought highly of Osborn and introduced him to his friend, P. R. Uhler, the hemipterist.

In 1890, Osborn was appointed Entomologist of the Iowa State Experiment Station, while serving as a field agent for the U.S. Division of Entomology, a service he performed from 1885 to 1894. In 1898 he became State Entomologist. That same year he moved to Ohio to assume the position of Head of the Department of Zoology and Entomology at Ohio State University, a position he held until 1916. He was also associate entomologist at the Ohio Agricultural Experiment Station and organized and directed the Ohio Biological Survey and the Lake Laboratory at Sandusky on Lake Erie.

In 1916 he was relieved of his administrative duties and elevated to a research professorship so he could advise and direct the research of a constantly increasing number of students in entomology. He continued this work until his retirement as professor emeritus in 1933. His reputation as an economic entomologist attracted numerous graduate students from many of the states as well as from foreign countries.

He was also a consultant for several of the state experiment stations and for the Tropical Plant Research Foundation; through his association with the Foundation he was enabled to travel in Mexico, Central America, and Panama. He also made several trips to Europe.

In 1898 Osborn became president of the American Association of Economic Entomologists, and in 1911 of the Entomological Society of America. He was a member of numerous scientific societies and received honorary doctorates from Iowa State College, the University of Pittsburgh, and Ohio State University. He also served as managing editor of the *Annals of the Entomological Society of America* from 1908 to 1928.

In entomology, Osborn had two main interests, the taxonomy of Homoptera, especially leafhoppers, and economic entomology. "Starting with publication of his first scientific paper in 1879, he produced more than three hundred papers for scientific journals. His books include *Insects Affecting Domestic Animals,* 1896; *Economic Żoology,* 1908; *Agricultural Entomology,* 1916; and (after retirement) *Fragments of Entomological History,* Part I, 1937; Part II, 1948; *Meadow and Pasture Insects,* 1939; and *A Brief History of Entomology* published in 1952." (Knulls)

Dr. E. P. Breakey has given me the following account of Osborn:

> I was most fortunate in being closely associated with Professor Osborn for a period of several years as his assistant on the Ohio Biological Survey. I had known Professor Osborn by reputation since my student days at the University of Kansas. We had exchanged letters for I had become interested in the Homoptera, an order in which he was a foremost authority. . . .
>
> Late in 1929, I decided to return to school and become a candidate for the Ph.D. degree. It was a happy circumstance that the American Association of Economic Entomologists and the Entomological Society of America were to hold their annual meetings in Des Moines, Iowa at the end of the year. There was my opportunity . . .

I told Professor Osborn of my plans and hopes in a letter and he agreed to meet me in Des Moines and talk things over. As a matter of fact, I became a passenger on the same train that was taking him to Des Moines, recognized him at once, introduced myself and was invited to share his seat (compartment). We had a very pleasant visit and he told me about an assistantship on the Ohio Biological Survey that would be open and suggested I make the necessary formal application for it, promising to send me whatever papers were required on his return to Columbus.

Professor Osborn impressed me as being modest, gentle and companionable. He gave me a most cordial welcome and listened attentively to what I had to say. It was only natural that I, a young man, quite inexperienced and with my whole future before me, should feel uncertain and ill at ease when, for the first time, I found myself in the presence of the great man. That feeling soon left me for Professor Osborn had a way of putting men at ease.

My appointment as assistant on the Ohio Biological Survey became effective October 1, 1930. Professor Osborn had a desk placed in his office beside his own for my use and so began a most stimulating and rewarding association. Moreover, he was chairman of the committee in charge of my graduate studies and in this capacity became my chief advisor. Seldom has a young man been so fortunate at this stage in his development. Professor Osborn was 76 and I was 30 years old. What a tragedy it would have been for a considerable number of young men, myself included, had Professor Osborn been arbitrarily retired at age 65.

Professor Osborn's office door was always ajar. He was never too busy to see a student. . . . He responded to each with kindness and a sympathetic understanding. Each was treated as a personality worthy of respect and consideration . . . Professor Osborn's kindly interest and understanding were a source of strength and encouragement to many students with problems. He helped many develop that degree of confidence and self-respect that is so vital to the success of a young man entering one of the professions. I often admired his patience and composure.

Professor Osborn would often accompany me on field trips. If there is a better way for two men to learn what each is like basically and fundamentally than to go fishing together, it would be to go on a field trip together and collect insects, particularly if they were entomologists and shared a common interest. Professor Osborn was temperate in his habits, conservative and persistent. He enjoyed good health, had a remarkable memory and superb eyesight. Though he was of an age when most men have impaired eyesight, he used a hand lens when most of us would have used a binocular microscope. His general knowledge of and interest in all phases of natural history, together with his enviable memory for insect species, made him a most stimulating companion on a field trip.

Professor Osborn had a fine sense of humor and a wholesome appreciation of the ludicrous. He was seldom critical, often deferring judgement with the plea that perhaps we didn't understand the motivation behind a man's behavior. It takes a big man to display such magnanimity. He was a pleasant and stimulating conversationalist and I like to recall some of the philosophical discussions we had, usually at the end of a day's work, when a little relaxation was in order. The philosophy that gave him motivation and guidance in his day-by-day living seemed to be based on his conviction that men of good will (intent) were personalities worthy of respect and consideration and when given the opportunity, some encouragement and perhaps a little advice, each would make a worthy contribution to the general welfare of the society to which he belonged.

We learn from other sources that Osborn was somewhat withdrawn and soft-spoken. He inspired "unusual enthusiasm" in his students and he only recommended students for a job when they were thoroughly ready for it. It is for this reason that so many of them were successful.

Professor Osborn gave his collection and his valuable library to the Ohio State University.

"Hale and hearty at 96, he reported twice weekly to his Ohio Biological Survey office on the campus." (Knulls) On September 20, 1954, at the age of ninety-eight he died as the result of a stroke.

EDWARD OLIVER ESSIG (1884–1964)

Essig was to entomology in the West what John Henry Comstock was in the East. Although Essig was noted as an economic entomologist, an author, and a taxonomist, he was, above all else, an inspiring teacher who, like Comstock, aroused in his students a real devotion to the study of insects. Michelbacher (1965), with several other of Essig's

Figure 40. E. O. Essig

students, wrote a biography of him, and the material herein is largely from this source.

Essig was born in Arcadia, Indiana, on September 29, 1884, his mother dying less than a year after his birth. He was cared for by an aunt until 1887, when his father moved to California. In this same year his father remarried, and Edward and his older brother were then reared by their stepmother. The relationship between the two boys and their new mother was an excellent one for she treated them "with tenderness and love as if they were her own." (Michelbacher *et al.*) For several years the Essigs lived in Calistoga, California, then in Florence, Oregon, and after that in Fortuna, California. The father was primarily a farmer, but had to eke out his living by doing other odd jobs.

Through the aid of a family minister, arrangements were made to send Essig to Pomona Preparatory School to prepare for his entrance to Pomona College. He planned to study for the ministry, but he was diverted from this by two expert entomologists, Professors A. J. Cook and C. F. Baker, who had a profound influence on Essig throughout his life. As Essig's family was of very modest means, it was necessary for him to work his way through college. He accomplished this by waiting on tables and sleeping in the College Bell Tower, where he was responsible for ringing the bell. During the summer he worked as a busboy and bellhop at the lodge in Yellowstone National Park. Essig got his B.S. from Pomona College in 1909 and his M.S. in 1912.

Since he had been reared in the citrus-growing area of Southern California, he was naturally conversant with the insect problems that constantly challenged the farmer there. It is for this reason that he was especially interested in aphids and scale insects, prime pests of the citrus grower. Under the tutelage of two such teachers as Cook and Baker, he became expert on the taxonomy and control of citrus pests even as a student.

Upon graduation from Pomona, Essig was appointed Horticultural Commissioner in Ventura County in Southern California and remained in this position during 1910 and 1911. There he undertook investigations that led to the control of the citrus mealy bug, an accomplishment that placed him in great favor with the farmer and with county and state agencies.

In 1911 Essig moved to Sacramento, where Cook had recently been appointed to the California State Commission of Horticulture. Essig served as Secretary to the Commission from 1911 to 1914 and was also editor-in-chief of the *Monthly Bulletin of the State Commission of Horticulture*. "The publication also afforded a vehicle in which he poured out his seemingly inexhaustible energies. His extremely valuable and timely publication 'Injurious and Beneficial Insects of California' appeared in Volume 2 and immediately became the entomological bible for the entire state. The demand for it was so great that for a while it became a collector's item until his revised and enlarged treatise was published in 1915. These publications furnished the foundation for his outstanding work 'Insects of Western North America' which was published in 1926 and became the most used entomological text in western North America." (Michelbacher *et al.*)

In 1912 Essig helped prepare and was largely responsible for writing the California State Plant Quarantine Law, a major achievement in economic entomology. This state law was soon followed by the National Plant Quarantine Act, which profited from the California law.

In 1914, he was appointed Instructor in Entomology at the University of California at Berkeley. Although Cook was very reluctant to see Essig go, he realized that the University offered excellent facilities and opportunities for a man of his talents. Essig became a full professor in 1928, at which time he also became Entomologist in the Experiment Station at the University. From 1942 to 1943 he served as Acting Chairman of the Division of Entomology and Parasitology, and from 1943 to 1951 he was Chairman, despite internal political opposition. The Division flourished under him and many young and able men were added to the staff. During his administration the California Insect Survey was started, the first course in insect pathology to be set up in any university was initiated, and a course in plant nematology was established.

There were few teachers who engendered such enthusiasm in his students as Essig did in his course on "Economic Entomology," which was taken by students from all disciplines. Michelbacher *et al.* wrote: "Of all Professor Essig's accomplishments, none stands out more vividly than his ability as a teacher. He was freely accessible to students, ready to listen to their problems, and he stimulated them to put forth their best efforts. He had the power to render encouragement, increase morale, and develop confidence when students were in greatest need of such help. . . . He was always ready to see that they received fair treatment, and represented them when such service was called for. Further, Professor Essig's lectures were well organized and outstanding because of the wealth of knowledge and experience he had at his disposal. Also, his command of language, and his contagious enthusiasm for the subject matter made his lectures very easy to assimilate. He furnished his graduate students with wise counsel, and where possible left the decision on conducting research up to them. Although always available for consultation, he was anxious to develop independent thinking on the part of the student."

While any professor can dote on a brilliant and accomplished student, the truly great teacher accomplishes miracles with a student who has lost his way, and many students who stumbled and faltered along the way finally succeeded through Essig's faith in them. During World War II, and even afterwards, he wrote hundreds of letters of encouragement, filled with news and kindly humor, to former students scattered in all parts of the world. He never forgot them and they never forgot him.

To accompany Essig on a field trip was a treat to any student, for he soon absorbed some of Essig's enthusiasm for the collection and study of insects. Essig was a first-rate botanist and had a profound knowledge of plants and plant breeding, as evidenced by his own garden. He bred peonies, fuchias, and iris, and was famous for his iris hybrids, for which he won many awards.

The race, color, or religion of a student was of no special interest to Essig. All he demanded of him was ability, sincerity and perseverance. Thus, foreign students who came under his wing were especially fortunate, for he went out of his way to help solve their problems and encourage them in their entomological pursuits. Many of these students have since become notable entomologists in their native lands and gladly acknowledge their debt to Essig.

For many years a student group called "Fitchia" met once or twice a month at night in his house for refreshments. Here they had informal presentations and discussed current literature and current problems. This was a social affair and for many of the students the only social affair they were able to attend.

Essig was a tireless worker. Smith's (1965) bibliography of his writings lists a total of 518 publications, many of them major works. This remarkable productivity was made possible by Essig's habit, like Comstock's, of arising early enough to complete several hours of work before going to his office at the University. Among his major publications were the 1,029-page *A History of Entomology* (1931), whose original, more exact title, changed by the publisher, was *A History of Entomology in California;* and *College Entomology* (1942), a textbook noted for its organization as well as its beautiful illustrations.

In the later years of his life, Essig returned to one of his first interests, the taxonomy of aphids, of which he described nine genera and 106 species.

In physical appearance Essig somewhat resembled Napoleon, short, bald, with blue eyes that penetrated like Roentgen rays. He was immaculate in dress.

He won many honors, both national and international. He was president of the Entomological Society of America in 1938, of the American Association of Economic Entomologists in 1944, and of the Pacific Coast Entomological Society for many years.

After his retirement in 1954 as professor emeritus, Essig continued his work on aphids while doing a revision of his *Insects of Western North America*. Unfortunately, his health began to fail when he was in his late seventies and he was unable to finish many projects that he had in mind. After a long illness, he died on November 23, 1964, at the age of eighty, survived by his second wife Marie, and a daughter by a previous marriage.

His valuable collection of scale insects was given to the California Academy of Sciences, his aphid collection and his large library of rare entomology books to the Department of Entomology and Parasitology of the University of California. His students and colleagues set up a fund to add rare books to this library. Few teachers in entomology left behind so many friends and so great a legacy of good will and kind memories.

ROBERT EVANS SNODGRASS (1875–1962)

One of the greatest insect anatomists and morphologists in any country, R. E. Snodgrass was also an artist, a philosopher, and a teacher "who was a source of inspiration to all scientists." He lived a long and productive life and left behind a legacy of many notable publications. Our information about his life is derived from E. B. Thurman (1959, 1962), R. H. Foote (1962) and J. B. Schmitt (1962, 1963).

Snodgrass was of English, Scotch, Welsh, and Irish ancestry; his parents were James Cathcart and Annie Elizabeth Evans Snodgrass. Robert, one of three children, was born on July 5, 1875, in St. Louis, where he lived until he was eight years old. The family moved to Kansas; when he was fifteen, they moved to Ontario, California, to a twenty-acre ranch of oranges, prunes, and grapes. Young Snodgrass attended a Methodist preparatory school. He shot and mounted or dissected birds and various animals. He was always an independent thinker and "his openly avowed belief in evolution estranged him at home and caused him to be expelled from Sunday School, much to his satisfaction" (Thurman, 1959).

When, in 1895, he entered Stanford University his notes and drawings of his dissections provided him with entrance credits in zoology, in which he majored. He studied fish under Dr. David Starr Jordan, president of Stanford University, and entomology under Professor V. L. Kellogg. Because of Snodgrass' interest in birds, Kellogg encouraged him to study their biting lice, and as a result his first two publications were *The Mouthparts of the Mallophaga* (1896) and *The Anatomy of the Mallophaga* (1899). While a student at Stanford, Snodgrass collected animals in the Pribilof and the Galapagos Islands, as well as other islands, and several publications resulted from these studies.

Figure 41(A).

Figure 41(B). (A) and (B). R. E. Snodgrass. (B) Courtesy of J. J. Davis

Upon receiving his B.A. in 1901, he took a teaching job at Washington State College (now University) in Pullman. After about two years the authorities concluded that some of his practical jokes were too much for them, whereupon he returned to Stanford as an instructor in entomology under Kellogg and began his well-known studies on the anatomy of honeybees. During a period when Kellogg was away in Europe, Snodgrass raised silkworms and stripped the campus mulberry trees of their leaves to feed his voracious charges. The undressed trees fared poorly and died, and once again he was out of a job.

Snodgrass was next reported as working in the art department of an advertising agency in San Francisco and attending art school at night. "He did some magazine covers, designs of clay-modeled animals, and was to be taken on the staff at the San Francisco Academy of Sciences. However, the morning after he had been accepted by the Academy, the earthquake of 1906 shook up things, and then the fire came." (Thurman, 1959) After camping out for several weeks amid the ruins of San Francisco, he returned to his home in Ontario, California. In the fall of 1906, he went to Washington, D.C., and was hired by L. O. Howard, Chief of the Bureau of Entomology, for $60 a month. This was eventually raised to $100. Snodgrass published five papers and continued his studies on the honeybee.

Then in 1910, dissatisfied with his salary at the Bureau, he moved to New York City, where he attended night classes at the Art Student's League, studied human anatomy, and contributed cartoons to *Life* and *Judge* magazines. When next heard of, Snodgrass was selling paintings to Indiana farmers. Then he was employed for two years as an artist by the State Entomologist of Indiana. In 1917 Howard hired him again to do art work, and at odd times Snodgrass was able to pursue his anatomical studies on insects. He also did entomological work for the Bureau in Connecticut and Maryland.

In later years, he devoted his time largely to the study of the anatomy and morphology of insects. He defined anatomy and morphology by saying that "anatomy is what you see with your eyes, morphology is what you *think* you see with your mind"; external characters could only be explained by knowing underlying structures. Thurman says that Snodgrass was "more interested in the evolutionary changes and relationships of anatomical structures . . . than [in] descriptive work."

From 1903 to 1909 he worked on the anatomy of insects, especially the thorax, and in 1910 he published his first works on the anatomy of the honeybee. In 1921 he published a paper on the mouthparts of the seventeen-year cicada and in 1924 studies on the anatomy and metamorphosis of the apple maggot.

In the latter year, he married Miss Ruth Mae Hansford, by whom he had two daughters. His wife was a musician, but understood his interests, encouraged him in his work, and helped type and edit his manuscripts.

In 1925 McGraw-Hill published his celebrated *Anatomy and Physiology of the Honeybee*. Other major works by him were *Insects, Their Ways and Means of Living* in 1930, *The Principles of Insect Morphology* in 1935, and *Textbook of Arthropod Anatomy* in 1952. All in all, besides these four books he wrote eighty scientific papers, which brought him scientific acclaim all over the world.

Snodgrass taught entomology at the University of Maryland from 1924 to 1947. He sketched very rapidly as he spoke. Graduate students under his supervision were guided very carefully. Schmitt (1962) writes, "I like to think of him as I knew him then and the simplicity of his equipment. His work table was a shelf supported on brackets by a window. The sunlight poured into the window. It was all he needed to see by, and the simplest of forceps and scissors served him better than the most remarkable devices we can get today for that purpose. . . . Who could talk of the work of Robert E. Snodgrass and not mention his exquisite and precise skill as an illustrator, his capacity for finding just the right choice of words, and his pleasing and yet discriminating style of writing."

Although Snodgrass retired from the Federal Government in 1945, he continued his own work in the U.S. National Museum and completed fifteen important studies.

Among the many honors he received were an honorary doctorate from a German university (his biographers do not say which) in 1953 and one from the University of Maryland in 1960.

Foote (1962) recalls how, in the National Museum, "Snodgrass would carry his coffee" from the pouring place across a public hall to his desk 50 feet away. In itself not significant, this insistence included the filling of his cup to within one-eighth inch of the brim and thereupon boasting of not spilling a drop the whole way. The first time, I didn't believe him and followed every movement of that cup—not a drop was spilled! I thought he couldn't possibly do it a second time, but this came to pass, and a third, and a fourth, and indeed many in the months to come.

"After much reflection I came to realize that this wonderful gift was but an outward sign of a deep, abiding inner tranquility. I also began to see other signs—his love of a good story, his ability to write without revision, his distaste for noise, his lack of the backward look so many men his age inherit, and his eagerness to become involved in a new adventure."

More than twenty years after he retired Snodgrass was physically fit and mentally alert, and Thurman (1959) described him as a "dignified, erect, gracious unassuming gentleman," and said his "thick shock of glistening white hair, a ruddy complexion, and alert blue eyes, which require the aid of glasses only for reading, denote unusual physical stamina. A phenomenal memory for facts and events, a wealth of basic knowledge at his ready command, thorough training in the use of the Classical and Romance languages, and an unlimited vocabulary in English and German, spiced with wit, which is ready but sometimes sharp, make conversation with him informative and memorable pleasures."

On September 4, 1962, Snodgrass died in his sleep at his home in Washington, D.C., at the age of eighty-seven.

GORDON FLOYD FERRIS (1893–1958)

G. F. Ferris was one of the "giants" of taxonomic entomology in this century, and Wiggins (1959), Usinger (1959), and McKenzie (1959) have done him justice in their fine obituaries. It is not easy to pigeonhole so accomplished an entomologist as G. F. Ferris, and I hope I will be forgiven if I place him in this section among the great teachers of entomology.

As a student of entomology at Berkeley, I often attended the monthly meetings of the Pan Pacific Entomological Society in the museum of the California Academy of Sciences in Golden Gate Park, San Francisco. Assembled here would be E. P. Van Duzee, E. O. Essig, E. C. Van Dyke, F. E. Blaisdell, and G. F. Ferris, among others. I still remember stocky Professor Ferris emphasizing a point with a lion-like roar; like most undergraduates, I held him in great awe.

Usinger obtained from Ferris shortly before his death the following sketch of the early years of his life: "Born January 2, 1893, at Bayard, Allen County, Kansas, a 'tank town' where his father was a gandy dancer on the railway. He was the fifth child and the fourth boy in a family of

Figure 42. G. F. Ferris

five. When he was between two and three years old the family moved to a 40-acre farm near Monticello, Cedar County, Missouri, where they lived in a one-room log cabin. When he was a little more than three years old his mother died after giving birth to another son, who also died while still young."

Gordon lived on a farm with his father until he was thirteen. Then he joined his brother Leslie at Ottawa, Kansas, where Leslie was enrolled as a student in the Academy there and acted as a circulation agent for the *Kansas City Star*. Gordon graduated from the Academy in 1909 and entered Ottawa University in Kansas that fall, but he did so badly he soon withdrew. He then sought to enter the Navy, but he was underage and unable to get his father's permission. His brother, who was then teaching in Telluride, Colorado, obtained a job for Gordon with the Telluride Power Company. Fortunately for Gordon, the company maintained an endowed institution, the Telluride Association, one of whose purposes was to encourage the education of its employees.

"The Telluride Association maintained a home at Cornell University," writes Usinger, "where most of its men who were selected for 'preferment' went. Gordon disapproved of the social emphasis at this house and asked instead to be sent to Stanford University. He had seen Kellogg's *American Insects* while he was at Ottawa, and Kellogg—then a professor at Stanford—was the only entomologist he had ever heard of. In the summer of 1912 he was granted $450 by the Telluride Association and came to Stanford. With the continued support of the Association—never amounting to over $500 a year—he finished his work for the degree of M.A. in 1917." Thereupon he was made a teaching assistant at Stanford, and this was the beginning of his long and distinguished career with the University.

In regard to Ferris' scientific studies, Usinger writes: "Ferris was primarily a taxonomist and in this, as in other activities, he brought a fresh approach and a plan for work. Starting with the lice under Kellogg and the Coccidae under [R. W.] Doane he soon set a pattern from which he deviated but little in later years. Foremost was his insistence on detailed drawings—doubtless as a reaction to the shocking state of knowledge of lice and coccids based largely on inadequate descriptions. In his first paper, with Vernon Kellogg, the drawings are not divided by a vertical line into ventral and dorsal halves, but he soon adopted this economical method and gave his philosophy on drawings in 1923 (*Science* 58: 266): 'A scientific illustration is not intended merely as a pretty picture and it has nothing to do with art. Its purpose is merely to present in the simplest and most accurate manner the things that it is desired to show and its production involves nothing more than good draftsmanship. If in addition to these qualifications it is also artistic—whatever that may mean—so much the better.'

"Once started on the course of drawing details of minute insects instead of describing them, Ferris was committed to a compound microscope and pen and ink for the rest of his life. With prodigious energy and usually with several projects running concurrently he turned out monumental works. He provided the foundations for our present knowledge of the Anoplura, Mallophaga, Coccoidea, Diptera-Pupipara, Cimicidae, and Polyctenidae. Probably no man has ever made so many original detailed drawings of insects."

Because of the cost of illustrations much of his work was reproduced in offset lithography, often with money from his own pocket. According to Usinger, he regarded much of the previous work in his fields to be "of dubious value and not worth the trouble to look up." And he quotes Ferris as saying, "I have noticed that entomologists, just as ordinary mortals, rarely care to do any more work than they are compelled to do. So it follows that those insects which can be taken care of merely by pushing a pin through them are relatively well known, while those minute forms that are useless as cabinet specimens and that require special methods of preparation are usually neglected. Consequently they form the most unexplored fields of entomology."

Ferris was not only an expert laboratory entomologist but an excellent field naturalist and often accompanied mammalogists on field trips so he could comb the ectoparasites from the fur of the animals collected by them. His field trips took him throughout the Southwest, including Lower California and Mexico, and also to Panama, China, Hong Kong, Taiwan, and other countries. He spent some time studying the Anoplura in Cambridge University and the British Museum of Natural History.

Ferris was an independent and forceful personality; he stated what was on his mind in energetic language, whether verbal or written. He had several quarrels with morphologists over his theories on comparative morphology. Again, since he believed that quarantine regulations were of little use, he engaged in various feuds with officials and colleagues who believed in the principles of quarantine.

Wiggins has recorded Ferris' academic career in some detail. Ferris taught at Stanford for forty-two years, starting as a laboratory assistant and eventually becoming a full professor. He vitalized any subject he taught and drove his students hard, demanding from them close attention to detail, good drawings, carefully prepared collections, and well-written reports. On one trip to Mexico in the early 1920's, reports Wiggins, "he made huge collections of both lice and scale insects in Mexico, walking many long weary miles with his meager field equipment packed on one burro and accompanied by a timid peon often greatly in fear of his life from 'bandits' or 'tigers.' " After these field trips his lectures and laboratory sessions reflected his enriched knowledge and his adventures. He presented his well-organized lectures without notes, and this was "a constant stimulus and source of admiration among his students," who were often fascinated by his lectures because they felt they were "sharing his experiences and his enthusiasm."

Ferris taught not only general entomology, histology, microtechnique, classification of insects, but specialized courses on aphids and economic entomology, with emphasis on the Coccidae. He also taught a course on "Insects and Man," known to the students as "Bugs and Bites," a very popular elective course under Ferris. "When he described the malarial menace in the tropics, he had personal knowledge of the disease and its vectors, for he had suffered severely from malaria after his Mexican trip in 1926. He had seen natives of several countries existing under the burden of numerous ecto- and endoparasites and gave graphic and accurate word pictures of his observations." (Wiggins)

Wiggins describes Ferris' technique in handling graduate students by means of the informal conference. "During such conferences he usually smoked incessantly, lighting one cigarette from the butt of the previous one. The student often had to supply the match with which to light the first one, and not infrequently, the subsequent cigarettes. The conference might last ten minutes, or it could continue for hours. Many, many times the discussion involving the student's research work, his problems in other courses, or general philosophical questions with numerous ramifications, would be interrupted long enough for them to drive to the Ferris home where the discussion could continue through the evening meal and well into the night. On such occasions the student was made to feel at ease and was given encouragement if that seemed to be called for. He could be, and sometimes was, subjected to rigorous verbal chastisement. But always he was urged to think his own thoughts, to approach every question as objectively as possible, and to keep an open, independent mind. . . . He did not pamper his students, but he brought out their better qualities, often helped them find unexpected capacities, and won their life-long loyalty and respect."

Students came from states outside California and foreign lands to study under Ferris. "Summer after summer, when he was under no legal or moral obligation to remain on campus he spent practically every week day in his office-laboratory in order to be available to graduate students trying to complete requirements for advanced degrees. He received no additional remuneration nor academic preferment from such extra services, but was content to know that his students were benefitting from his presence."

The names of a few of the most important of Ferris' 217 publications are the "Catalogue and Host List of the Anoplura," *Proceedings of the California Academy of Sciences* (1916); *Principles of Systematic Entomology,* Stanford University Publications (1928); *Atlas of the Scale Insects of North America,* in 4 volumes 1950–1953); *The Sucking Lice, Memoirs of the Pacific Coast Entomological Society* (1951).

Ferris died on May 21, 1958, at the age of sixty-five.

6 Notable Neuropterists

I wonder in what fields today
He chases dragonflies in play
My little boy—who ran away.
Kaga No Chiyo

Lacewings, dobsonflies, mayflies, dragonflies, ant lions, damselflies, and stone flies are some of the Neuroptera of the early entomologists. Anyone who has observed or collected them in streams, ponds, and swamps can understand their fascination for these delicate and beautiful insects.

JAMES GEORGE NEEDHAM (1868–1957)

Entomologists are indebted to Dr. Needham for his outstanding contributions to the study of aquatic insects. It is surprising that I was able to find so little printed autobiographical material on a man who has played so prominent a part in American entomology. Dr. S. W. Frost, one of Needham's most prominent students, has given me an account of his late teacher. There is also an obituary by H. H. Schwardt (1959).

Needham was born in Virginia, Illinois, on March 18, 1868, and attended the public schools there. He received his B.S. and M.S. degrees from Knox College in Illinois and while teaching there from 1894 to 1896 wrote *Elementary Lessons in Zoology*. This brought him to the attention of Professor J. H. Comstock of Cornell. Needham was granted a scholarship to Cornell, and while there he and Comstock wrote their well-known *The Wings of Insects,* which was published in 1899 by the Comstock Publishing Company.

From 1898 until 1907 Needham was Professor of Biology at Lake Forest University, Illinois. During the summers he worked for the New York State Conservation Department, studying the role of aquatic life in fish food. He was then invited to teach limnology at Cornell, to become "a world leader in this branch of biology, and to found a tradition of excellence in this field that still flourishes at Cornell." He also

Figure 43. J. G. Needham

gave there a biology course for non-biology majors, which won him great acclaim. For years he studied fresh water biology in a field station at the head of Cayuga Lake.

Schwardt writes that "Ecology entered the teaching curriculum of the Department of Entomology because Professor Needham developed it along with limnology and biology. Any student who had the privilege of going into the field with him has a lasting memory of a great naturalist at work. Nature unfolded its intricacies around him and with warmth and enthusiasm he made one see the life in a pond, in a stream, on an alder bush, or a goldenrod plant as one had never dreamed it existed. Professor Needham was so much at home with all of his friends in the plant and animal world that students sensed his inspiration and shared his enthusiasm for nature."

Upon the retirement of Professor Comstock in 1914, Needham was appointed head of the Department of Entomology and as such served with distinction until his own retirement in 1935. He then continued his research and published an additional 150 papers.

He was widely traveled, and he taught and did research on insects in China. With an international reputation for his teaching, greatly admired and loved by his students, he attracted young people to Cornell from all parts of the United States and the world.

One of his students, Dr. S. W. Frost, wrote me his recollections of his teacher:

I knew Dr. James G. Needham for many years. When I entered Cornell University in the fall of 1911 he was in the early days of his professorship.

Although he wrote many books on biology, he always maintained that nature should be studied outdoors not from books. He was a master in directing field courses and the students always looked forward to these trips. I well remember journeys to nearby woods and swamps to study plant and animal life. In his old clothes, sometimes wearing boots, he looked little like a professor. His dignity and enthusiasm always overshadowed his awkward posture. He dealt with common things, the soil, the flowers, the wind and the weather. The common dandelion took real shape and color when he put it in its proper setting. It became a gorgeous rosette of leaves closely pressed to the ground trying to avoid other plants that towered above it.

He was fond of young people, especially children, and they liked him equally well. He was instrumental in persuading many young people to come to Cornell. Naturally he was most interested in promising biologists and encouraged many to continue their studies in this field. I had only one course with him in general biology; however he took as much interest in me and my work as though I were one of his special students. He inspired me and guided me in my studies. The joint production of *Leaf-Mining Insects* is evidence of this.

His dialect seemed foreign, and to many of us it was difficult at first to understand him. The better students took seats towards the front of the classroom to grasp every word. Even to the last I found myself listening closely to catch some of his expressions.

In summer he came to class without a coat. This was somewhat unusual among professors in those days. I remember one morning he was wearing jeans with suspenders. It was a warm pleasant morning and he apparently had been working in his garden until the last minute. One of the loops of his suspenders in the back bothered him and he reached back cautiously to try and button it. Without success he turned about and remarked to the class, "I told my wife to sew that button on."

He shunned scientific names wherever he could avoid them and preferred to speak in common terms. This is probably the secret of his popularity in lecturing to many groups. When describing the appendages of the alimentary canal of an insect he likened it to a clothes' line and proceeded to hang the organs from head to tail, illustrating each with one of his inimitable sketches.

For many years he refused to obtain an automobile. He really liked walking, especially from his home to the University, which was a rather long steep climb. Frequently he stopped to listen to a bird or pluck a weed that he might use in class. As far as the auto was concerned he finally succumbed to the demands of the family and purchased a car. However he was never much of a driver. Even when he was staying at the Archbold Biological Station in Florida, which was somewhat isolated and ten miles from the nearest town, he preferred to walk and depended on others for transportation when he wanted to extend his collecting areas.

The Needhams often entertained students in their home. Members of the faculty frequently attended. These meetings were always informal. I remember a game that was played on a large table. The contestants took sides on opposite ends of the table. The object of the game was to blow empty egg shells across the middle line. It was a curious sight to see dignified professors crouching and puffing at these unruly objects. Ofttimes interesting remarks or demonstrations added intellectual touch. A student interested in Indian lore demonstrated the art of chipping arrow heads. These meetings certainly made a student feel at home and must have been relaxing for the older members.

The last winters of his life, after he had retired, were spent in Florida collecting dragonflies. He spent some time at Englewood but finally found comfortable and profitable hours at the Archbold Biological Station located about ten miles south of Lake Placid, Florida. The staff and visitors spoke of him in the highest terms and he made friends wherever he went. I spent many winters there myself and everywhere I traveled people would ask about Dr. Needham. He spent most of his time out of doors studying the unusual plants as well as the dragonflies of this area. He equipped everyone with nets and they aided him in collecting dragonflies. Even the clerk in the office tells me how she gathered specimens for him.

Those at the station remember how, almost every clear evening, he climbed the hundreds of steps to the top of the water tower, singing as he ascended. There he relaxed and let his thoughts wander over the extensive pine barrens stretching in all directions to the horizon. Occasionally he would capture some interesting insect flying at this relatively high elevation.

It is not generally known that he was somewhat of a poet. In 1959 he published a pamphlet, "Ontario and Other Verses." From these one learns much of his philosophy of life, his love for his family and the great out of doors.

I am glad to have known this outstanding biologist, especially during the years when he was most active. During his latter years his hearing was poor and he used a hearing aid. His humor never dwindled. Once he remarked to me, "Frost, when will we resort to a 'smelling aid'?"

Schwardt comments: "One of the highest and most gratifying tributes that a college professor can receive is the respect and admiration of his graduate students, and this Dr. Needham enjoyed in full measure. Those of us who were present will always remember his appearance at a Cornell luncheon at our Richmond meetings a few years ago. Each of the more than 100 present was asked to introduce himself, identify the years spent at Cornell, and the professor with whom he studied. By chance Dr. Needham was the last to rise, and his eyes were moist, as he smiled and acknowledged the long, standing ovation that followed. No words were spoken. None were needed."

Needham died in Ithaca on July 24, 1957. He was eighty-nine years old and his mind was fully alert to the end.

He was the author of approximately 350 publications, a number of which were major contributions to the identification, behavior and ecology of water-inhabiting insects. Some of his more important publications are "Aquatic Insects in the Adirondacks," with C. Betten (*New York State Museum Bulletin* 47, 1901); *A Monograph of the Plecoptera or Stoneflies of America North of Mexico,* with P. W. Claasen

(1925); *Leaf-Mining Insects,* with S. W. Frost and B. H. Tothill (1928); *Handbook of Dragonflies of North America,* with H. B. Heywood (1929); *The Life of Inland Waters,* with T. J. Lloyd (2nd Edition, 1930); *The Biology of Mayflies,* with J. R. Traver and Yin-Chi Hsu (1935); and *A Manual of the Dragonflies of North America* (*Anisoptera*), with M. J. Westfall, Jr. (1955).

PHILIP POWELL CALVERT (1871–1961)

At the 1960 annual meeting of the Entomological Society of America in Atlantic City, I saw a distinguished-looking old gentleman hovering in the antechamber of the main hall, apparently seeking in vain some familiar face. Later I learned that this was the eminent entomologist Philip P. Calvert, who, at the ripe age of eighty-nine, had outlived almost all of his contemporaries except his good friend, J. A. G. Rehn. Of Calvert, Rehn writes "He was a counselor, a friend, and a colleague of some sixty-odd years," and the following is from Rehn's (1962) tribute to him.

Calvert was the son of the Philadelphia lawyer, Graham Calvert, and Mary Sophia Powell Calvert and was born in Philadelphia on January 29, 1871. He attended the Friends' and public schools of Philadelphia, finishing at Central High in 1888. In 1892 he graduated from the University of Pennsylvania with a certificate in biology and received a Ph.D. in 1895. At the University of Pennsylvania he was exposed to the influence of such scientists as Joseph Leidy and Edward Drinker Cope.

Figure 44. P. P. Calvert

Calvert continued his studies in 1895–1896 at the Universities of Berlin and Jena in Germany.

He began to teach zoology at the University of Pennsylvania in 1907, where he remained until he retired in 1939 as Emeritus Professor of Zoology. According to Rehn he "was outstanding in sparing neither time nor effort to give his best to both undergraduates and graduates in his lectures and his more personal contacts with students in their work and problems. His students were fellow-workers, and his attitude toward them was that of a fellow-scientist. This encouraged the development of originality and responsibility on the part of the student, tempered, as the teacher's suggestions always were, with the utmost consideration for the student and the latter's comprehension of his immediate problem. This combined with the teacher's personal interest, and his kindly and gentle sense of humor, endeared him to a host of students in his years of teaching, many of whom recall the value to them of the help of this kindly scholar."

In 1901 Calvert married Amelia Smith, who had received her B.S. in biology from the University of Pennsylvania. "Their home life was one of those congenial ones of highly cultured people, fully alive to the world about them, its people and its problems. A visit to their delightful modernized old Pennsylvania farm house was always something one long remembered, not only for the setting but for the brilliant and kindly pair who made it their home."

Calvert was a great authority on the Odonata, not only in America but in the world, and he wrote more than three hundred articles on this order, starting in 1889 and completing his last in 1961, a few days before his death. All this work was accomplished "in the occasional intervals of a very busy teaching schedule." A few of his major works are "Catalogue of the Odonata (Dragonflies) of the Vicinity of Philadelphia, with an Introduction to the Study of This Group of Insects" (*Transactions of the American Entomological Society,* 1893); "Insects Neuroptera Odonata" (*Biologia Centrali-Americana,* 1901–1908); "Contribution to a Knowledge of the Odonata of the Neotropical Region, Exclusive of Mexico and Central America" (*Annals of Carnegie Museum of Pittsburgh,* 1909); and "The Rates of Growth, Larval Development and Seasonal Distribution of the Genus *Anax*" (*Proceedings of the American Philosophical Society,* 1944).

During the year 1909–1910, while on sabbatical leave, he and his wife studied the flora and fauna of Costa Rica and wrote a semipopular volume. *A Year of Costa Rican Natural History.* Rehn reports that in Cartago the authors experienced the earthquake "which, on April 15, 1910, destroyed a large part of the city. They fortunately escaped injury, although their friends in Philadelphia were without news of them for some days. They reached New York safely two weeks after their harrowing experience."

Besides taxonomy, Calvert also wrote on anatomy, distribution, paleontology, and ecology. His taxonomic studies encompassed not only the United States and Costa Rica but Guatemala, Paraguay, British Guiana, and other Central and South American countries.

Calvert was one of the pillars of the oldest entomological society in the United States, The American Entomological Society, serving as president from 1901 to 1915 and as editor of *Entomological News* from 1911 to 1944. He never received any compensation for his efforts in behalf of the Society, and as a matter of fact, at times he dipped into his own pocket to help it. He was also one of the founders of the Entomological Society of America.

A bibliography of Calvert's writings appeared in *Entomological News* for 1951, Volume 62, pages 3 to 40. This bibliography encompassed all his publications from 1889–1950; upon his death in 1961, there appeared in *Entomological News* another bibliography of publications subsequent to 1950.

Shortly after retiring in 1939, he suffered a stroke but recovered and lived until August 23, 1961. Rehn bade farewell to him with these words, "a gentle and kindly nature, a touch of whimsy, and an utter lack of self-laudation . . . made Philip Calvert beloved by all who knew him well."

NATHAN BANKS (1868–1953)

Nathan Banks was an expert in systematic entomology, a world authority on spiders, scorpions, solpugids, ticks, mites, the Neuroptera (broad sense), Isoptera, Psocoptera, and Diptera.

Banks summed up his life in one paragraph (Gunder, 1930):

"The 13th has always been unlucky for all spiders and bugs which get in my way. Perhaps it is because I was born at Roslyn, New York, on the 13th of April, 1868; however, be that as it may, Roslyn still remains a good town! Like all boys of a kind, I collected and my first book was [J. G.] Wood's *Insects at Home*. Graduated from Cornell in 1889 and thought so much of the school and studying with Prof. Comstock that I took the postgraduate course the year following. Was employed in the old Division of Entomology at Washington under Riley from July, 1890, till September, 1892, when the Democratic Congress (bless their free-trade on insects) reduced appropriations and the young men were fired, or rather kissed goodbye. Went home to Sea Cliff, New York, where I carried on my insect studies, collected and began to publish largely on spiders. In 1896 was again appointed to the Government's

Figure 45. Nathan Banks

Division of Entomology under the orderly regime of Dr. Howard with work on bibliography, ticks, mites, dipterous larvae, etc., until 1916, when I left to come up here [Museum of Comparative Zoology, Harvard]. In the meantime had built up a good private collection of Arachnids and also Neuroptera. Have eight children and one helps as preparator in the Museum. Live twenty-five miles out of Cambridge at Holliston, on a ten-acre place, where the collecting is good and I sometimes find new spiders in the back yard. Don't know which of my published articles to recommend now, can't find that out till after I'm dead!"

F. M. Carpenter and P. J. Darlington (1954) note that he received his M.S. in 1890, and they report that although he became dissatisfied with his Washington job as early as 1904, when he applied for a position in the Museum of Comparative Zoology, where Samuel Henshaw had recently been appointed director, he had to wait until 1916 before an opening was available there. In the interim he was offered another museum job, but he turned it down because he would have been limited to working with the Thysanura, Arachnida, and Neuroptera, and "he rather fancied himself an entomologist."

When he moved to Cambridge, he presented his collection of more than one hundred twenty thousand specimens with about eighteen hundred types to the Museum and also his library of "about a thousand pamphlets and books not in the museum library." In 1928 he was appointed Associate Professor of Zoology. He gave no formal courses but guided advanced students in their research. In 1941 he became Head Curator of Insects, a title he held until his retirement in 1945 at the age of seventy-six. He remained active for a few years after retirement until his strength gradually failed. He died at the age of eighty-four on January 24, 1953, in his home at Holliston, Massachusetts, survived by his wife and eight children.

Although Banks was a good field collector, his curatorial duties afforded him little time for extensive field trips, but he made a number to the Black Mountains of North Carolina, to the Great Smoky Mountains "accompanied by his son Gilbert, and P. J. Darlington, Jr., and F. M. Carpenter," as well as to Barro Colorado Island in the Canal Zone. Of course he collected widely in Massachusetts and nearby states.

"Banks greatest contribution to entomology was through his service and devotion as curator of the entomological collections. Having virtually no assistance at the museum he did the routine curatorial work on the collection, compiled catalogues of types and genera, and prepared specimens for shipment to investigators in other institutions. The continued high rank of Harvard's entomological collection is chiefly the result of his efforts. The collection itself is indeed a memorial to the devotion which he had for entomology. . . . [He] had a great capacity for work, a single-mindedness of purpose, and a good memory. It is a misfortune that his knowledge of the structural diversity and adaptations of insects was not more available to biologists. This was partly his fault. He had not learned the modern vocabularies of genetics and evolution. Moreover, Banks, feeling that he was too busy, did not join the smoking and conversation group on the Museum steps in mid-morning and mid-afternoon, further restricting the circulation of his knowledge. Nevertheless he liked people and liked to talk, and was kind, helpful, and had a sense of humor." (Carpenter and Darlington.)

According to Harry K. Clench, Nathan Banks in his advanced years was a somewhat grim-looking gentleman but not quite as grim as he looked. For instance, he began each of the type catalogues in the Museum of Comparative Zoology with some doggerel of which the following is typical:

> With Joy I add each number
> Another type will be
> Eternally in slumber
> In the good old MCZ.

His bibliography consists of 440 technical papers, at first mostly on arachnids, and later on the Trichoptera, Mecoptera, Neuroptera, Perlaria, Isoptera, and Psocoptera.

A few of Banks's more important publications include *Bibliography of the More Important Contributions to American Economic Entomol-*

ogy (1898, 1901, 1905); *Catalogue of the Neuropteroid Insects (except Odonata) of the United States* (American Entomological Society, 1907); *A Revision of the Ixodoidea or Ticks of the United States* (Technical Series, U.S. Division of Entomology No. 15, 1908); *A List of Works on North American Entomology* (Bulletin U.S. Division of Entomology No. 81, 1910); *The Acarina or Mites* (U.S. Department of Agriculture Report No. 108, 1915); *Revision of Nearctic Termites,* with T. E. Snyder (Bulletin 108, U.S. National Museum, 1920); and *Revision of the Nearctic Myrmeleonidae* (Bulletin 68, Museum of Comparative Zoology, 1927).

EDWARD BRUCE WILLIAMSON (1877–1933)

Williamson was born on July 10, 1877, in Marion, Indiana. His father, who was president of the Wells County Bank in nearby Bluffton, Indiana, enjoyed collecting dragonflies and encouraged his son in this pursuit. Edward studied at Ohio State University and received his B.S. in 1898. From 1898 to 1899 he was an assistant curator of insects at the Carnegie Museum under Dr. W. J. Holland, with whom he apparently did not get on very well. For a time Edward thought of studying medicine at the University of Pennsylvania, but an impairment of his hearing put an end to that idea.

Williamson began to work in the Wells County Bank in 1902 and became president in 1918. He had taken a job at the bank with the thought that eventually he could obtain a curatorship in one of the

Figure 46. E. B. Williamson. Courtesy of J. J. Davis

entomological museums in the country, but in this he was unsuccessful. Nevertheless he continued to collect and study dragonflies with great enthusiasm and was in constant correspondence with P. P. Calvert, who was also working on this order. Williamson collected widely throughout the United States, in Guatemala, and in Columbia. In 1899 his "Dragonflies of Indiana" was published in a report by the Department of Geology of Indiana. All in all, he published more than a hundred papers on the Odonata. Besides dragonflies, he collected birds and bird eggs. In 1929 he continued his study of the Odonata as a research student in the Museum of Zoology at the University of Michigan. He died in Ann Arbor, Michigan, on February 23, 1933. His collection, which contained many types, was turned over to the University Museum.

Calvert (1935) writes: "That which especially distinguished E. B. Williamson, it seems to me, was the combination of indefatigability as a collector, and minuteness of observation as a student and author. This resulted in his knowing the Odonata, both in the field and in the laboratory, with probably a greater thoroughness than any of his predecessors or contemporaries."

7 Notable Orthopterists

Of all the beasts that ever were born
Your locust most delights in corn
And though his body be but small
To fatten him takes the devil and all!

Grasshoppers and locusts are among the most familiar insects to man. They have been the bane of his existence from the earliest times, and it is not surprising that the older entomologists showed a deep interest in them.

SAMUEL HUBBARD SCUDDER (1837–1911)

It is a moot point whether Scudder should be classed as a lepidopterist, an orthopterist, or a paleoentomologist, for he was outstanding in the field of each of these. If we classify him as an orthopterist, at least we have the support of J. A. G. Rehn, who said, "He was the greatest orthopterist America has produced."

Figure 47. S. H. Scudder

The Scudders were of Puritan origin, and they followed the sea, his father being Charles Scudder, a merchant, and his mother, Sarah Lathrop Coit. Samuel was born in Boston on April 13, 1837. The family moved shortly thereafter to a pleasant country home three miles from Boston. "Here among the woods and fields of a 30-acre estate young Scudder spent his early years, his only known adventure being a successful attempt to jump over a cow, which resulted, however, in a broken arm on his part, but no recorded injury to the cow." (Mayor, 1924)

Samuel Scudder attended the Boston Latin School, and when he was sixteen, was sent to Williams College so that he might be under the guidance of Mark Hopkins the educator, just as F. H. Snow was a few years later. Williams College had a number of great teachers, such as Ebenezer Emmons in geology and Albert Hopkins in natural history, who had a profound influence on Samuel. A glass case of butterflies on a friend's wall aroused his first interest in insects. He then constructed a net and began to collect in the vicinity of his home. In his junior year in college he decided to make the study of insects his life's work. The Lyceum of Natural History in Williams College, with its growing library and museum, helped to shape his future.

In 1857 Scudder graduated from Williams at the head of his class; he received his A.M. in 1860 and his D.S. in 1890. His brother, David Coit Scudder, of the class of 1855, became a missionary to India, only to be drowned a few years later. Horace Elisha Scudder, another brother, of the class of 1858, edited the *Atlantic Monthly* for some time, and a third brother, Charles, was a prominent business man in Boston and treasurer for many years of the Massachusetts Institute of Technology and the Boston Society of Natural History.

Samuel loved the mountains of New England, especially the Berkshire Hills, and, according to Mayor, "became the leading spirit of the 'Alpine Club of Williamstown.'" Later he was a founder and first vice president of the Appalachian Club, and then its second president. The journal of the club, *Appalachia,* published many articles by Scudder such as "A Climb on Mount Adams in Winter," "A Winter Excursion to Tuckerman's Ravine," "The White Mountains as a Home for Butterflies," and "The Alpine Orthoptera of North America."

After graduating from Williams College, Scudder entered the Lawrence Scientific School of Harvard to study under Professor Louis Agassiz. A. G. Mayor (1924) writes, "So to Agassiz he went with the statement that he intended to devote his entire life to the study of insects. The great master shook his head, and drawing a very dead and discolored fish out of a bottle of alcohol he deposited it in the hands of young Scudder telling him to observe and report. For ten minutes he studied this unattractive object, and then endeavored to report, but fortunately the professor was away. For three whole days he gazed at that single fish before he could satisfy the professor upon the important point that it possessed 'symmetrical sides with paired organs.' Then followed eight months entirely devoted to the study of somewhat similar fishes, all *Haemulous,* until the pupil saw *Haemulous* in his dreams, and grew to associate the odor of preserving fluids with pleasant memories. One thing seemed thereafter to have been burned into his very nature; devotion to all but infinite detail. Indeed throughout his scientific career one wonders not so much at the great bulk of his writings, as at the vast mass of minute and accurate details of observation therein presented."

After studying with Agassiz for four years he received a B.S. in 1862. He remained as an assistant to Agassiz until 1864.

From 1862 on, Scudder was associated with the Boston Society of Natural History in a variety of capacities, as recording secretary, librarian, custodian, vice president, and from 1880 to 1887 as president. From 1879 to 1882 he was also an assistant librarian at Harvard University. In 1879 he helped found the Cambridge Entomological Club, and was one of the founders of *Psyche,* and from 1883 to 1885 its editor.

A. P. Morse (1911) describes him as follows:

Personally Mr. Scudder was the highest type of a scholarly gentleman: a broad-minded, dignified, cultivated, courteous savant, in whom were united the finest attributes of the scholar and man of science; yet genial withal, and most kind and helpful to the inquiring student. Well do I remember the cordial welcome he extended to me, an unknown quantity, in response to the rat-a-tat of his laboratory knocker—that quaint conceit, a knocker in the form of a locust, beating upon the door with its hind legs!—when I first called upon him, as well as the many delightful hours spent there afterward in the study of his collection. His unrivaled library, rich in everything entomological and as complete as possible in his specialty; his collection unequaled in America, containing specimens from the ends of the earth; and most of all the man himself, well-versed in many branches of the science, made his laboratory the Mecca of every entomologist, resident or migrant, native or foreign.

In those days (the 90's) the Cambridge Entomological Club met [in his house], its members few but determined to keep the lamp alive and maintain the high traditions of an earlier time. Mr. Scudder was a host in himself, Roland Hayward, now with the great majority, was very regular in attendance; Mr. [Samuel] Henshaw came frequently, less often in the later years; Messrs. Bowditch and [J. H.] Emerton, still with us, occasionally appeared; rarely, birds of passage visiting the Museum of Comparative Zoology or Mr. Scudder himself, among them Dr. Geo. H. Horn, Prof. Lawrence Bruner, and other entomologists of note; and among the younger men, while resident in Cambridge, I recall especially J. W. Folsom and W. L. Tower.

This period was at the flood tide of Mr. Scudder's productiveness on the orthoptera. Never a meeting passed but that he had something to communicate;

—additional or newly worked material, new discoveries based on his studies, or notes of interest gleaned from his wide reading of entomological literature.

Scudder's first work on insects, a report on T. W. Harris' collection, appeared in the *Journal of the Boston Society for Natural History* for July 1859; his *Catalogue of Scientific Serials* appeared in 1879; and his *Nomenclator Zoologicus* in 1882. For forty or more years a flood of scientific papers and popular works flowed from his prolific pen.

J. S. Kingsley (1911) thinks that "Especial emphasis should be laid on his connection with the periodical 'Science.' There had been a struggling and jejune journal with that name but Scudder was able to interest some wealthy men in the project of a weekly scientific newspaper which should adequately represent all departments of science. So the old journal was bought, so as to control the name, and the new one was started, with Scudder as chief editor, in 1883. It was ably edited and rejoiced the hearts of scientific men of the day. It began by paying for all contributions and soon exhausted its guarantee fund; there were not subscribers enough to pay the expenses and no one thought of the later expedient of making it the organ of some large association. So after two years in the editorial chair, Mr. Scudder dropped out. The scientific public was not large enough and the general public would not support the journal, so, after lingering along for a few years it died."

According to Morse, Scudder's first paper on the Orthoptera appeared in 1861, and he continued working with this order until 1902, when physical infirmities terminated his labors. He was especially active in his studies on the Orthoptera from 1891 to 1901. A few of his many important papers on this group are the *Catalogue of the Orthoptera of North America Described Previous to 1867* (1868), *Guide to the Genera and Classification of the North American Orthoptera Found North of Mexico* (1897), *Revision of the Orthoptera Group Melanopli (Acridiidae) with Special Reference to North American Forms* (1897), *Index to North American Orthoptera* (1901), and *Catalogue of Described Orthoptera of U.S. and Canada* (1900).

Scudder became a world authority on the Orthoptera. Most of his field work on them was done in his youth; the New England Orthoptera were largely collected by himself, and the rest he got by exchange or purchase or received from Government surveys of the West. Besides the classification of the Orthoptera, he was very much interested in their biology and distribution. Unlike some of his work on butterflies, his work on the Orthoptera was of a technical nature, and very little was written for beginners.

"Mr. Scudder worked rapidly," wrote Morse, "too rapidly for accurate results sometimes; but in no other way could he have accomplished so much. Once, on remarking to him in connection with some of my own studies that if one worked rapidly he was liable to make errors, he rejoined with emphasis 'Sure to, sure to!' In the early period of his activity he was perhaps prone to draw conclusions from too limited a supply of material but that was the custom of the time quite as much as of the man. In later years, with long series before him, he often worked with selected examples of the species he had tentatively discriminated, instead of with the series as a whole. This method, though economizing time, provided the possibility of numerous errors, from which he did not wholly escape. No one, however, was ever more ready to acknowledge and rectify mistakes when called to his attention." In his work on the Orthoptera, he described 106 genera and 630 species.

Scudder's monumental work in three volumes, *The Butterflies of the Eastern United States and Canada* (1889), was the culmination of more than thirty years' effort. After its publication he concerned himself largely with Orthoptera and fossil insects, but still found time to write several popular books on butterflies. He intended to write a manual of the butterflies of North America, but unfortunately this was one task fate did not permit him to complete. But W. L. W. Field (1911) has written: "Doctor Scudder's essays upon migration, geographical distribution, protective coloration, dimorphism, and other evolutionary aspects of butterfly life, pointed the way to broad fields for research among our native insects. These essays, most of them placed as 'excursi' between the accounts of different families and genera in his great monograph, are perhaps the most widely read of his writings." He was greatly interested in morphology, feeding habits and phylogeny; and he was a pioneer in the study of the genitalia of butterflies for the purposes of classification. He described thirty species of butterflies and "was ever revisiting the haunts of particular butterflies, and amending or confirming, by patient observation, his accumulated data."

According to Mayor, "Scudder believed that generic names should be used to indicate differences rather than to show relationships. He was thus one of the type of systematists known as 'splitters,' and nowhere does this tendency appear in his works in a more accentuated degree than in his own treatment of the generic names of butterflies."

Butterflies were the subject of most of his popular works on insects, and he wrote of them with a contagious enthusiasm. In the preface to his book, *Everyday Butterflies,* he says, "Butterflies have often been compared fancifully to flowers. They are certainly like them in that each kind has its own season for appearing in perfect bloom; and thus they variegate the landscape in the open season of the year." Two of his other popular writings on butterflies were *Frail Children of the Air* and *The Life of a Butterfly in a New Region,* the latter showing the gradual invasion of North America by the Old World cabbage butterfly. Many butterflies owe their common names to him.

A third avenue for Scudder's tremendous ability and energy was the study of fossil insects. He instigated American insect paleontology and contributed a remarkable number of papers on the subject, his first appearing in 1865. Mayor remarks that Scudder as a pioneer made many mistakes in the study of fossil insects: he often had only fragments of them to work with. From 1886 to 1892 he was paleontologist for the U.S. Geological Survey, and named 838 North American insects of the Tertiary period, most of them from Florissant, Colorado. He also named 300 other fossil insects, and in 1891 the U.S. Geological Survey published his 744-page *Index to the Known Fossil Insects of the World, Including Myriapods and Arachnids.*

Although he had little ability in drawing, his works contain many fine illustrations. Kingsley, who drew fossil cockroaches for him, says, "When my drawings came to him for criticism, I was astonished again and again, at his extreme accuracy; a vein a hundredth of an inch from its proper position was always noted."

According to T. D. A. Cockerell, a noted entomologist and naturalist in his own right, "He was perhaps, the greatest entomologist of his time. Whether we regard the mere mass of his work or its excellence or the breadth of view shown, we who belong to this later generation must stand amazed and humbled. . . . I think of two especially prominent characteristics, his *enthusiasm* and his *kindness.*"

In addition to all his other work, Scudder managed to go on a number of field trips. In 1903, he wrote a description of one of these: "Fully thirty years ago, the last week of July found my companion and myself in a railway town in Wyoming, camping on the floor of the storage-room of a Western post office and 'store' combined, frequented alike by Indians, half-breeds, and whites. We had just room to lay ourselves down at night on buffalo robes in the narrow passage between barrels of molasses on one side and cheeses and firkins of pretty strong butter on the other, while skins and furs dangled from the rafters overhead. Sometimes cats entered by the one open window and actually fought on our prostrate bodies, awakening us from profound sleep by squalling in our very ears."

Scudder's personal life was a tragic one. He married Ethelinda Jane Blatchford in 1867. She died in 1872, leaving a son, Gardiner Hubbard Scudder, born in 1869. The boy was the focus of all his hopes, and Scudder took him on many of his early field trips. Gardiner graduated with honors from Harvard and entered Harvard Medical School. Very shortly afterwards, he became seriously ill of tuberculosis; he died in 1896. This was a terrible blow to the long-widowed Scudder, and in the same year he developed paralysis agitans, which "was to gradually undermine his physical and mental powers and close the avenues of his busy life."

Scudder thereupon gave his collection to the Museum of Comparative Zoology and his library to the Boston Society of Natural History, with duplicates to Williams College. Day by day and year by year the disease increased his invalidism, and only the loving care of his sister-in-law, a Miss Blatchford, sustained him.

Cockerell reports: "I only saw Scudder once after he was stricken with paralysis, and his work was done. This was in 1907, at the time of the Zoological Congress in Boston. I was allowed to talk with him for three minutes only, but in those minutes he enquired after various old friends in the west, and looked at some new fossils from his old-time hunting ground at Florissant. . . . I mention these facts to show that his mind was still active, although he was physically unable to work and mentally incapable of any continuous strain. It is one of the most pathetic facts in the history of science that for seven years this great naturalist remained paralyzed and helpless, with so much of the work he had planned to do still unfinished."

On May 17, 1911, Samuel Hubbard Scudder was released from his suffering. Perhaps the most poetic tribute to him comes from Field: "So the breath of outdoors is in his writings, and the eagerness of the explorer was in his spoken words; and the sight of the autumnal flocks of the Monarch, or the feeble flight of *Oeneis* about leeward ledges, must always recall to us their greatly gifted interpreter."

LAWRENCE BRUNER (1856–1937)

Lawrence Bruner, famous as an entomologist, ornithologist, and teacher, was one of Nebraska's favorite sons, and we are told that the State legislature elected him "as the State's representative to the San Francisco Exposition [1915] against such competition as William Jennings Bryan."

It is fortunate that one of his students, M. H. Swenk (1937), later his successor in entomology at the University of Nebraska, prepared a memoriam that presents Bruner's scientific activities in great detail. Two of his former students, W. Dwight Pierce and C. E. Mickel, kindly supplemented Swenk's biographical notes through correspondence with me.

Bruner was born on March 2, 1856, in Catasauqua, Lehigh County, Pennsylvania. He was one of seven children of Uriah and Amelia Bruner. When Lawrence was six weeks old, Uriah moved to Nebraska with his

Figure 48. Lawrence Bruner

family, and the trip from Iowa City, Iowa, to Omaha, Nebraska, was made by stagecoach. The family settled on a farm near Omaha, and it was here that young Lawrence began to study the birds, mammals, and insects that were everywhere around him.

The Bruners later moved to West Point, a newly-founded town in Nebraska, and this was Lawrence's home from 1869 until 1888, when he entered the University of Nebraska. He received his early education in his own home and in a local academy.

From 1873 to 1876 Nebraska and some of the nearby states were devastated by plagues of Rocky Mountain grasshoppers, and naturally these disastrous invasions made a great impression on Bruner. He began seriously to collect grasshoppers and to study their classification, behavior, life histories, and control. In 1876 he published his first technical paper, entitled "New Species of Nebraska Acrididae." Two of the three species described were the same species called by different names, not surprising in the case of a beginner isolated from entomological literature. The following year he published a list of the grasshoppers of Nebraska, and then for a period of seven years ceased describing new species and instead sent them to Samuel H. Scudder, the great American authority on grasshoppers.

In 1877 Congress approved the formation of the United States Entomological Commission, consisting of three eminent entomologists, C. V. Riley, Alpheus S. Packard, Jr., and Cyrus Thomas, who were appointed to study and recommend measures for the control of the Rocky Mountain grasshoppers. The Commission almost immediately selected Bruner to assist in these studies; he began to work on a temporary basis in 1878, and in 1880 he was formally appointed as an assistant to the Commission. For the next eight years he worked for the Commission; later he worked for the U.S. Department of Agriculture.

From 1880 through 1883 Bruner studied the habits of the Rocky Mountain grasshopper throughout much of the rugged and primitive areas of the mountain states in the northwest.

He married Miss Marcia Dewell of Little Sioux, Iowa, on Christmas Day, 1881. L. O. Howard (1930) recalls this anecdote: "An interesting incident connected with Professor Bruner may be told. He came to Washington on his wedding trip. He had married a charming Nebraska [?] girl. Some time later I received a telegram from him announcing the birth of a daughter. I wired a reply of congratulations, and added that if he would name the daughter Psyche the Division of Entomology would stand as godfather. This dispatch was not answered, but I learned a year later that his little girl bore the unusual name of Psyche; whereupon the entomological force in Washington sent her a silver cup. It may be interesting to know that this little girl eventually became the wife of Harry S. Smith, so well known to all American entomologists."

During the summer of 1884 C. V. Riley appointed Bruner as Special Agent for the Division of Entomology to study the breeding habits of grasshoppers. Bruner did this for several years and also studied the habits of other pests, such as the chinch bug, cutworm, Colorado potato beetle, cabbage worm, and snowy tree-cricket.

In the 1870's Bruner became acquainted with the work of Professor Samuel Aughey on birds as predators of grasshoppers and other noxious pests, and from this association he developed a life-long interest in birds as destroyers of pests.

In 1888 he was appointed entomologist of the Nebraska State Agricultural Experiment Station in Lincoln but continued to act as Special Field Agent for the U.S. Division of Entomology for the next six years, during which time he prepared annual reports for the U.S.D.A. on birds and pests. From 1893 on he also published scientific papers and popular articles on grasshoppers and a variety of insect pests as well as on birds.

Myron Swenk (1957) records how he obtained a copy of Bruner's extensive report for 1896 entitled "Some Notes on Nebraska Birds": "From this time on the ornithological enthusiasm of the writer waxed greatly, and before the close of 1898 he had started an oölogical collection and began to mount birds, following instructions found in Apgar's *Birds of the United States*. But he still desired a copy of Professor Bruner's report, and most of all, to meet the man himself. So, in the spring of 1899, he mustered courage to write Professor Bruner about some birds that he had seen, receiving a prompt and most kindly reply, as well as an invitation to call and to become a charter member of the Nebraska Ornithologist's Union, which, as just stated, was then under process of organization. The actual meeting, which occurred the following September in Professor Bruner's office, has always been a red-letter day to the writer. The forty-three year old man, of much superior knowledge and experience about birds, and on a busy day, gave up a large part of his afternoon telling the sixteen year old lad about his recent trip to South America, showing him the specimens of bird skins and eggs brought back, as well as the insects collected, giving him a copy of the 1896 report and encouraging him to enter the University upon graduation."

In 1890 he began to offer a regular course in entomology at the University of Nebraska, and in 1895 a Department of Entomology and Ornithology was established with Bruner as the only professor. In this post he was assisted by his "boys," many of whom later became prominent entomologists—Harry G. Barber, Walter D. Hunter, J. C. Crawford, W. Dwight Pierce, Harry S. Smith, and C. E. Mickel, to name just a few. Swenk writes that "from 1900 to 1910, Professor Bruner's home at 2314 South 17th Street in Lincoln became largely the leisure-time rendezvous of all of these 'boys,' and not only his 'den,' in which his collections and library were kept, was open to them, but his entire home . . ."

At the request of the Argentine government, Bruner sailed for Argentina in the spring of 1897 and, on a year's leave of absence from the University, studied the grasshopper problem there. As a result of this trip he wrote several technical papers. His scientific studies were given due recognition when, in 1897, he received an honorary B.S. from the University of Nebraska.

In 1898 Bruner was made State Entomologist, while retaining his other jobs, and in 1899 was elected president of the American Association of Economic Entomologists.

In the winter of 1901–1902 Bruner and some of his students collected insects, birds, and mammals in Costa Rica, and, in the ensuing years, extensively throughout the West.

Bruner published numerous papers on the Orthoptera, and from 1900 to 1908 he completed most of the second volume on Orthoptera of the *Biologia Centrali-Americana*. When, in 1911, Bruner decided to devote his taxonomic efforts exclusively to exotic Orthoptera, he sold his collection of North American Orthoptera to Morgan Hebard of Philadelphia; it is now part of the collection of the Philadelphia Academy of Natural Sciences. As a result of his South American studies, he issued a number of monographs on crickets and locusts from many localities.

In 1913 Bruner collected in Japan, China, and especially in the

Philippines and subsequently published a catalogue of Philippine Orthoptera. And in 1915 he was designated "The most distinguished citizen" of the State of Nebraska in a celebration of Nebraska Day at the Panama-Pacific International Exposition at San Francisco.

Mickel writes: "It is 43 years since I last saw Professor Bruner and memory is sometimes tricky . . .

"My recollection of Prof. Bruner is [of] a kindly, patient man, with an absorbing interest in entomology, which he somehow communicated to his students. His lecture notes, research notes, and often his first draft of research manuscripts were written on the backs of envelopes, which he systematically saved for the purpose. He had a wealth of stories drawn from his long experience as State Entomologist of Nebraska and his travels abroad. I remember his lectures as stimulative and informative. He was a good field collector, and I learned considerable entomology from him while on collecting trips with him in the field."

W. Dwight Pierce has even earlier recollections, having begun to attend a series of lectures by Professor Bruner when a high school senior in Omaha. In 1901 Pierce began to study under Bruner at the University of Nebraska. Pierce notes:

"The Professor was a short, broad set man with a pompadour hair style. He spoke with a soft voice and a quaint sense of humor and had many interesting little stories about the insects.

"He was of a Quaker family opposed to titles, and while he had been to college he would not take a graduation degree. As a boy . . . he was entranced by the grasshoppers, and when supposed to be plowing he was often found lying on the ground watching at close vision the intimate activities of the wonderful grasshoppers."

Having been made chairman of the Department of Entomology and Ornithology and entomologist for the Experiment Station, Bruner's responsibilities began to wear on him by 1919 and he asked to be relieved of them. Once freed of these duties, he again began to collect extensively and to enlarge the University insect collection. From that time on he spent a good deal of his time in California and eventually moved to Berkeley. In 1931 the University of Nebraska made him Professor of Entomology Emeritus. In 1933 Bruner presented his famous collection of Orthoptera and his splendid library to the University of Nebraska. He died on January 30, 1937, about two years after the death of his wife.

OTTO LUGGER (1844–1901)

L. O. Howard (1901) provides the major source of information available for Otto Lugger. He writes that Lugger "was one of the most widely known of the many Americans of German birth who have

Figure 49. Otto Lugger. Courtesy of E. I. McDaniel

obtained high scientific reputation in this country. He was born at Hagen, Westphalia, September 16, 1844. His father was a professor of chemistry in a Prussian university. Lugger was educated in Hagen, and in 1864 he came with his parents to the United States and secured a position with the engineer corps of the army, and for two years was engaged in the survey of the Great Lakes. He had always been interested in entomology, and collected specimens while engaged in his engineering work. He became acquainted with the late C. V. Riley, who at that time was occupied in newspaper work in Chicago, and when, in 1868, Riley was appointed State Entomologist of Missouri, Lugger went with him as his assistant. During the years 1868 to 1875, when Riley established his great reputation as an economic entomologist and published eight of the nine annual reports that brought him lasting fame, Lugger remained his quiet, unassuming, self-sacrificing and devoted helper. In 1875 he married Lina Krokmann and went to Baltimore, where he became the curator of the Maryland Academy of Sciences and naturalist of the city parks. In 1885 he was appointed assistant in the Division of Entomology of the U.S. Department of Agriculture, remaining in Washington until 1888."

Lugger left Washington in 1888 to serve as State Entomologist of Minnesota, a post he retained until his death. He soon won the farmers' confidence and gratitude by helping them fight off the ravenous swarms of Rocky Mountain locusts that were then attacking their crops in Ottertail County. His publications on the Orthoptera, Lepidoptera, Coleoptera, and Hemiptera of Minnesota augmented an already established reputation as a fine economic entomologist.

"Aside from his scientific ability," Howard continues, "Lugger was a man of admirable qualities. His wide information, his agreeable personality and his keen sense of humor made him one of the most delightful companions I have ever known. Many of his stories and humorous sayings are current among entomologists all over the United States."

Lugger died from pneumonia on May 21, 1901, after a very brief illness, leaving a widow and two children.

ALBERT PITTS MORSE (1863–1936)

Albert Morse was not only an eminent orthopterist, with sixty or more highly regarded papers on the Orthoptera to his credit, but also an expert on the Odonata.

He was born on February 10, 1863, in Sherborn, Massachusetts, where his father was prominent in town affairs. He attended the local schools

Figure 50. A. P. Morse

and graduated from Sawin Academy in Sherborn in 1879. This was the only formal schooling he received, because he was needed to help with the work at home. Also he missed many days of school because of ill health. From his boyhood on he was greatly interested in natural history, collected specimens, and studied taxidermy. He was encouraged in his interest in wild life by local naturalists.

According to Dow (1937), he abandoned farming as a livelihood when he was twenty-three years old and took up drafting, a vocation which he followed for several years. "After the death of his parents in 1886 and 1888, he accepted a position as assistant in the Zoological Department of Wellesley College, with which institution he was connected in different capacities for more than 45 years (until 1933). As collector and instructor, he served the students and teachers in various ways, developed the museum, and lectured on elementary and systematic zoology and entomology. During the first part of this period he attended the summer school of the Marine Biological Laboratory at Woods Hole, took a long summer course in entomology at Cornell University under Professor J. H. Comstock, and made extensive collections of New England insects, paying particular attention to the Orthoptera and Odonata, in which orders he discovered and described many new species."

During the summer of 1897, encouraged by S. H. Scudder, he left for a collecting trip on the West Coast. Morse returned with several thousand specimens of the Orthoptera, many of the new species of which were described by Scudder. In 1903 and 1905 he was a Research Assistant at the Carnegie Institution of Washington, where he studied the Orthoptera of the southern United States and wrote two reports on the identification and ecology of grasshoppers. At some point he taught zoology and entomology at Teachers' School of Science in Boston and during the summers lectured on natural history to youngsters and teachers in Vermont. In his spare time he worked on a monograph on Orthoptera, which appeared in 1920 as a *Manual of the Orthoptera of New England* under the aegis of the Boston Society of Natural History.

In 1911 Morse was appointed Curator of Natural History at the Peabody Museum in Salem. While there he worked on the *Orthoptera of Maine* and also on the food habits of grasshoppers with reference to their attacks on binder-twine.

The Museum of Comparative Zoology at Harvard acquired his collection of more than fifty thousand specimens in the winter of 1920–1921.

Morse's health began to fail in 1934, and he died on April 29, 1936, survived by his wife and two children.

In Dow's words, Morse "was above all a lover of nature with a remarkable knowledge of natural history. The members of the Cambridge Entomological Club, at whose meetings he was a constant and welcome attendant, greatly miss his presence and his interesting, often humorous contributions. He was an accurate observer and meticulous in attention to detail. Few men with as little formal training are able to command so much respect for their scientific work."

ANDREW NELSON CAUDELL (1872–1936)

Caudell was born in Indianapolis on August 18, 1872, and moved with his parents to Oklahoma when a young boy, where he was brought

Figure 51. A. N. Caudell

up on a farm. Howard and Busck (1936) note that "his first contact with entomology took place in an old building where he took refuge from a heavy rain; there he found a copy of one of the early Yearbooks of the U.S. Department of Agriculture and became absorbed in the possibility [of using it to identify] some of the insects he had noticed and collected in his boyhood. He wrote for help in this task to the Department of Agriculture and was rewarded by an encouraging and understanding letter from the Chief of the Division of Insects (C. V. Riley) whose encouragement caused Andrew to become a student and later an assistant at the Oklahoma Agricultural College, 1895–98. After that, he was for a short time connected with the Gypsy Moth work and with the Agricultural College in Amherst, Massachusetts, and in 1898, he obtained a position in the Division of Insects of the U.S. Department of Agriculture. Then he chose the study of Orthoptera as his specialty and was made custodian of this group of insects in the U.S. National Museum, in which capacity he remained until his death, early becoming a recognized authority in his field and publishing many important papers on the classification of these insects."

Caudell published papers on numerous families of the Orthoptera order as well as on insects in other orders such as the Zoraptera.

He was well known for his catalogues and card indices, especially his card index of the food plants of North American Lepidoptera and one of the literature on Lepidoptera.

"Indexing was in fact one of Caudell's most characteristic habits. His card indices which are left with the collection of Orthoptera in the

Museum are most exacting and complete, even the smallest item relating to Orthoptera being included. This tendency to record found an expression also in his personal diary from which he was able to state in detail what happened to him every day for nearly forty years.

"The methodical indexing habit was a phase of Caudell's love for order, his desire for having a special place for each object, which was one of the qualities that made him such a good systematic entomologist; he devoted much time to tabulating the species and genera, which he studied, and he could always put his finger on the particular pigeonhole for any specimen, book or note, that was required in his studies." (Howard and Busck)

Because of the methodical way he worked, he accumulated in time a large amount of information on taxonomic nomenclature and, in 1912, with Nathan Banks as joint author, published the widely used work, *The Entomological Code. A Code of Nomenclature for Use in Entomology.*

Caudell accompanied Harrison G. Dyar, a lepidopterist on a number of trips to collect Lepidoptera, Orthoptera, and other insects.

In 1915 he was president of the Entomological Society of Washington. Howard and Busck note that his "quiet native humor and genial kindness made him friends wherever he went—he probably had no enemy. In our Society, he was not a frequent speaker but whenever Caudell got up with a communication everybody was smilingly attentive, because we knew he had something worth while to say and that he would present it in his own droll original manner, which invariably caused friendly merriment."

Caudell had married Penelope Lee Cundiff in 1900 and they had one daughter. He died in Washington, D.C., on March 1, 1936.

MORGAN HEBARD (1887–1946)

J. A. G. Rehn was a close friend of Morgan Hebard's, and the following is chiefly from Rehn's (1948) obituary of him.

Hebard was born in Cleveland, Ohio, on February 23, 1887. His family derived its wealth from the lumber business, owning a cypress tract of almost five hundred square miles in southeastern Georgia. The family had homes in Georgia, Michigan, and Miami, and a permanent residence in Chestnut Hill near Philadelphia. Hebard received his early education from tutors who moved with the family to their various seasonal homes; after graduating from a private school at Acton, North Carolina, he entered Yale University, where he received his A.B. in 1910.

"His boyhood collections covered a wide range, from historical relics to postage stamps," Rehn writes, "but the interest in nature was always

Figure 52. Morgan Hebard

paramount, accompanied as it was by the determination to find out as much as he could about a matter of particular interest, even if this entailed wading in swamps (without boots!) or excavating gopher holes. One characteristic early developed was a meticulous exactness and neatness in labelling his possessions, an appreciation of orderly arrangement which remained a marked trait throughout his life."

Rehn, who was five years older than Hebard, met him through Dr. Henry Skinner, then Curator of Insects at the Philadelphia Academy of Natural Sciences, and was responsible for getting Hebard interested in Orthoptera. Rehn spent a month in 1904 collecting insects with Hebard at the latter's home in Thomasville, Georgia. They put out a joint paper in 1905, the first of many.

At his father's request, after graduating from Yale, Hebard worked for a year in a Philadelphia bank. From 1911 until 1930, when illness curtailed his activities, he devoted his full time (except for a period in the armed forces during World War I) to collecting Orthoptera in the field and then studying them in the laboratory. He and Rehn collected over a hundred thousand specimens and wrote hundreds of papers of field observations. For many years the two men worked as a team, both in the field and in the laboratory. Their aim was to prepare a monograph on the Orthoptera of North America, a work which Rehn was to continue after Hebard's death. They published numerous joint papers, and in later years Hebard alone wrote many papers on the Orthoptera. He financed publications and the art work that went in them, as well as collecting expeditions from 1905 to 1928. He and Rehn made several trips together to Jamaica, Panama, and Colombia; Hebard also collected in Cuba, the Bahamas, Europe, and North Africa. Hebard was an authority on North American cockroaches; and his monograph, *The Blattidae of North America North of the Mexican Boundary,* was published by the American Entomological Society in 1917.

Rehn reports that Hebard had a keen incisive mind, and when working in the laboratory was oblivious to outside distractions, always completing what he had started. He was well versed in French, German, Spanish, and Greek. He typed his own manuscripts and preferred to mount his own specimens. Although quick-tempered and outspoken, he was actually kind and sympathetic, with a good sense of humor. Despite his wealthy upbringing, he made friends with people of all kinds, and on field trips "soon won the friendship, respect and cooperation of those who cooked the meals and wrangled the horses or drove the truck." He was a great outdoorsman and loved hunting, fishing, golf, tennis, and trapshooting; he was a good shot with a rifle, revolver, or shotgun.

Hebard served as Curator of Insects in the Philadelphia Academy of Natural Sciences and accepted no salary. He rarely attended formal meetings of any sort because he considered them a waste of time. In 1945 he presented his collection of Dermaptera and Orthoptera to the Academy; these included the famous Lawrence Bruner and J. L. Hancock collections. The total collection (both exotic and native) consisted of 1,369 types and 2,000 paratypes and contained 250,000 specimens from North America alone. "The whole series filled 2,400 Academy standard glass-top cabinet drawers, contained in 147 metal cases . . ." (Rehn).

In his later years Hebard suffered so severely from arthritis that he had to have a nurse assigned to him to set up his slides and focus the microscope for him. He died of a heart attack on December 28, 1946, leaving as his widow, Margaret Claxton Hebard, whom he had married in 1913. She was a granddaughter of the artist, John La Farge.

JAMES ABRAM GARFIELD REHN (1881–1965)

The following information is taken from obituaries for Rehn written by Phillips (1965) and Gurney (1965).

Rehn was born in Philadelphia in 1881. He attended local schools there, including the Public Industrial Art School and the Pennsylvania Academy of Fine Arts. He was a member of a boys' group whose interest in natural history was directed by the well-known entomologist C. W. Johnson, then Curator of the Wagner Free Institute of Science. For a number of years Rehn was interested in both mammals and insects, and on June 1, 1900, he joined the staff of the Academy of Natural

Figure 53. J. A. G. Rehn

Sciences as a Jessup Fund Student, a specially selected student in the field of natural history.

Shortly after joining the staff he became an associate in the Entomological Section of the Academy; in 1910 he became a resident member of the American Entomological Society. Throughout the years he held many important positions in both institutions. E. T. Cresson, the hymenopterist, Henry Skinner, the lepidopterist, and C. W. Johnson, the dipterist, all directed Rehn's interest toward entomology, and from 1900 onwards he began to publish steadily on what was to be his life's work, the study of Orthoptera.

Phillips writes that Rehn was a field naturalist, who "explored the mountains and plains, the rain forests and the deserts to study the ways of his favorite form of animal life. He has studied their habits, their ecology, their distribution and interrelationships in great detail. Few men have explored this country more thoroughly and with greater care in a search for all the species in their special field." Rehn accumulated over twenty volumes of meticulous field notes covering his collections and described over one thousand new species.

He made many collecting expeditions to foreign countries, including, besides those countries he visited with Morgan Hebard, Africa, Costa Rica, Honduras, and Brazil. Through their efforts and those of other dedicated staff members they gradually accumulated one of the finest collections of Orthoptera in the world.

Gurney writes: "Trips by Rehn and Hebard to the Southwest during the early years before motor vehicles and good highways were generally available sometimes included rugged camp living, with a helper to manage the horses and to cook, with tiring days in the field followed by long, sometimes cold, evenings preparing the day's catch. . . .

"Personally, he was loyal to his friends, devoted to principles which he considered sound, outspoken in conversation, pithy in long letters on matters which he felt deserved censure, and not one to waste time on efforts which he regarded as trivial. My large file of his letters, together with memories of his colorful and instant recall of experiences in numerous interesting places and of his associations with biologists, are among my own personal treasures."

According to Phillips, "There can be no doubt that at the time of his death Mr. Rehn stood in a very small and select group of the world's great Orthopterists. On the basis of his published work, he was perhaps the 'best.' " He wrote in all about twenty papers on mammals and over three hundred on the Orthoptera. Of his many works, two begun late in his life are especially notable: *The Grasshoppers and Locusts (Acrididae) of Australia,* in three volumes, begun in 1949; and, together with Harold J. Grant, Jr., *A Monograph of the Orthoptera of North America (North of Mexico)*, Volume I, in 1961.

Rehn was the recipient of many honors. In 1945 he was president of the Entomological Society of America. He was research systematist and member of the Academy of Natural Sciences of Philadelphia and the American Entomological Society for almost sixty-five years, and the present prominence of both of these organizations can be attributed in large part to Rehn's interest and activities.

Upon the death of his wife in 1964, Rehn's health began to decline, and his death followed on January 25, 1965. He was survived by his son, J. W. H. Rehn, also an entomologist.

8 Notable Homopterists-Hemipterists

The June bug hath a gaudy wing
The lightning bug a flame
The bedbug hath no wings at all
But he gets there just the same
Anon.

Plant lice, scale insects, and the true bugs are among our most injurious plant pests. Thaddeus W. Harris and Asa Fitch, two early economic entomologists I have already described, devoted a great deal of time to them. For a variety of reasons, insects in these orders have attracted taxonomists and applied entomologists the way roses attract aphids.

PHILIP REESE UHLER (1835–1913)

The man E. O. Essig, in his *History of Entomology* (1931), calls "America's greatest hemipterist" has had surprisingly little written about him. Uhler was one of America's pioneer hemipterists and he helped and encouraged many others, including Herbert Osborn and E. P. Van Duzee, in the study of the Hemiptera.

Uhler was born in Baltimore, Maryland, on June 3, 1835, and died there on October 21, 1913. Howard (1913) describes his father George Washington Uhler, as "a well-to-do and philanthropic merchant of that city." His great grandfather had served in the Revolutionary War. Philip received a good education in Latin and German at private schools. When he was about ten years old his father acquired a farm near Reisterstown, where the family spent their vacations, and Philip began to collect butterflies and moths. In this pursuit he was encouraged by the Rev. J. G. Morris, Librarian of the Peabody Institute in Baltimore, and a German entomologist named J. F. Wild.

Although his father placed him in business, Philip spent most of his

Figure 54. P. R. Uhler

time studying geology, botany, zoology, and insects. In 1863 he was appointed assistant librarian at Peabody, under the Rev. J. G. Morris, and in 1864 Louis Agassiz placed him in charge of the insect collection and library in the Museum of Comparative Zoology at Harvard, where he assisted Professor Agassiz and taught entomology to undergraduate students. He also attended the Lawrence Scientific School at Harvard and studied with Agassiz, Asa Gray, and other noted scientists. Uhler remained at Harvard for three years; then according to Howard, he was called away from Cambridge rather suddenly (Howard does not say why), at a time when Professor Louis Agassiz "was in a very helpless condition and unable to give Uhler the degree of Bachelor of Science, which he was entirely qualified to receive. The degree, however, was later given to him by the University. Among his papers is a most appreciative note received from Dr. Oliver Wendell Holmes just before he retired from the work at Harvard."

Howard notes that, before he went to Harvard, "Uhler had become a member of the Academy of Natural Sciences of Philadelphia (1858), and of the Entomological Society of Philadelphia (1859). He had published a number of systematic papers on the Coleoptera, Neuroptera, and Hemiptera, and had translated for the Smithsonian Institution and edited (with the assistance of Osten Sacken) Hagen's elaborate *Synopsis of the Neuroptera of North America,* published by the Smithsonian in 1861. It was this work which attracted the attention of the elder Agassiz."

After leaving Harvard, Uhler returned to Baltimore and was reappointed assistant librarian at the Peabody Institute; in 1870 he was made librarian; he served as provost from 1880 to 1911. He continued working with insects, especially the Homoptera and Hemiptera, and often published on other aspects of science, including archaeology and geology. Some of his earliest work on insects was concerned with their injuries to agriculture. Later he occupied many of his spare hours with taxonomic studies of Homoptera and Hemiptera and corresponded with collectors all over the United States about their identification. He described many of the Hemiptera and Homoptera gathered by various expeditions to the West. His "List of Hemiptera of the Region West of the Mississippi River, Including Those Collected by the Hayden Explorations of 1873," published by the U.S. Geological Survey in 1876 and 1877, contained monographs of the Cydnidae and Saldidae families. He also published on the Hemiptera of Lower California and many other localities.

According to Howard: "He gave much help at the time of the forming of the Johns Hopkins University, and was the first associate professor appointed in the University, and in this capacity was connected with the institution until the time of his death. His life was the quiet and uneventful one of the student; his profound modesty kept him in the background, and he disliked what he termed 'cheap notoriety'. . . . No worker appealed to him in vain, and to many he was of the greatest help. He was broadly read, and possessed an astonishing memory. Mrs. Uhler tells me that in the summer of 1893 he went abroad and purchased for the Peabody Library about twenty thousand dollars' worth of books. They were bought without the aid of lists, since he trusted to his memory of the books already in the library, and when the accessions were finally catalogued it was found that he had bought but three duplicates of those previously possessed." On this trip he studied the Hemiptera in some of the European museums.

Herbert Osborn (1937), who found Uhler most "cordial and generous with his time," says that his early descriptions, "which were numerous, were extremes of careful noting of details, many of them occupying a full page or more of text and some of them may be characterized as descriptions of individuals rather than of 'species.'" Although his descriptions were often elaborate they were rarely illustrated. Uhler was of medium height and build with a congenial disposition. He was a good speaker and presented his material in a lucid manner.

After 1890, Uhler's activities in the Peabody Institute and failing eyesight forced him to curtail his systematic studies. From 1905 on he was afflicted with glaucoma. At the meetings of the American Association for the Advancement of Science in Baltimore in 1908, Osborn noted that he "took part in some of the exercises, but not in the entomological program, and he was doing little or nothing at this time with his collections of Hemiptera. I walked with him one morning from his home to the Peabody Institution and was especially impressed with his remark as we left his home that he regretted it was such a cloudy day and he feared it might rain. In fact it was a clear sunny morning and I realized as I had not before that his eyesight was greatly impaired. Nevertheless he walked with scarcely any hesitation the familiar route from his house to his office, a route traveled I suppose thousands of times in the course of his busy life."

Part of Uhler's collection is in the Museum of Comparative Zoology at Harvard and part is in the National Museum. When he died in Baltimore on October 21, 1913, he left behind his second wife, a daughter, and a son, the latter a professor of physics at Yale; he also, regrettably, left much of his work on the Hemiptera unwritten.

THEODORE PERGANDE (1840–1916) *

When L. O. Howard first began to work with C. V. Riley in the U.S. Department of Agriculture in 1878, he was greeted by Riley, he says (1933), "with sufficient cordiality" and introduced . . . to "his only other assistant, Theodore Pergande, a little German of forty, with a heavy brown beard, who spoke fluent but rather ungrammatical English, and who had charge of the rearing of insects and the making of notes."

Figure 55. Theodore Pergande

* This sketch is adapted from "An Early Federal Entomologist," which I published in *Entomological News* 78(5): 113–116, 1967.

Thus, when Professor and Mrs. J. H. Comstock came to Washington in 1879, Howard and Pergande were there to greet them. Mrs. Comstock (1953) recalls Pergande as "a small, delicate-featured, bearded German with a gentle manner and lovable character. He was about thirty-nine years old and had come to America before the Civil War. A rich man in town where he was born in Germany had wanted him to become a Catholic, marry his daughter, and go into business with him. Pergande told me that he would have liked the partnership with the man but that he could not stand either the church or the daughter, so he came to America." Howard, however, states that Pergande came to this country just at the "outset" of the Civil War, and adds that he had been a mechanic in Germany and was a man of "slight" education.

"He landed in New York," Howard continues,

> with no English and very little money. He did not know where to go. He found his way to Grand Central Station, fell into line at the ticket office and noticed that the man ahead of him bought a ticket for Syracuse. The word Syracuse sounded familiar, so he, too, bought a ticket to that point. He arrived in Syracuse early in the evening, wandered about the streets, homesick for the German tongue, and presently found himself before a chapel where he heard German spoken by the people passing in. So he, also, entered the church. Behold! he was back in the religious atmosphere. He spoke (in German, of course) to the young man who sat next to him, found him agreeable and went home to spend the night with him at his boarding-house. The young man had no job, and when, on the next morning, while walking together through the streets, they saw a recruiting station, they both volunteered, adding their names to the first three hundred thousand recruited for the war. So, by the irony of fate, twenty-four hours after he landed, he found himself back in both the religious and warlike atmospheres.
>
> At the end of the first three months, the enlistment having expired, Pergande's new found friend went back to New York, but the immigrant, with characteristic perseverance stuck to the army for the full four years of fighting. At the end of the war he was discharged at St. Louis, and having no other trade, went into the big gun works there. He had always been an amateur entomologist, collecting butterflies and beetles and such things, and on one of his Sunday afternoon collecting rambles he met Otto Lugger, then Riley's assistant. Lugger was about to resign and recommended his friend for the job. So Pergande stayed with Riley, and came with him from Missouri to Washington in 1878, where he remained until the time of his death in the early 1900's.

Mrs. Comstock (1953) says that after the Civil War "he married a pleasant, thrifty little German woman who took good care of her husband and their daughter. He was a tireless worker, faithful to his task and to his fellow workers; he wrote an exquisitely fine hand, as legible as print. His notes on the insects he studied were of the greatest value because of their accuracy and careful descriptions. He was ambitious to write in perfect English, so he began studying Shakespeare. Mr. Comstock and Leland Howard had many a secret chuckle over notes on some minute insect, written in true Shakespearian diction. Pergande had discovered the male form, never before observed of a scale insect. When Dr. A. S. Packard of Brown University, visited our offices, he remarked, 'You are fortunate to have so many of these rare insects.' Pergande answered with a smile, 'Fortunate? No, not fortunate! We hoont for them.' Pergande was not fitted for independent scientific work, but his knowledge of insects was great, and as an observer in a scientific laboratory he was invaluable. He could mount the most minute insects to perfection; his slender hands could manipulate, with exquisite precision, the wings of the smallest Tineid moth. He loved his work and loved to discover new things. When my mother wished him a long life, he answered: 'Jes, jes, I hope so too, dare are so many tings to find out and I hope I live to fine dem.' "

Howard continues in the above vein and enlarges on his abilities by noting that, though he was "not too careful about his personal appearance," he was "a positive genius in his work on the life history of insects. He was invaluable to Riley and invaluable to the entomological service at Washington. For many years he kept the main insectary notes of the service; and the great bulk of the life history work published in the many entomological publications of the Department for many years was based upon his careful notes and observations."

It was Pergande who in the summer of 1886 showed that the plum tree was an alternate host for the hop-plant louse; by spraying the aphids on the plums, the hop plants were freed from this destructive pest. He also worked out the life history of the destructive clover-seed midge and many other economic pests. He was an outstanding authority on the Aphididae and did much work on scale insects and thrips. Osborn (1937) notes that Pergande was "a very exact and keen observer and a great deal of his time was devoted to the preparation of material used in the preparation of papers by others, so that his published papers do not represent in any degree the results of his research on insects." In time he began greatly to resent Riley's use of his work without any accreditation.

As he grew older he became somewhat cantankerous and difficult to get along with, and Howard writes of his last years: "Pergande had many friends and admirers who estimated him at his true worth. No one who worked with him will ever forget him. He received little public credit for his work, but his very few published papers show his great knowledge and keen ability. He had a delightful sense of humor, and told fascinating stories of his experiences. He had strong likes and dislikes as to persons, and was very outspoken. His mind began to fail toward the end, and he had a number of curious hallucinations."

Pergande died in Washington, D.C., on March 23, 1916.

OTTO HEIDEMANN (1842–1916)

Heidemann is remembered by entomologists for his entomological illustrations and as an early specialist on certain families of the Hemiptera.

He was born in Magdeburg, Germany, on September 1, 1842. He came to Baltimore at the age of seventeen. There, having learned wood engraving in Leipzig, he entered the wood-engraving business. In 1876 he moved to Washington, D.C., where he prepared illustrations for Government publications. He became interested in insects through his engravings of them. Albert Koebele, E. A. Schwarz, and Theodore Pergande helped and encouraged him in his studies. In 1898 he was employed by the U.S. Bureau of Entomology as an assistant and a specialist on the Hemiptera. In time he won respect for his knowledge of the families Aradidae, Miridae, and Tingidae. In 1907 he became Honorary Custodian of Hemiptera in the U.S. National Museum. Although he took up entomology when he was past fifty, he published thirty-four papers on the subject, mostly on the Hemiptera.

Howard *et al.* (1916) say of him that he was a cheerful man of unimpeachable character and broad culture, "a peer among the leading contemporary Hemipterists, a writer of plays in both German and English,

Figure 56. Otto Heidemann

several of which have achieved public performance, a scientific artist of the very first rank and an earnest student of the social problems of our day." He was active in the Entomological Society of Washington and was its president for two terms (1909–1910).

Heidemann died in Washington, D.C., on November 17, 1916. He was survived by his wife Mica, who was known as "a sculptress and maker of insect models." Cornell University purchased his fine collection of insects.

EDWARD PAYSON VAN DUZEE (1861–1940)

"Without a doubt the greatest [hemipterist] in this country and one of the few great world-hemipterists," wrote De la Torre Bueno (1941), Van Duzee was an outstanding example of a man who, "without previous training, without a college degree, was able by his own effort to make himself a master of his science."

In the obituary Van Duzee wrote in 1934 for his brother, Millard Carr Van Duzee, a collector of Diptera, he reveals the environment that nurtured their love for natural history and insects.

> Our father was a naturalist of the old school, and had gathered extensive collections in Mineralogy, Geology, Botany, Conchology and Ornithology, sufficient to fill a three story annex to his home in Buffalo, the upper floor of which was an astronomical observatory for which he had purchased an acromatic refracting telescope with an 18-inch objective, at that time the largest privately

Figure 57. E. P. Van Duzee

owned telescope in the United States. Here and at father's farm at Lancaster, N.Y., where our summers were spent, we acquired the habit of observation as well as of collecting, both of which were encouraged by our father in every way possible. Some of the happiest days of our lives were spent on hunting trips through the woods with father while he was adding to his large collection of birds. Millard and I devoted our efforts mostly to the insects and together we built up our collection which we had in common. The paths about our country home in Lancaster were lined with flower borders and about these flowers we would hunt with our nets during the evening twilight for moths, or take them as they came to the lighted windows. About 1876 Mr. A. R. Grote, then Director of the Buffalo Society of Natural Science, taught us how to sugar for moths and how to dig for pupae and to raise larvae, and at the close of the season he determined our material for us; he also encouraged us to work in other orders of insects, so we formed a fairly good general collection of insects.

Grote also taught the boys how to mount their specimens and advised Edward to study the Hemiptera.

E. P. adds that his father's collections and observatory were open to the public several days each month. There, also, gathered those members of the community with a serious interest in natural history; these gatherings resulted in the organization of the Buffalo Society of Natural History.

Essig and Usinger (1940) provide the material for much of the following account of E. P. Van Duzee. He was born in New York City on April 6, 1861. His father, William Sanford Van Duzee, was a talented scientist, and had been a missionary to the Hawaiian Islands, a teacher, then a dentist, and later a building contractor. Of his large family, two sons, Millard and Edward, became naturalists and entomologists.

The boys were educated in private schools or by private tutors. Naturally, considering their father's interests and collections, they soon became immersed in botany, geology, and later astronomy. Through nightly sessions with the telescope, Edward became well acquainted with the planets, stars, sun spots, and other astronomical phenomena, and when students visited the observatory, it was he who would explain the objects they saw through the telescope.

When the elder Van Duzee died in 1883, Millard also became a building contractor, and a prominent dipterist in his spare time. Edward studied for a short time with Professor A. E. Verrill at Yale University, where he also met and worked with the famous dipterist, S. W. Williston.

E. P. Van Duzee became assistant librarian of the Grosvenor Library of Buffalo in 1885. "This was exclusively a research library," write Essig and Usinger, "and hence provided ideal surroundings for the scientific work which was to occupy most of his waking hours during the next half century." With a medium-power monocular microscope and his collection within reach, he did much of his early work on the Homoptera. His evenings and Sundays at home were also busily occupied with his taxonomic studies of the Hemiptera.

Besides working on the Hemiptera, he collected and published lists of the Lepidoptera and dragonflies in and around Buffalo. After ten years at the Grosvenor Library, he became head librarian, a position he held for seventeen years. During his years at Grosvenor "he acquired a knowledge of books and bibliographical methods which made him one of the foremost Hemiptera bibliographers of his time. Moreover, on yearly trips to library conventions and while purchasing books for the Grosvenor Library, he was able to visit the entomologists at the U.S. National Museum and, on at least two, in 1885 and 1888, a fellow librarian and the foremost hemipterologist of America, Philip Reese Uhler of Baltimore. Dr. Uhler was a constant source of inspiration through correspondence and must have welcomed such able assistance because of his own failing eyesight." (Essig and Usinger)

Van Duzee did a great deal of field collecting even during the winter, when he concentrated upon the hibernating insects in Buffalo and vicinity. This field work definitely contributed to his health by offsetting the sedentary nature of his library work. In later years he collected throughout much of the United States.

At the beginning of this century he began his great monograph on the Hemiptera, a work that took ten years to complete. According to Essig and Usinger, "The excellent library facilities of the eastern United States were all available to supplement his own very complete library so the resulting catalogue proved to be an exhaustive treatise and a model or standard of excellence for such bibliographical work." When the Grosvenor Library was enlarged, and the staff increased, the added work and responsibility affected his health, inducing a nervous condition that made it necessary for him to leave his job in 1912. He visited friends in Colorado and then went to San Diego, where six months of collecting restored his health. In 1913 and 1914 he worked in the Scripps Institute of Oceanography at La Jolla, where he completed a rough draft of his catalogue.

While in San Diego, Van Duzee wrote the following letter to Dr. W. J. Holland, Director of the Carnegie Museum in Pittsburgh:

> For some time past I have been spending all my spare moments and a good deal of my private means in the preparation of a catalogue of the North American Hemiptera and the work is now well advanced. On account of poor health I have been obliged to sever my connection with the Grosvenor Library of Buffalo where I have been employed for 27 years and now consider myself

a resident of California. Before taking up any regular line of work here I am anxious to push the catalogue to completion and to make this possible am anxious to dispose of some of my duplicate Hemiptera of which I have a large amount. My plan is to sell carefully determined North American Hemiptera at 10 cts. per species for an average of two specimens to a species. Do not you think the Carnegie Museum can take a set of these? I am sending you the first offer and with the first set I will be able to include quite a number of co-types. I will also be able to include a number of rare western forms. In any case you would get the best set I can now furnish. Such a sale would be a real help to me just now in the completion of the Catalogue and I am sure you would be pleased with the material.

I am anxious to have this Catalogue published here on the coast but if I fail in this may have to send it east; perhaps the National Museum may undertake it. My reason for wanting it published here is so I can attend to reading the proof in person.

Holland agreed to purchase the specimens.

In 1914 C. W. Woodworth, head of the Division of Entomology at the University of California, Berkeley, offered Van Duzee a job as an instructor in entomology at the University. Van Duzee accepted. He then completed his *Catalogue of the Hemiptera of America North of Mexico, Excepting the Aphididae, Coccidae, and Aleyrodidae,* which was printed by the University of California in 1917.

However, Van Duzee did not like teaching, and in 1916 he accepted the position of assistant librarian and curator of entomology at the California Academy of Sciences. This position was open because Dr. E. C. Van Dyke had just left the Academy to teach at the University of California. Van Duzee was assistant librarian for eleven years, until replaced by a full-time librarian.

Van Duzee was fifty-five when he started with the Academy, and in the twenty-four years he was connected with it the collection of insects increased from 30,000 to over one million. In his curatorial duties he was assisted by Drs. F. E. Blaisdell and E. C. Van Dyke, and the collection was one of the neatest and best organized then extant. He was also the editor of the *Pan-Pacific Entomologist* for the first fourteen years of its existence.

It was my good fortune to attend some of the meetings of the Pacific Entomological Society in the Academy in San Francisco. There I casually met Mr. Van Duzee, who seemed to be a quiet man, small in physical stature, and as much a part of the entomological section as his insects. I shall never forget the large collar that floated loosely around his thin neck.

When he died on June 2, 1940, he had more than two-hundred-and-fifty publications to his credit.

WILLIAM THOMPSON DAVIS (1862–1945)

Davis is that rare entomologist who has a biography in book form. It is a most readable account, but is unfortunately out of print. The following sketch is largely taken from this book, entitled *The Life of William T. Davis* by Mabel Abbott (1949) and from the writings of his friends, Teale (1942), Parshley (1951), Cleaves (1942), and Wade and Barber (1945).

Parshley recalls Davis as "the eccentric, old-fashioned, keen-minded, simple, sincere, thrifty, generous, and lovable character that hundreds of people—from country boys and New York business men to John Kieran and Edwin Way Teale—remember with affection. . . . Encountered in the open, he was a strange apparition: small and spare, dressed in dusty black, wearing a battered stiff straw hat; burdened with an old army knapsack, a black umbrella, and an insect net which was often stored under his coat behind; his face adorned with a large, red nose and a straggly mustache, his eyes shining through rimless glasses—an odd figure to be sure, but somehow radiating kindliness, intelligence, and an infectious joy in every aspect of nature, including people—except perhaps those who set forest fires and 'improve' landscapes."

The son of George B., Jr., and Elizabeth Thompson Davis, William was born on Staten Island on October 12, 1862. When he was ten his parents were divorced; thereafter he lived with his mother's family and was educated in two private schools on the Island.

Even as a youngster he was quite proficient in natural history and he

Figure 58. W. T. Davis

chose friends with similar tastes. One man who had a tremendous influence on him was Augustus Radcliffe Grote, who was twenty-one years his senior. Grote, in time, became an eminent entomologist, naturalist, composer, and poet. Davis came to know him in 1880, when Grote returned to Staten Island, where he had lived briefly as a child. Miss Abbott writes, "Grote kept in his house his splendid collection of moths (which afterward was sold to the British Museum), and this the boy studied. Grote drew for him on a scrap of paper, with a fine pen, the wings of a butterfly, showing the nerves clearly, and placed a numeral at the outer end of each nerve. That scrap of paper was found after Davis' death, among his carefully preserved scientific effects. When Grote found a female *Eacles imperiales* alive he dropped a postcard in the mail inviting Davis to come and get it. 'It will lay eggs,' he wrote. 'Try your luck rearing a brood.' Davis did so, and the moth lived up to the prophecy, producing more than a hundred eggs.

"Many long and happy hours he spent in the older naturalist's little parlor, listening to expert information and advice. Grote definitely—and naturally—threw the weight of his influence toward entomology as against all the rest of natural history, which at that period was beckoning to the youth from every bird's nest, muskrat track, and cast-off snakeskin, as well as from every wild flower, ant hill and cocoon."

Davis did not go to college but at the age of twenty went to work as a three-dollar-a-week clerk in the New York Produce Exchange—not because of poverty but because it was a family principle that each member be self-supporting and learn to earn and save. It has been said that Davis earned his degrees outdoors with the birds, insects, trees, and flowers, and from this school he graduated *magna cum laude*.

In 1879 he had another significant encounter when he met Charles W. Leng; for the remainder of their long lives "Charlie" the beetle man and "Billie" the cicada man were as close as brothers. The two belonged to the Brooklyn Entomological Society when its meetings were held in the room of Franz G. Schaupp, above a saloon in Williamsburg, Brooklyn. After a meeting was over, Leng noted (Abbott, 1949), "there was rarely any delay in adjourning to Schaeffer's Saloon downstairs, where with the help of beer, coffee, cakes and sandwiches, the informal meeting was prolonged to a late hour. I had to watch the clock for a Roosevelt Street ferryboat that would enable me by running to Whitehall Street to catch the last boat for Staten Island and then walk four and a half miles to my home."

The young men attended the meetings together whenever they could and Davis, living near the ferry got home ahead of Leng—at 2:00 A.M. More often their meetings were in the fields and woods, where they collected and studied the flowers and trees and shrubs, the birds and other animals, and, especially, the insects. Leng, said Abbott, "brought Davis into the Brooklyn Entomological Society; Davis proposed Leng for membership in the New York Entomological Society. Leng found cicadas for Davis, and Davis collected beetles for Leng. It was a quiet friendship and a deep one."

After his day as a clerk in an insurance office in New York, Davis would wander through the woods and fields, often until late in the night. According to Teale (1942), Davis was an all-around naturalist of the old school. "He was the first scientist to report that the sex of land tortoises can be determined by the color of the eyes. Females have dark eyes, males reddish eyes. He discovered and named the devious Eastern walking stick, *Manomera atlantica,* a species in which no male has ever been discovered. He recorded the first corn snake found in the state of New Jersey. He discovered a new type of hybrid oak, *Quercus brittoni.* At other times he added new species of dragonflies and grasshoppers to the lists of science. And, in the field in which he was pre-eminent, as America's leading authority on the *Cicadidae,* he named and described more than 100 of the approximately 170 species of cicadas known to North America."

Many of his scientific papers appeared in the *Journal of the New York Entomological Society* or in the *Proceedings of the Staten Island Association of Arts & Sciences.* In 1892 he published 310 copies of his work, *Days Afield on Staten Island,* which he sold at ninety cents a copy.

On November 7, 1900, he married Bertha Mary Fillingham. She became fatally ill shortly after their marriage and died on December 17, 1901. It was a tragic blow to him.

After twenty-six years' employment as a clerk, he resigned his job in 1909 at forty-six years of age to devote full time to his avocations, the flora and fauna, as well as people, of Staten Island.

Naturally, only a man who was well-to-do could retire at this early age. When he was twenty-two, he noted in his journal that he had saved a little over $1,300 "to aid my desired occupation as a future tramp." He managed to do this, and more. Abbott writes: "A woman struggling to pay off a mortgage might find it canceled; a boy needing money for his education would receive it, and more later, if necessary; youngsters all over the country were paid, and overpaid, for collecting insects of which Davis already had too many; more than one friend trying to build a home received a loan with no time limit; a widow was liberally paid for her husband's collections. Perhaps he himself, methodical as he was about accounts, did not know exactly how much he disbursed in these ways. Yet he spent almost nothing on himself; he filed his immense accumulation of clippings and extracts in pasteboard shirt boxes obtained without cost from a friend who kept a haberdashery; he used piled up orange crates for bookcases; and when he travelled, unless he was with companions who desired comfort, he chose the cheapest lodging he could find."

Cleaves (1942), in adding to the above, says, "Nobody can say how many times he has sent five, ten or fifteen dollars to some hillbilly correspondent in Mexico, Texas or California for collecting a few specimens of cicadas, not that the insects are worth that, but because 'The poor devil probably needs the money.' "

Davis was the "guiding spirit and chief financial support" (Parshley) of the Staten Island Association of Arts and Sciences and honorary curator of the Department of Zoology in the Museum of the Staten Island Institute of Arts and Sciences.

Miss Abbott describes him as he looked in his attic office: "Day and night the genius of the attic was the small man who wore garments which could be said to fit him only because they had in the course of time adjusted themselves to the peculiarities of his person, who looked up over his spectacles with bright, interested eyes at a visitor, and was always ready to drop his work to show his famous collection either to fellow naturalists or to children, or to ignoramuses to whom he must explain patiently matters which were less than ABC to him."

His interests and that of his friend for sixty years, Charlie Leng, were not limited to natural history, but also encompassed ancient houses, old graves, and the history of Staten Island. Together they wrote *Legends, Stories, and Folklore of Old Staten Island,* published in 1926, and the four-volume *Staten Island and Its People: A History 1609–1929,* which appeared in 1930.

There are many stories about Davis. The following is one that Cleaves tells. Davis was on one of his infrequent trips from Staten Island when "the train came to a halt which promised to be of considerable duration. The surrounding country held promise, so Mr. Davis left the train to do a bit of collecting. He had just caught a desirable looking and lively beetle when the locomotive whistle gave several toots. There was no time to kill or bottle the insect so Mr. Davis popped it into his mouth and sprinted to the train."

"Once when I visited him on Staten Island," says Teale, "I found him picking up cicadas from the grass of his yard and putting them in trees where they would be beyond the reach of prowling tomcats. There is something delightfully appealing in stories about how he piled brush before an exposed towhee's nest to shield it from the eyes of enemies, and raked dry leaves away from the trunk of a favorite woodland tree to lessen its danger from fire. He went about, a kind of St. Francis of Staten Island, unobtrusively doing good deeds not only to those of his own kind but also to the small and often-unconsidered creatures of the earth. Yet his kindness was not weakness. Armed only with a butterfly net, he once put to flight three holdup hoodlums armed with a gun."

As Davis grew older he developed an anxiety, after several financial setbacks, regarding his monetary affairs, which affected his normally happy temperament, and after a long illness, he died in Staten Island on January 22, 1945; he was buried in the Moravian Cemetery. His types were deposited in the American Museum of Natural History.

Miss Abbott writes: "His fear of poverty was entirely groundless. Certainly he had lost a good deal of money, but he was able to leave $20,000 to the Historical Society, a few small bequests to close friends, and a residue to the Staten Island Institute of Arts and Sciences which amounted in gross figures to nearly $200,000. His worries were just a tragic trick that the years had played on him. The fortune he left supports the Institute's work on natural science."

But more than money, Davis left a long list of devoted friends.

ALEXANDER DYER MacGILLIVRAY (1868–1924)

A teacher of entomology and a specialist in coccids, MacGillivray was born on July 15, 1868, at Inverness, Ohio. In 1889 he entered Cornell University. According to W. A. Riley (1924), "With characteristic singleness of purpose, he turned to this study [insects], which was destined to be his life work, and neglected some of the formal requirements of the undergraduate course. He registered sporadically and followed much his

Figure 59. A. D. MacGillivray

own bent in his work. The result was that when I first came to know him in the fall of 1898 he was technically still classed as a sophomore, although apart from [M. V.] Slingerland, who gave a brief course in Economic Entomology, he was Professor Comstock's sole assistant in the strenuous teaching work of the Cornell Department of Entomology." He finally realized the error of his ways and began to work for his degree, taking German, French, and Italian as well as other subjects. He received a Ph.D. from Cornell in 1904. From 1900 to 1906 he was an instructor in entomology and invertebrate zoology and then became an assistant professor. In 1911 he joined the faculty of the University of Illinois, where he eventually became a full professor.

MacGillivray made a name for himself through his studies of the taxonomy of silverfish, springtails, sawflies, and scale insects. He wrote two important books, *External Insect Anatomy, A Guide to the Study of Insect Anatomy and an Introduction to Systematic Entomology* and *The Coccidae.*

According to MacGillivray himself, he was full of "Scotch stubbornness." Riley claims he was a "termagant" in his courses and did not believe in "sugar-coating" his information. "Sensitive to a degree, he valued friends and friendship to an extent that many did not realize. He was keenly interested in his students personally and his home was always open to them."

After he died, on March 24, 1924, the University of Illinois acquired his collection of 5,000 specimens and 400 holotypes of Tenthredinidae.

CHARLES FULLER BAKER (1872–1927)

E. O. Essig was one of Baker's favorite students and Essig's book, *Insects of Western North America,* was dedicated to Baker and A. J. Cook, both of whom had taught him. The following is from Essig's tribute to Baker in 1959, when Essig was seventy-five years old.

Baker was born on March 22, 1872, in Lansing Michigan, one of ten children of Major Joseph S. and Alice P. Baker. Two of Baker's brothers achieved fame in pursuits far afield from Charles's: Ray Stannard Baker, who wrote under the *nom de plume* of David Grayson, and Hugh P. Baker, forester, educator, and college president.

Charles attended Michigan State College, where he was taught by the famous entomologist, Professor Albert John Cook, and he found Cook's enthusiasm contagious. "Cook once told me that Baker, as a student, spent nearly all of his cash for insect boxes, much to the embarrassment of his father, and that by the time he graduated he had several hundred boxes of specimens—a larger and more representative collection than was

Figure 60. C. F. Baker

possessed by the college at that time." Cook was greatly taken by his star pupil, on whom he had a great influence and in whom he never lost interest. Charles graduated from Michigan in 1892.

Cook recommended Baker as a laboratory assistant to C. P. Gillette of the Colorado Agricultural and Mechanical College at Fort Collins, and Baker collected insects and plants in this relatively unknown Rocky Mountain region. He soon began to publish his work on the material he found here.

His first paper, written in collaboration with Professor Gillette in 1895, was "A Preliminary List of the Hemiptera of Colorado." In this paper he described the beet leafhopper, *Circulifer tenellus* (Baker), which was to become infamous as the carrier of the "curly top" virus of sugar beets.

Baker was a zoologist at the Alabama Polytechnic Institute and an entomologist in the Agricultural Experiment Station from 1897 to 1899; he worked on such insects as the San Jose scale, the fruit-tree bark beetles, and the peach-tree borer.

In 1898 and 1899 he was a botanist on the H. H. Smith expedition to Colombia. There he got his first taste of tropical collecting, an experience that eventually enhanced his fame and destroyed his health. It is reported that, upon the completion of this expedition, he presented a collection of 60,000 Hemiptera and Hymenoptera to the U.S. National Museum.

From 1899 to 1901 Baker was the highest-ranking teacher of biology at the Central High School in St. Louis, and in 1902–1903 he studied with Professor Vernon L. Kellogg at Stanford University.

In 1903, at the request of Professor A. J. Cook, then at Pomona College, Claremont, California, Baker became assistant professor of biology there. He began to study insects in the West with his usual assiduity and to publish his findings in the serial, *Invertebrata Pacifica,* which he published and financed out of his own pocket. During this year he also wrote two of his important entomological papers, which were published by the U.S. National Museum, *A Revision of American Siphonaptera or Fleas* in 1904, and *The Classification of the American Siphonaptera* in 1906.

He left Pomona College in 1904 to become chief of the Department of Botany at the new Cuban Experiment Station in Santiago de las Vegas, Cuba, which was established in 1904, where he began the study of tropical economic botany and soon established a botanical garden.

Baker told Essig that "while he was in Cuba a large American tobacco company offered him an unusually high salary if he would undertake a program of plant breeding to improve the strains of tobacco. Because Baker believed that tobacco was not good for young people, and since he never smoked or otherwise used tobacco, he refused to accept the position."

In Cuba he befriended a young native boy who "had a rare native ability as a naturalist." The young boy was Julian Valdez, who "scoured the valleys and the mountains, the grasslands and the tropical jungles, in search of new things that kept Baker pinning, pressing, and labeling far into the night and early morning. When Baker returned to Pomona College and, later still, went to the Philippines, young Valdez accompanied him." Essig says Valdez became probably "one of the most remarkable entomological and botanical field collectors of all time."

Baker left Cuba in 1907 to become curator of the botanical gardens and herbarium at the Museum Goeldi in Pará, Brazil, where people were dying like flies from yellow fever. He stayed one year, then returned to teach at Pomona College. He took with him thousands of insect and botanical specimens, which he presented to Pomona College.

"It was at the beginning of my junior year," writes Essig, "that I came under his singular guidance. Baker was a handsome man of medium height, slightly built, and very active and graceful. His personality was dynamic and he had unbounded energy and alertness. I never heard him say a harsh or unkind word to anyone, and I never left him without a heightened feeling of strength and enthusiasm. For his students and associates he set a pattern of friendliness, honesty, frankness, and industry unexcelled. He fashioned and attained all his objectives with confidence and optimism, and these qualities made him a great and inspiring teacher."

Baker and Cook did a remarkable job at Pomona. Their students investigated the systematics and life histories of many citrus pests and other insects in the vicinity. They organized orchard inspection, which resulted in excellent field experience and good financial returns to advanced students, as well as beneficial results for the growers. They began a *Journal of Entomology* and a *Journal of Economic Botany,* and the Pomona College Laguna Marine Laboratory. But most of all, they turned out many first-rate biologists.

When Cook was appointed California State Horticultural Commissioner in 1911, the brilliant team was split up. In 1912 Baker accepted an important professorship in tropical agronomy at the University of the Philippines' College of Agriculture at Los Banos. Here for the last fifteen years of his life he contributed greatly to the development of scientific agriculture in the Philippines. His brilliance as a teacher was such that he "often took students who were thought incapable of learning and made of them worthy scholars." He helped edit the *Philippine Journal of Science,* the *Agricultural Review,* and the *Philippine Agriculturist,* and he co-operated with numerous Governmental agencies. Throughout the Philippines, Straits Settlements, and Borneo, either by his own efforts or that of his emissaries, he constantly acquired insects and plants, especially fungi.

During World War I he spent six months in Singapore as assistant director of the botanical gardens. In 1919 he was appointed dean of the College of Agriculture at Los Banos. He had a personal laboratory and storage rooms for his immense collection in his own home. Out of his salary he constantly aided impoverished students, and from the end of the war until his own untimely death "he gave half of his salary to fellow scientists in Europe who had been reduced to poverty as a result of the war."

Through his efforts hundreds of entomologists and botanists were constantly supplied with insects and botanical specimens, including fungi, so that their activities opened the eyes of the scientific world to many of the natural wonders of the Western Pacific. He published papers on the taxonomy of Homoptera and Hymenoptera of the Philippines, Malaya, Papua, and even Australia, and worked on the cercopids, jassids, and fulgorids in the Homoptera, and the braconid parasites in the Hymenoptera.

Baker had been separated from his wife for many years and had a romantic attachment to his cook, Tome San, a comparatively young widow. He intended to marry her as soon as he could get a divorce from his wife. Then one day in 1922 his Japanese gardener went berserk and tried to kill Baker with an axe. Tome San ran between Baker and the gardener and was struck by the axe, leaving Baker unscratched. She died five days later. Her death was a blow from which he never recovered.

The University's decision to replace the white faculty with Filipinos and Baker's rapidly deteriorating health soon made it imperative for him to think of other climes. He made overtures to the California Academy of Sciences, but these were complicated by the extent of his collections, valued at $50,000, in which he had invested his life's savings, and arrangements with the Academy were never completed.

David L. Crawford, president of the University of Hawaii and a former student of Baker's, offered him a position which combined the resources of the Sugar Planters' Experiment Station, the Bishop Museum, the University of Hawaii, and other groups. His duties were to conduct an entomological survey of the Pacific Islands in the South Seas. The arrangements suited Baker and he made plans to resign from the University of the Philippines in November 1927. The University of the Philippines, well aware of his tremendous contributions to Philippine science, voted him Dean Emeritus of the College of Agriculture and Director Emeritus of the Experiment Station, to take effect on December 1, 1927. But this was an honor Dean Baker never realized.

In June 1927, previously weakened by numerous tropical diseases and consumed by an attack of chronic amoebic dysentery, Baker was rushed to a hospital, where he died on July 21, 1927, a man who had done "the life work of ten men."

Baker's will asked that he be cremated and that his ashes be buried in a Buddhist Temple in Kobe, Japan, close to Tome San's ashes, and this was done. His will also provided that his collections should go to the Smithsonian Institution and the National Museum. It took Dr. [R. A.] Cushman, a representative of the National Museum, five months to repin, pack, and ship about one-thousand-four-hundred Schmidt boxes of insects and 35,000 index cards on Malaysian insects.

A stone on Baker's grave in Japan sums up his life in these words, "He gave his life for the education of other people."

CLARENCE PRESTON GILLETTE (1859–1941)

C. P. Gillette was born, according to List (1942), on April 7, 1859, at Maple Corners, Ionia County, Michigan. He studied under A. J. Cook at Michigan State Agricultural College, where he received his B.S. in 1886 and his M.S. in 1887, and in 1916 he received an honorary degree of Doctor of Science. He was employed for a year as assistant entomologist by the Zoology Department of Michigan State Agricultural College and from 1888 to 1891 in the Experiment Station of Iowa State College. In 1891 he was placed in charge of the Department of Zoology,

Figure 61. C. P. Gillette

Entomology, and Physiology at the Colorado Agriculture College (now Colorado State University), and for more than forty years was associated with the growth of the College in numerous capacities. In 1907 he became Colorado's first state entomologist, a position he held for twenty-four years, while still carrying on his work at the College; and in 1910 he became Director of the Colorado Experiment Station. During the last few years of his service before his retirement in 1932, he was vice president of Colorado State.

Gillette was an excellent and popular teacher of entomology, genetics, eugenics, and other subjects and was in great demand as a writer and speaker. In taxonomy he was best known for his work on the Cynipidae, Cicadellidae, and Aphidae. He and Miss Miriam A. Palmer were co-authors of the "Aphidae of Colorado" published in the *Annals of the Entomological Society of America* between 1931 and 1936. He was said to be a tireless collector and to have a remarkable memory as to just when and where he had taken a certain species.

A very capable administrator, he was nevertheless a modest and retiring individual, who took a kindly interest in the members of his staff. The Experiment Station grew and prospered under his administration. For most of his life he was an active member of the Presbyterian Church and his Sunday school class was well attended.

Toward the end of his life his eyesight began to fail and he courageously adjusted himself to cope with this condition. Except for his eyes, he was in reasonably good health until he was felled by a stroke and died at his home in Fort Collins, Colorado, on January 4, 1941.

JOSÉ ROLLIN DE LA TORRE-BUENO (1871–1948)

J. R. de la Torre-Bueno was born, according to Sherman (1948), in Lima, Peru, on October 6, 1871, and his family moved to the United States when José was fourteen years old; he was "fully acquainted with our language having studied under English tutors in Peru."

He attended the School of Mines at Columbia University and graduated in the class of 1894. He was employed by the General Chemical Company of New York, where his duties consisted of editorial and other work.

Torre-Bueno was one of the stalwarts of the Brooklyn Entomological Society for many years and "the prime mover in the revival, after 27 years, of the Bulletin of the Society in 1912 (New Series, Vol. 8) and in 1926 of the New Series (Vol. 7) of *Entomologica Americana* of which the final volume 6 of the First Series was published in 1890. He was the editor of both series until he died." The Brooklyn Entomological Society, to which Sherman refers, once one of the liveliest entomological societies in the United States, was closed several years ago. Undoubtedly trees grow in Brooklyn, but they and the rolling fields of old, with their hordes of engrossing insects, have largely disappeared. Like the Indians of the plains, who vanished with the buffalo, only the ghosts of amateur entomologists now haunt the paved streets of the Borough. And Torre-Bueno (1948) reminds us of the enthusiastic and thirsty members of his day, "who adjourned to a German biergarten nearby the place of meeting, going into the back room by The Family Entrance, where they

Figure 62. J. R. de la Torre-Bueno

were served sauerfleisch and other hearty Teutonic food and delicacies washed down with foaming steins of 'echt bier'—none of the feeble latter-day imitations or 'ersatz.'"

Torre-Bueno was primarily interested in members of the order of the Hemiptera and of course published his papers in the journals he edited. One of his most important studies was his "Synopsis of the North American Hemiptera Heteroptera," which appeared in three parts in 1939, 1941, and 1946 in *Entomologica Americana*. In 1937 the Brooklyn Entomological Society published the entomologist's *Glossary of Entomology*, a revision by Torre-Bueno of J. B. Smith's *An Explanation of Terms Used in Entomology*.

Olsen (1948) remembers Torre-Bueno as an enthusiastic entomologist and collector, a writer and a linguist: "It is many years ago, although it seems as if it were a short time back that I first made the acquaintance of Bueno. It was when Bueno resided at number 14 Duzenbury Place, White Plains, New York. It was at the time when his children, now grown up men and women, were babies and our own girls were mere tots in rompers. The picture that most strongly comes to mind from those early days is that of Mr. Bueno, youthful, alert, generally mild-mannered, very courteous and attentive, surrounded by a growing family of children, all of whom he adored and felt proud of. . . . aside from all his other abilities, Bueno had a beautiful speaking voice, with a rare, clear and bell-like tone that, together with his perfect diction, accompanied with a slight Spanish accent, made his talks interesting, convincing and unforgettable." Although of mild disposition, he was a man of independent spirit who would not tolerate anything unjust or unscrupulous.

His wife was a Miss Lillian Reinhardt of Brooklyn and they had four sons and three daughters. In September 1934 he moved to Tucson, Arizona, and died there May 3, 1948.

The University of Kansas purchased the collection which he had amassed over forty-seven years and which was especially rich in aquatic Hemiptera.

EDITH MARION PATCH (1876–1954)

Although female professional entomologists are relatively few, Dr. Edith Patch was an uncommonly good one and yielded to no one as an authority on aphids.

She was born on July 27, 1876, the youngest of six children, in Worcester, Massachusetts, where she lived until 1884, when her family moved to Minnesota. For two years she lived in Minneapolis; then her

Figure 63(A).

Figure 63(B). (A) E. M. Patch; courtesy of Mrs. J. B. Adams. (B) E. M. Patch, center; J. R. Parker, left; F. B. Paddock, right; courtesy of J. J. Davis

father purchased ten acres of prairie land close by, about a mile from the Mississippi River. The farm was within a short distance of a lake, swamp, and woodland area rich in flowers, birds, animals, and insects—all of which were a constant delight to a child with an innate love of the life around her.

This period is described in some of her unpublished notes:

Among all the intermingled interests, the Monarch Butterfly fluttered through the vista like a beckoning sprite. I followed its metamorphic fortunes with fascination: first a pale egg with reticulated engravings; next a larva ringed with yellow and black and white stripes, gay of garb as a court jester; then a noble chrysalid, a jade green jewel touched with gold and jet black; and last the majestic butterfly taking a leisurely flight with wings richly tawny above and bordered with black velvet.

This gracious insect undoubtedly influenced my life. It brought me a gift of $25 by functioning as the subject of my essay in a prize contest during my senior year in the South High School of Minneapolis, the spring of 1896. With part of this welcome money, I purchased the book I most desired—*Manual for the Study of Insects*. As I looked at the title page of that book by John Henry Comstock, Professor of Entomology in Cornell University, and Anna Botsford Comstock, member of the Society of American Wood-Engravers, I wished I might know the authors. Because of them, Cornell University became a goal. (I could not predict the happy time to come fifteen years later, when I should be offered my choice of the artist-proof of certain famous engravings bearing the inscription, "With the cordial regards of the artist and engraver, Anna Botsford Comstock." I could not foretell that in 1924 I should receive a gift copy of the 1,044-page book, *Introduction to Entomology,* with the penned inscription by the author, "To Professor Edith M. Patch, with the affectionate regards of her old friend, J. H. Comstock.") I received my Ph.D. degree at Cornell University in 1911, after studying there during leaves-of-absence from Maine; and Professor Comstock's book, just mentioned, includes material from my doctorate thesis. . . .

My first impression of Entomological Cornell was that it was sort of a family with the faculty acting as older brothers to the graduate students and everybody loving the Professor and Mrs. Comstock better than they did anybody else and that Cornell was the friendliest group of people in the world.

Edith entered the University of Minnesota in 1897 and received her B.S. in 1901. At this university, Professor O. W. Oestlund first introduced her to the study of aphids. She taught for two years in high schools in Minnesota, "while hopefully prospecting for a chance to enter my chosen field." Finally, after being turned down by many experiment stations because entomology was "not a work for women," she was invited in 1903 to organize a department of entomology at the University of Maine, in Orono. She was head of that department until her retirement in June 1937, when she became Entomologist Emeritus and was given an honorary D.S.

During her thirty-four years at the University of Maine she wrote about eighty technical articles on insects, most of them from an economic or taxonomic viewpoint. The brown-tail moth, the strawberry crown borer, the potato plant louse, the woolly apple aphid, and many other insects were investigated by her as important farm pests. Her studies in taxonomy encompassed aphids and psyllids, and some of her important papers were *Homologies of the Wing Veins of the Aphididae, Psyllidae, Aleurodidae and Coccidae* in 1911, *Psyllidae of Connecticut* and *Aphididae of Connecticut* in 1923, and *Food-plant Catalogue of the Aphids of the World* in 1938.

Dr. Patch's contribution to the knowledge of American aphids is evident in her voluminous correspondence, which shows how she helped entomologists in every state of the Union and in many parts of the world by determining aphids for them, describing aphid habits, and encouraging and assisting other workers in their study of aphids. She was usually behind in her correspondence, and her work bench was overcrowded with aphids awaiting her determinations. This was a service she rendered to state, Federal, and other institutions, in addition to her own many and varied duties at the University of Maine.

A typical letter, dated September 1, 1916, to Mortimer Leonard at Cornell gives the flavor of her personality:

> Young man,—the next aphids you send drop alive into *thin* balsam on a slide, clap on a cover glass and transport *whole*. It has taken my assistant half a day to collect the fragments from the debris which arrived, and me half an hour to examine the different portions of wrecked antennae and select one with the sixth segment entire. An aphid is a delicate animal and swishing it around in a large vial of fluid works havoc. Alcoholic material of the larger species does very well if you crowd a smooth plug of absorbent cotton down so close as to prevent swishing. If you hear of anybody thinking of sending me material,—just pass my advice along if you don't mind. It makes me savage to see perfectly good aphids broken to bits!
>
> After which tirade I will express my appreciation for the privilege of seeing this interesting species which is new to me, and is apparently undescribed for America at least.

In her "off hours" Dr. Patch wrote more than forty articles on popular science and nature subjects in various magazines and about one-hundred nature stories for children in juvenile magazines. She published some seventeen books on natural history for younger readers, a number of them with Carrol Lane Fenton as co-author—such as *Hexapod Stories, Holiday Pond, Outdoor Visits,* and *Desert Neighbors.*

Dr. Frank H. Lathrop, a State and Federal entomologist, was kind enough to record his memories of her:

"She was ambitious to do her best at whatever she undertook. This was manifest in the care that she used in her professional work and also in her daily life.

"She was persistent. . . . In any argument or contest she usually won her way. In winning, however, she used gentle persuasion rather than the more forceful methods attempted by so many people.

"She was unhurried. She always took time to do well whatever activity was at hand.

"She was helpful to anyone who came to her in need. She was generous with sound advice, and when necessary, with financial assistance. Sometimes she seemed overgenerous to the extent that she fell prey to dishonest graspers.

"To crown it all, she was friendly, and she had a pleasing sense of humor, not to ridicule, but to laugh with friends."

Mrs. Jean Burnham Adams was her last graduate student, and she writes of Edith Patch with reverence. "I remember my graduation day so well. She took me to lunch in Bangor; in the midst of general conversation she paused and said, 'Jean, on your graduation day, I have a thought to leave with you.' I awaited a profound statement—'Never in the years ahead be afraid to say "I don't know," but try not to have to say it twice about the same thing.' At the time I was rather let down; soon, however, I began to realize how very basic this wise counsel was, and as years have gone on, how often she has saved me from pretentiousness!" To Jean Adams, Edith Patch typified "whimsy, love of beauty and of contrast; a gentle, courageous, audacious, highly intelligent, beauty-loving woman; whether the beauty lay in daffodils or pure, fine observational work mattered not. She had an impish quality that made her most lovable and sometimes unpredictable, as when the title of a serious survey of aphids in potatoes appeared in press as 'Marooned in a Potato Field.' "

Dr. Patch described her home in Orono: "My home, with my sister Alice Patch (who taught in Minneapolis until she came here in 1918), is a typical 100-year-old New England white house situated about a mile from the University of Maine campus. The 50-acre home lot is a combination of river front, meadow, and woodland. Among the plants in my wild-flower garden I grow milkweeds—as an invitation to Monarch Butterflies."

She won many honors during her career. In 1930 she was the first woman president of the then Entomological Society of America. She belonged to a number of honorary societies, including Phi Beta Kappa. She travelled and collected through much of the United States and parts of Canada and Mexico and in 1927 spent six months at the Rothamsted Experimental Station in Harpenden, England. She was a recognized world authority on aphids.

She died at Orono on September 27, 1954, and was buried in Worcester, Massachusetts, her childhood home. Adams and Simpson (1955) wrote her obituary.

HARRY GARDNER BARBER (1871–1960)

H. G. Barber was born on April 20, 1871, in Hiram, Ohio, his father being a veteran of the Union Army and a teacher of classical languages. In 1881 his parents moved to Lincoln, Nebraska, where his father taught Latin and Greek in the newly organized University of Nebraska. Harry graduated from this university in 1893 and then assisted Professor Bruner in the Department of Entomology, receiving his master's degree in 1895. In 1896 he began to teach at Nebraska City High School and the same year married Blanche E. Davis. In 1897 and 1898 he attended Bussey Institute at Harvard, where he received another M.S. Despite the fact that he studied under William Morton Wheeler, the noted ant authority, he maintained his interest in the taxonomy of the true bugs.

In 1898 he was appointed teacher of biology at DeWitt Clinton High School in New York City, and he taught there until his retirement in 1930. Although he was, according to Leonard and Sailer (1960), "a first-class teacher, the daily routine of teaching biology to large classes of boys, city-born and bred, and most of whom had little interest in the subject, must at times have been trying. Undoubtedly he found relaxation and intellectual stimulation from his week-end and long summer collecting trips as well as from his correspondence with hemipterists in all parts of the world."

He was an active member of the New York Entomological Society and its president in 1916–1918; he was an associate of such entomologists as

Figure 64. H. G. Barber

C. W. Leng, Wm. T. Davis, J. R. Torre-Bueno, F. E. Lutz, A. J. Mutchler, W. Beutenmüller, C. Olsen, H. B. Weiss and others. Leonard and Sailer (1960) write, "When the senior author was in high school, he well remembers Mr. Barber leaving school with his suitcase just as soon as he could get away on Friday afternoons during spring and fall to go on a collecting trip with some of these men." During the summer vacations he travelled and collected throughout the United States with Mrs. Barber, who encouraged him in his avocation.

After his retirement as a teacher he became a specialist in charge of the true bugs in the USDA Bureau of Entomology. He resigned in 1942 to return to his home in Roselle, New Jersey. Upon his wife's death in 1949, he was made Collaborator at the U.S. National Museum, a job he had held previously.

Barber published about one-hundred-and-ten papers, many on the Lygaeidae, and in 1950 gave his fine collection and library to the National Museum. He continued to work actively until two weeks before he died in Doctor's Hospital, Washington, D.C., on January 27, 1960.

HOWARD MADISON PARSHLEY (1884–1953)

H. M. Parshley was described by Usinger (1954) as "one of the most versatile hemipterists of our time." He was born in Hallowell, Maine, on August 7, 1884, and grew up on a farm in New York State. He graduated from Harvard in 1909 and attended the New England Con-

Figure 65. H. M. Parshley

servatory of Music from 1906–1909; he received his M.S. from Harvard in 1910 and his doctorate in 1917.

Parshley was appointed to the zoology department of Smith College in 1917. From 1914 to 1925 he published fifty-three papers on the Hemiptera, and his well-known *Bibliography of the North American Hemiptera-Heteroptera* appeared in 1925. He was also managing editor of the *General Catalogue of Hemiptera,* which appeared in twelve parts between 1929 and 1949. His "penetrating mind" exerted great influence over other hemipterists, but after 1925 he devoted his scholarly studies largely to ethics, evolution, psychology, eugenics, and the biology of sex. He wrote several books on these.

Parshley was an accomplished musician and played in the bass section of the Springfield Symphony Orchestra.

He died on May 19, 1953.

ZENO PAYNE METCALF (1885–1956)

Z. P. Metcalf was born in Lakeville, Ohio, on May 1, 1885. He earned his A.B. degree in 1907 at Ohio State University and his doctorate at Harvard in 1925. He taught at Michigan State for one year and then joined the staff of the North Carolina Department of Agriculture. In 1912 he became a member of the North Carolina State College and rose to be head of the Department of Zoology and Entomology while also serving in other administrative positions, including that of Associate

Figure 66. Z. P. Metcalf

Dean of the Graduate School. He retired from his administrative duties in 1950 and for the remainder of his life taught, did research, and wrote. He was the author of nine books and numerous other publications and was engaged in the preparation of a forty-two-volume catalogue of the Homoptera of the world. Of these, fifteen volumes were printed or ready to go to press at the time of his death.

Metcalf served as president of the American Microscopical Society in 1927, of the Entomological Society of America in 1949, and the Ecological Society of America in the same year.

Clyde F. Smith, a former associate of his, was kind enough to present, in a letter, his impression of him:

"Dr. Metcalf was one of the kindest, most gentle men that I have ever known. At the same time he could be extremely stern and positive in dealing with students and faculty alike. He was a great biologist in the truest sense of the word. Because of this, he realized that the man who was interested in his work, and had freedom to follow his interests, accomplished the most in the long run. He was, therefore, against regimentation. This was brought to my attention most forcibly. The first week I was on the job, I was assigned the responsibility of working on fruit insects. After outlining many things which I thought should be done, and which would take many men to do, I went to Dr. Metcalf to ask which I should tackle first. His comment was: 'He didn't care what I did as long as I worked.' 'Z. P.,' as he was affectionately known, was a tireless worker. He expected everyone else to put in his full quota of work. Many people complained that they could never get to see Dr. Metcalf; however, I found that it was always possible to see him within twenty-four hours after making a request. I say after twenty-four hours because it was often difficult to see him during the regular working hours of 8:00 A.M. to 5:00 P.M.; however, he was always willing to stay and see you any time after 5:00 P.M. or before 8:00 A.M. While he was devoted to his work as an entomologist, he also found time for his family, his church, and his community. Those of us who had the opportunity of knowing Dr. Metcalf are living a richer and fuller life because of this association."

Professor Metcalf died suddenly in his home in Raleigh, North Carolina, on January 5, 1956, and was survived by his wife and a daughter. Smith (1956) wrote his obituary.

HERBERT BARKER HUNGERFORD (1885–1963)

According to Woodruff (1956, 1963) Hungerford was born on August 30, 1885, in Mahaska, Kansas, the son of a pioneer Kansas family. He

Figure 67. H. B. Hungerford

attended Kansas State Normal School (Kansas State Teachers College) in Emporia and was a teacher, principal, and superintendent of schools there from 1904 to 1909.

He married Mary Frances Kenney in 1905, by whom he had one daughter. In 1909 he enrolled at the University of Kansas as a full-time student and received his bachelor's degree in 1911, when he became a member of the faculty. An outstanding student, he had been elected to Phi Beta Kappa in his senior year.

In 1924 the University appointed him Head of the Department of Entomology, and the same year he also assumed the position of Kansas State Entomologist, a position he held for twenty-five years.

He was a great student of the ecology and biology of water bugs as well as of the taxonomy of these insects, especially the familiar Corixidae and Notonectidae. In time he accumulated one of the largest collections in the world of aquatic and semiaquatic Hemiptera and became a world authority on the subject.

Woodruff writes that Hungerford was "a dedicated teacher with an inspirational flavor. He . . . was endowed with a tenacious memory and a great ambition to excel, coupled with an integrity of mind and purpose which his students could expect only to emulate. His energy and persistence are legend and a constant example to his many students and colleagues." He was a friendly man, but could be aloof when preoccupied. "Quick to respond to friendship, he has a subtle wit when appropriate. Many students through the years have been aided financially through his concern and benevolence."

Entomologists and biologists had a very high regard for Hungerford and he was president of the Entomological Society of America in 1936 and the Society of Systematic Zoology in 1953. He was instrumental in the formation of the Kansas Entomological Society. He was the recipient of many honors, including a Leidy Medal from the Philadelphia Academy of Sciences for being an outstanding biologist.

After retiring from the University in 1956 he continued his research on "water bugs"; he named hundreds of new species of them. He died on May 13, 1963 at the age of seventy-eight.

CARL JOHN DRAKE (1885–1965)

According to Froeschner (1966) and Gurney *et al.* (1966), Drake was born on a farm in Eaglesville, Ohio, on July 28, 1885. He attended public schools and Heidelberg Academy in Tiffin, Ohio, and received his B.S. and Bachelor of Pedagogy from Baldwin-Wallace College in Berea, Ohio, where in 1912 he participated in athletics, especially basketball and baseball.

He undertook graduate work at Ohio State University from 1913–1917, where he studied entomology under Herbert Osborn and became a taxonomic specialist on lacebugs and semiaquatic Hemiptera. For almost fifty years he devoted himself to the study of these and related Hemiptera. Ohio State University gave him an M.S. in 1914 and a Ph.D. in 1921, and he taught there between 1913 and 1917. From 1917–1922 he was a specialist in entomology in the School of Forestry, Syracuse University.

Figure 68. C. J. Drake

From 1922 to 1946 he was head of the Department of Zoology and Entomology at Iowa State University, where he established a reputation as educator, entomologist, conservationist, and administrator. Under his direction the Department acquired a reputation for its broad training in the sciences and for its strong graduate program. Many outstanding students of entomology received their training under his enthusiastic tutelage.

While he served as head of the Department, he was also head of the Entomology Section of the Agricultural Experiment Station as well as State Entomologist. His farm background helped him understand the problems of the farmer and was invaluable in enabling him to find practical solutions to grasshopper, chinch bug, Hessian fly, and European corn borer plagues.

Drake was a member of many important organizations, including the Central States Plant Board, the National Plant Board, and the American Association of Economic Entomologists; he headed the Tucura (Grasshopper) Commission to the Argentine in 1938–1939.

Froeschner reports that in 1946 he "grudgingly relinquished his three official responsibilities and began concentrating on his taxonomic interests." Never having married, he lived alone in a frugal fashion, investing his income so as to acquire the independent means that permitted him to pursue his interests. In time he gathered the world's best collection of lacebugs and a very fine collection of semiaquatic Hemiptera of the Western Hemisphere.

In 1957, he became an Honorary Research Associate of the National Museum in Washington, D.C., and moved his collection and important library there. Froeschner adds that "His associations with the many staff scientists and with those who visited the Museum catalyzed his enthusiasm into even greater taxonomic efforts. He worked about six and a half days nearly every week and was able to surpass his lifetime goal of 500 scientific papers. But more importantly, he brought to a monumental close his half century of tingid studies with an outstanding monograph on the morphology of these insects and a unique catalog of the tingids of the world."

Dr. Drake died on October 2, 1965, in Washington, D.C., from a diabetic and circulatory illness, leaving his collection and library to the National Museum.

RAYMOND HILL BEAMER (1889–1957)

Beamer was born on a farm near Hallowell, Kansas, on October 20, 1889. He received his early schooling in nearby towns and entered the

Figure 69. R. H. Beamer

University of Kansas in 1909. Initially he majored in geology but later switched to entomology. He received his A.B. in 1913 and became assistant curator of the Francis Huntington Snow Collections in 1914 while doing graduate work. He obtained his M.A. in 1917 and then for five years, during and after the war, helped on his wife's farm near Hallowell, Kansas. In 1922 he returned to the University of Kansas as Assistant Curator of Collections and Assistant Professor of Entomology. Continuing his graduate work, he received his Ph.D. in 1927, becoming full Professor in 1939 and Curator of the Snow Collections in 1949.

At the University he taught systematic entomology and guided and inspired many students who have since become well-known taxonomists. He was a friendly, helpful man, quite skilled in the making of all kinds of equipment. He published 107 papers on the Homoptera, especially on the Cicadellidae and Fulgoridae, and described 563 new species.

He died on November 21, 1957, after a long illness. Hungerford (1958) and Hungerford and Oman (1958) wrote his obituary.

HAROLD MORRISON (1890–1963)

Morrison was an important contributor to our knowledge of scale insects. He was born on a farm in McCordsville, Indiana, on May 24, 1890. He studied with Vernon L. Kellogg at Stanford University and then with John Henry Comstock at Cornell University. Morrison received his B.A. from Cornell in 1914, his M.A. from Stanford in

Figure 70. Harold Morrison

1915, and his Ph.D. from Harvard in 1927. He was a star high jumper at both Stanford and Cornell.

From 1911 to 1915 Morrison was a part-time employee in the office of the State Entomologist of Indiana, where he and Harry F. Dietz wrote "The Coccidae or Scale Insects of Indiana," in the *Report of the Indiana State Entomologist* for 1916. R. E. Snodgrass, who later became a famous student of insect anatomy, illustrated this article.

In 1916 Morrison joined the U.S.D.A. as a plant quarantine inspector for the Federal Horticultural Board, and in 1919 he became an entomological explorer in the U.S. Bureau of Entomology, travelling to the West Indies, where he collected coccids and other insects. In 1920 he was appointed a specialist on scale insects for the Bureau, a position he retained until he retired in May 1960. There he was in charge of taxonomic investigations from 1928 to 1935.

Morrison prepared important studies of scale insects in several subfamilies. A number of his papers were written in collaboration with his wife Emily. According to Russell (1963), "He pioneered in the preparation of specimens for microscope examination, and contributed to the improvement of mounting techniques of scale insects and other microscopic arthropods."

Russell tells us that he had an excellent memory and that temperamentally, "He was direct, critical, thorough, cautious and uncompromising."

Morrison died from a stroke in Washington, D.C., on March 11, 1963.

9 Notable Coleopterists

The lightning bug seems brilliant
But he has not any mind
For he stumbles through existence
With his head light on behind.
Entomological News, 16(3):88, 1905

Beetles in their profusion of species and numbers and in their multiplicity of forms and colors have always attracted collectors. Even the layman with no special interest in them is aware of lady beetles and the May and June beetles. Some, such as the scarab beetles, have been objects of worship. Others, such as the cotton boll weevil, the Japanese beetle, and the alfalfa weevil have not been held in such high esteem. In any event, American entomologists have devoted a great deal of study to beetles.

JOHN LAWRENCE LeCONTE (1825–1883)

S. H. Scudder, no mean entomologist himself, wrote in 1886, "LeConte was the greatest entomologist this country has yet produced." LeConte is indeed our greatest coleopterist, not because he named almost five thousand species of beetles, but because he showed their systematic relationships and pointed the way to the scientific classification of American insects.

Lesley (1883), Riley (1883), Horn (1884), Scudder (1884, 1886), Dow (1914), and others have sketched the life of Dr. LeConte and the portrait of him presented here is drawn from their accounts.

As Lesley said in his sketch of Dr. LeConte, "A memorial of the life of our fellow member and friend would be incomplete without a personal description of old Major LeConte, to whose vigorous intellect, excellent common sense, and great experience in zoological studies, John owed not only his extraordinary abilities, his aptitude for mathematics, his eye for form and color, his exactness, his imagination, his love of

Figure 71. J. L. LeConte. Courtesy of the Academy of Natural Sciences

the study of languages, his taste for historical metaphysics, and especially mythology, and his pronounced capacity for practically putting things in order and managing affairs, but also the opportunity for cultivating and displaying all these various, and as many people vainly imagine, contradictory mental powers."

The LeContes, a wealthy and prominent family in France of Huguenot ancestry, fled to the New World to escape the intolerance of Louis XIV and the revocation of the Edict of Nantes. According to Dow (1914), LeConte's father, Major John Eatton LeConte "entered the army from pride, not necessity, and retired long before he was forty." He was thirty-seven when he married a Miss Lawrence and settled in New York City. Their first two children died in infancy. His wife died a few months after their third child was born. The child was raised in the care of his father, who passionately pursued his life-long interest in natural history. According to Dow, he worked day after day on his beetles with the "little toddler on his knee."

Dow says that the Major, being a man of simple tastes, lived economically, for he was determined to provide his son with "the broadest foundation of general knowledge" and to see that he should have the leisure to pursue his interests.

About twenty years before John's birth, the Major had published some papers on the Coleoptera; later he took a greater interest in the previous stages of Lepidoptera, according to Horn (1884). "The well-known monograph which he published together with Dr. [J. A.] Boisduval from Paris, France, was the first of these studies. But when the son decided upon the line of his studies, his father returned also to his former favorites, and published a monograph of the Histeridae of the United States, for which the son had drawn some excellent plates. These plates evince a prominent talent for entomological drawing, and it is not easy to understand why he did not follow up this remarkable talent."

In addition, the Major compiled catalogues on the plants of New York City and wrote papers on mammals, reptiles, batrachians, and crustacea as well as systematic papers on butterflies and beetles.

When John once sent back 10,000 beetles in alcohol from San Francisco the Major mounted them, identified all he could, and jotted down the most important characteristics of all of them.

In 1852 the LeContes moved to Philadelphia. Professor S. S. Haldeman was then at the University of Pennsylvania and Dr. Frederick E. Melsheimer had become president of the Entomological Society of Pennsylvania at York, formed in 1842. Haldeman and John LeConte had volunteered to edit and bring up to date for the Smithsonian the Melsheimer check list. The American Entomological Society (formerly the Entomological Society of Philadelphia) was formed in 1859 by Ezra T. Cresson, George Newman, and James Ridings. The Major used to attend its meetings, "a little bent from his earlier carriage, one hand bearing heavily on his cane, the other on the shoulder of his boy," writes Dow, the "boy" being then in his thirties. Among the visitors to the LeConte house at 321 West Locust Street were August Sallé, a French coleopterist, and Victor Ivanovich Motschulsky, a Russian coleopterist, who was using the LeConte collection, which then consisted of about seven thousand carefully arranged species, to identify some beetles he had collected in the South.

Major LeConte died in Philadelphia in 1862 at the age of seventy-eight. Besides his collection he left behind a series of water colors of native insects and plants that he had painted. We can only wonder, with Dow, what happened to the some two or three thousand water colors of insects done by John Abbot, which the Major had accumulated over many years.

John showed a natural aptitude for mathematics and language, and with his retentive memory, rapidly progressed through school, graduating from Mount St. Mary's College in Maryland in 1842. His fellow students in the College used to ridicule him for his interest in natural history, and his teacher once tried to discourage his interest in this field until the Major intervened. Lesley (1883) recounts this incident in the strictly disciplined class:

"One day silence reigned in the school-room. Everybody was conning his task at his seat. The tutor was silently reading at his desk. Suddenly there was a great fracas—John LeConte was seen starting from his seat and scrambling on the floor in the middle of the room. He was called up to the tutor's desk to give an account of himself. He held in his hand two beetles. He explained that they were rare, that he could not help trying to catch them, that he had to be quick about it, that he did not know he would make such a noise, etc. The others in great excitement sat expecting dolorous consequences for John. But they were disappointed. The tutor remembered the Major, or perhaps had received orders from the upper region. He merely sent the boy back to his seat with his beetles, and a warning not to make so much noise another time."

After graduating from St. Mary's, LeConte entered the College of Physicians and Surgeons in New York, and in 1846 was granted a medical degree. However, he never entered private practice. Dow wonders how he got his medical degree: "He made a journey to the Far West in 1843. In 1844 he visited Lake Superior, working his way along the entire south shore and crossing the country to the sources of the Mississippi River, and this trip was soon repeated. In 1845 he went up to the Platte River to Fort Laramie, thence to the foot of the Rocky Mountains. He followed the Santa Fé trail to New Mexico to turn up once more the insects known only from Say's descriptions of his own types. There was little time left for the study of medicine." Dow also refers to "the ten years when young LeConte was hurrying from Superior to Florida, from Nova Scotia to San Diego, from Coney Island to South Orange, losing 20,000 specimens in the San Francisco fire of 1852, robbed of his horses by the Indians near the Gila River and having to walk to camp thirty miles over the desert, constantly amassing the actual material from which he constituted his classification."

In 1844 LeConte's first paper on beetles, "Descriptions of Some New and Interesting Insects Inhabiting the United States," appeared in Volume 5, No. 2, of the *Boston Journal of Natural History* of the Boston Society of Natural History.

In his third paper published the same year, while he was still a student in medical college and all of nineteen, John voiced this vigorous challenge:

"The indolence of our entomological observers is the more deplorable, as we are few in number, and therefore more is to be expected from each individual. The field of research is still open, and anyone who travels in it, with even ordinary care and attention, will not fail, under the numerous stones scattered on its surface, and the weeds which apparently obstruct his path, to discover as fine insects as have ever graced the cabinet of a [F. W.] Hope or a [P. F. M. A.] Dejean. I trust that the day is past when our insects must be sent to Europe for determination. Are we to be bound by the mere dictum of some European entomologist, of equal indolence with ourselves, who chooses to *name* the insect which we have discovered? Where should our insects be better known than in the country which gave them birth; but in what civilized land are they less studied?"

These early papers dealt with ground beetles, tiger beetles, long-horned beetles, and dytiscids from the Eastern United States. Other early papers by him were devoted to faunal and ecological relationships among insects and animals. In 1851 he read a paper before the American Association for the Advancement of Science on the distribution of California beetles.

With the aid of Baron Osten-Sacken on the Diptera and P. R. Uhler on the Hemiptera, LeConte edited *The Complete Writings of Thomas Say on the Entomology of North America*. This important collection appeared in two volumes in 1859 and has proven invaluable to American students.

In 1861, shortly before his father died, LeConte married Helen C. Grier, the daughter of Judge Robert C. Grier, and they had one son, Robert.

The first parts of a *Classification of the Coleoptera of North America* appeared in 1861 and 1862, but the Civil War temporarily halted LeConte's entomological pursuits. He resumed publication in 1873; the work was never completed, however. Meanwhile, he entered the Army Medical Corps as a volunteer surgeon and soon became medical inspector with the rank of lieutenant colonel. LeConte's organizational abilities proved to be of great value to the Medical Corps. With the end of the war, LeConte once again was able to immerse himself in his entomological, geological, paleontological, and ethnological studies. In 1867 he served as geologist on a railroad survey throughout Kansas and New Mexico, and the beetles of these two states undoubtedly received much-needed recognition as a result of this trek.

From 1869 through 1872, LeConte and his family lived in Europe, where he studied many important collections. During this period he also travelled in Algeria and Egypt.

One of his important works was his *Species of Rhyncophora*, published in 1876 as a volume in the *Proceedings* of the Academy of Natural Sciences, Philadelphia. And with the collaboration of his friend and pupil, Dr. George H. Horn, he completed the monumental *Classification of the Coleoptera of North America,* Volume 26 in the *Smithsonian Institute Miscellaneous Collection,* 1883. The latter work was based on 11,000 species of beetles from the collections of LeConte and Horn. All in all, LeConte was author of about one hundred and eighty publications, among which were many important monographs.

He was a founder and president of the American Entomological Society, which celebrated its centennial in 1959. He was also the first president of the Entomological Club of the American Association for the Advancement of Science, a forerunner of the American Association of Economic Entomologists. The Entomological Club met for the first time in Detroit in 1875. He was elected president of the A.A.A.S. in 1874.

Besides being a person of remarkable intellect, an authority in several fields of natural science, and a seminal influence in American entomology, LeConte was a most amiable and approachable person. F. G. Schaupp (1883) describes a typical encounter:

"I shall ever remember his kind reception when I visited him during the Christmas weeks of latter years. After his colored porter had opened the door for me and had taken him my card, the Doctor shouted from the top of the stairs: 'Welcome! Very glad to see you! Please come up stairs!' He talked with me for hours, and gave me all the information I desired, and then left me alone with his collection. He presented me many typical specimens and I shall never forget his kindness."

And Scudder (1886) writes, "I remember well with what timidity I, an utter stranger, a mere boy, first ventured to seek him, a man but twelve years my senior, yet clothed with all the garb of learning—and with what kindness I was received and counseled. The pains he took for others, the time he has given, the immense labor he has undertaken, in determining series of beetles for a hundred correspondents all over the country can never be known." Yet, strangely enough, Dr. Asa Fitch complained that despite numerous "importunities," Dr. LeConte would not determine his beetles for him.

According to Scudder, entomologists from Europe on their first visit to him, were "charmed with his learning and affability, and the freedom with which he communicated his rich stores of knowledge, [and] always spoke with enthusiasm of his erudition and his generous and simple spirit. It was the same on his visits abroad. His minute familiarity with all the details of structure through long series of various degrees of complexity, his wonderful retentive memory, his quick and accurate judgement, his courage and self-reliance, all gave his words weight . . ."

Riley notes that LeConte was greatly interested in the applied aspects of entomology and he wrote "Hints for the Promotion of Economic Entomology in the United States" in 1873 and "Method of Subduing Insects Injurious to Agriculture."

With the possible exception of LeConte's wife and son, no one knew him as intimately as George H. Horn (1883), who wrote of him: "For nearly twenty-five years our association had been of the most intimate nature. I sought his advice and instruction as a neophyte in entomology, finding a welcome which I had no reason to expect. Our friendship ripened to an intimacy never shadowed by the slightest cloud. . . .

"In a general review of LeConte's writings, we find them remarkably free from controversial tendencies. He gave to science the results of careful study, knowing that in time whatever was worthy would be adopted. His dissent from the views of another was always couched in the mildest terms. He was above . . . those petty jealousies which too often prevail between those working in the same field."

In 1878 LeConte was appointed Assistant Director of the United States Mint in Philadelphia, a position he held for the remainder of his life.

For the last few years of his life he suffered from ill health. He died in Philadelphia on November 15, 1883. His widow lived in her son's home there until her death in 1917.

LeConte's collection went to Harvard; both Louis Agassiz and Agassiz's son, Alexander, had been his life-long friends.

Horn sums up Dr. LeConte's contribution to American entomology with these words: "The results of LeConte's work in Coleopterology in America are plainly marked. He entered the field ten years after the death of Say, who seems to have had no higher ambition, if indeed capacity, then the description of species which he collected. LeConte on the other hand, began the framework of a systematic structure which he lived to see completed in all its parts. He reduced chaos to order."

GEORGE HENRY HORN (1840–1897)

Horn was born in Philadelphia April 7, 1840, his parents being of German extraction, and his father a drugstore proprietor. George graduated in 1858 from Central High School, where his teachers had encouraged his interest in natural science. In 1861 he received his degree in medicine from the University of Pennsylvania. From 1862 to 1866 he served as a surgeon with the California Volunteers in the infantry. His service in California enabled him to collect insects throughout much of California and in parts of Arizona and Nevada.

When he returned to Philadelphia in 1866 he resumed private practice and in time became an expert obstetrician, frequently called on by other doctors for consultation in difficult cases. He also resumed his association with the Academy of Natural Sciences and was elected president of the American Entomological Society.

He began his systematic studies of animals while still a student in medical college. His early papers were on the Coelenterates and the Bryozoa, and his first paper on Coleoptera appeared in 1860. His description of

Figure 72. G. H. Horn. Courtesy of the Academy of Natural Sciences

a beetle attracted the attention of Dr. J. L. LeConte, who then sought him out and this resulted in continuing friendship and collaboration. J. B. Smith (1897) writes of this association as follows:

"It is difficult to speak of Dr. Horn without referring also to Dr. John L. LeConte:—first and always his teacher, afterward also his co-laborer. And it was wonderful how these two men supplemented each other! Dr. LeConte was the broader student of nature; his grasp was wider and he saw the Coleoptera more truly in their relation to other orders, and the insects in their relation to the rest of the animal kingdom. Dr. Horn was narrower, but his knowledge of detail was greater and more accurate. The result of combining these two characteristics may be seen in the *Classification* of the North American Coleoptera . . ."

C. V. Riley (1883), writing about LeConte, notes that he "must have felt proud of the excellent work done by the younger naturalist, and the manner in which Horn's more advanced views and often more thorough labors—made possible by accumulated knowledge and material—were accepted, even where they undid much of his [LeConte's] previous descriptive work, was one of the truest marks of greatness in LeConte."

Horn sought more than one character to distinguish one species from another; he used the insect as a whole. He did not care to describe a single species; instead he prepared monographs to show the relationships of a number of species, particularly their anatomical relationships. According to Smith (1897) Horn "always claimed that there was no evidence that an insect was really suffering for the want of a name, and that no wrong would be done to it by postponing the christening for a brief period." H. B. Weiss (1936) called Horn "a born systematist, whose genius for arranging things, for discovering relationships and for consistent unification was developed to a remarkable degree."

J. B. Smith (1910), who was well acquainted with Horn, describes his room, "Not unusually, the bed was filled with papers, and everything was in such condition that scientific work could be resumed at a moment's notice whenever the doctor came in from a round of calls or had a few moments to spare during office hours." Smith continues, "and many an hour did I spend in his room among his boxes, while he was on his rounds; for the doctor had a large practice and entomology was his recreation. I regret that I can not give a picture of that room. There was a cot in one corner which was often the only available place to sit; there was a huge table or desk occupying most of the floor and, during the many years that I knew that room, this table was cleared only once. Occasionally the cigarette stumps would be gathered together and thrown out; but the dust and dirt were never otherwise disturbed. Cabinets and book-shelves were about the walls and books were everywhere—on the floor, the chairs and often even on the bed. It was strictly a work-room and the doctor was an indefatigable worker."

Horn's only recreation, apart from an occasional fishing expedition, was beetles. As Noland (1897) wrote:

"Dr. Horn's devotion to science was singularly undivided. Although every obligation of his profession was regarded by him as binding, he practised medicine merely as a means to an end. He found at once his relaxation and intellectual profit when, after an exhausting day of attendance on the sick, he was at leisure to pore over his cherished insect-cases until far into the night. Not having married, he was not distracted by domestic ties from his favorite occupation, and for social engagements he cared little. Art and literature were to him outside issues, very well in their way, but to be left to the cultivation of others. As a contributor to knowledge, his function was well-defined, and recognition of his success as an entomologist was valued by him the more because of the singleness of his interest."

Horn visited Europe on three occasions, when he studied insect collections and became acquainted with the entomologists there. He preferred to work with dead insects; J. B. Smith (1897) remarked that he was what might be called a "closet naturalist," although he had considerable experience in the field. He published 265 papers, in which he described 154 new genera and 1,682 species. Many of these descriptions were illustrated by his own drawings.

Calvert (1898) says Horn's sister described him as being five feet, eight-

and-a-half inches tall; "he was slender and rather delicate in build, of fair, pale complexion, with dark brown hair. Of nervous temperament, his energy was boundless, enduring fatigue and loss of rest, which was apparently unnoticed by him and resulted as you know. He had a remarkable retentive memory, was always studious from childhood, quick to learn and ready to retain, and capable of imparting his knowledge to others. In all matters of judgement he was very independent, and adhered to his opinions. In regard to character he had marked originality. His fondness for children was so great that one might almost say a little child could lead him. His mechanical talents were quite marked. He had good practical business habits and was good at figures. He was fond of music without any particular talent therefor."

He was not interested in extended argument or in engaging in personalities, as were so many of his contemporaries. C. V. Riley, who had a genius for entomology and controversy, once goaded Dr. Horn to reply in a private printing. The argument was over who had first proved that the unusual larva of *Platypsyllus* was a beetle larva. Horn said:

"No literary work is more distasteful to me than controversy especially when there is a personal element. . . .

"In publishing my reply to Dr. Riley privately I wish to express my disapproval of the use of the pages of scientific periodicals for the ventilation of personal grievances to the exclusion of more useful matter."

In a letter dated October 23, 1886, Horn replied to W. J. Holland regarding some spirited competition they had had in acquiring a beetle collection, and reveals his sense of humor:

I saw a letter you wrote to Mr. Wilt about his father's collection and was a little surprised, probably a little amused also, at its sense.

I suppose I am possibly at fault in the matter. He told me he had corresponded with you but I did not suppose that there was anything more than a flirtation. I at once urged him to write discontinuing negotiations and I offered to make up a large part of the amount necessary to reimburse him for the cost to him of obtaining the collection. As I had suggested making the Coleoptera collection a memorial to his father I thought no better thing could be done than to aggregate the whole lot. I do not suppose the collection would have increased materially the value of your own any more than the Coleops would have increased mine, bulk & beyond that nothing.

However the material can always be made accessible to you & if I have been partly the cause (not knowingly) of rubbing you the wrong way I am sorry, but when you come to the city do not fail to call and I have no doubt we can fight it out in the most approved style. I wish I could get out there to see you and Dr. Hamilton and I have no doubt but that he would stand by us both as long as we had a button on our coats.

The entire Coleoptera collection was presented to our Society and we thus get a magnificent foundation for a collection, in fact more than a foundation. I am now at work reviewing it and to make it more complete add from my cabinet where I have to spare.

Horn was an honorary member of many European as well as American entomological societies. He was librarian and one of the secretaries of the American Philosophical Society. Upon the death of LeConte, he was elected president of the American Entomological Society. The University of Pennsylvania conferred on him the honorary title of Professor of Entomology, but he never actually taught.

Horn's feelings for his former mentor and friend, LeConte, can be sensed in these remarks (Horn, 1896) published years after LeConte's demise:

Some months after the death of Dr. LeConte I considered it a duty to assist in fulfilling his will by suitably preparing his cabinet and transporting it to the Museum at Cambridge. Annually, since, I have made one or two visits for the more accurate study of its types after a thorough study of my own material had been completed. In that collection I find not only the bare facts, for which I seek, but much besides. In the more than thirty years of our association there is not a box which has not been before us the topic of discussion or for consultation. Every one recalls its memories, and even particular specimens recall incidents of interest. To me such a visit is therefore more than the comparison of specimens, it puts me again in touch with a friend.

In 1896 Horn suffered a stroke, from which he never fully recovered, and he died at Beesley's Point, New Jersey, on November 24, 1897. He was buried in Central Laurel Hill Cemetery in Philadelphia. The Reverend Henry C. McCook, an entomologist of note, delivered the funeral address, although Horn himself had not been a member of any church. His collection, his library, and the sum of $5,000 were bequeathed to the American Entomological Society; he also willed $1,000 to the Academy of Natural Sciences and $500 to the American Philosophical Society.

EUGENE AMANDUS SCHWARZ (1844–1928)

Schwarz, one of the best-liked and most erudite men in the history of American entomology, was an entomologist and beetle specialist in the U.S. Department of Agriculture for more than fifty years. L. O. Howard, H. S. Barber, and A. Busck were all close friends of Schwarz's, and it is

Figure 73. E. A. Schwarz

from their obituary (1928) that we learn many of the following details about him.

He was born in Liegnitz, Silesia (Prussia), on April 21, 1844. He received his education at the Liegnitz Gymnasium and the universities of Breslau and Leipzig. He studied zoology and entomology under such notable European entomologists as E. Bugnion, R. Leuckart, and K. Letzner. It is believed that his defective eyesight kept him out of the war between Prussia and Austria in 1866.

According to Howard *et al.* (1928) Dr. Horn explained Schwarz's departure from Germany as follows: "Schwarz wished to become a zoologist and specialize in entomology, but his parents did not like the plan. They compelled him to study philology, with the idea of the career of a high-school master. Schwarz followed his parents' directions for a few years only; then, without their knowledge he devoted himself to zoology. The time came when he could go no further without his parents' knowledge, and, rather than explain to them he had neglected their plans and disappointed them in their ideas, he left for the United States." According to Horn, "had Schwarz mustered up his courage and explained the situation fully to his father, the elder Schwarz would have swallowed his disappointment and consented to the change of career. What a loss it would have been to American entomology had our friend faced the situation in that way!"

Schwarz joined the Museum of Comparative Zoology in 1872 as a preparator with G. R. Crotch under H. A. Hagen and attended lectures by Hagen and Louis Agassiz. After the death of Agassiz in 1873, the Museum had financial difficulties, and Schwarz sought employment elsewhere. Apparently his parting from the Museum resulted in strained relations between him and Hagen, for years later Hagen refused to speak to Schwarz while visiting the U.S.D.A. entomology laboratory in Washington.

It was at Harvard that Schwarz met H. G. Hubbard, who later became Schwarz's pupil, fellow collector, and faithful friend. The entomological accomplishments of the two came to be so closely linked together that it is not possible to get a complete picture of the one without reference to the other. In 1874 Schwarz accompanied Hubbard to the home of his father, Bela P. Hubbard, a wealthy and prominent citizen of Detroit, with a pronounced interest in science. There Hubbard and Schwarz founded the Detroit Scientific Association and began their collaboration on the notable Hubbard-Schwarz collection of Coleoptera.

When the American Association for the Advancement of Science met in Detroit in August 1875, Schwarz encountered J. A. Lintner, C. V. Riley, A. R. Grote, C. R. Osten Sacken, William Saunders, J. L. LeConte, and others, who were Mr. Bela Hubbard's guests. There Schwarz and Hubbard exhibited the insects they had collected in Florida. LeConte was so taken with this collection of beetles that he became a staunch admirer of the two collectors. Howard *et al.* report that "At this time Schwarz was so afflicted with ague [malaria?] from the Florida trip that he had difficulty in answering questions."

In 1876 Schwarz collected alone in Florida; he then returned to Michigan and made a trip by schooner all around Lake Superior with Hubbard. The two men returned to this territory the following year to collect again, and they and LeConte published *The Coleoptera of Michigan.* "At Doctor LeConte's instigation, Schwarz went to Colorado early in 1878, LeConte employing him to collect beetles there. But the season had hardly started when he received a telegram from Riley offering him a position in the Department of Agriculture. This he accepted, and spent the following winter investigating the cotton worm from Texas through the Southern States to the Bahamas. Riley's bulletin (No. 3, U.S. Entomological Commission) utilized this work." (Howard *et al.*)

L. O. Howard joined the Department of Agriculture in November 1878; Riley and T. Pergande were there, and Schwarz was in the South working on the cotton caterpillar. When Riley resigned from the Department in 1879, Schwarz, who had just returned from the South,

went with him to the offices of the U.S. Entomological Commission which Riley opened in his new house at 1701 13th street. Pergande and Howard remained with the Department, under J. H. Comstock, who succeeded Riley as chief of the service and who remained until March, 1881. Schwarz and W. H.

Patton worked with Riley on R. Street, and as there was a certain amount of bickering between the heads of the two offices, the assistants rarely met during these two years.

When Riley returned to the Department in March, 1881, he brought Schwarz with him, added B. P. Mann, W. S. Barnard and George Marx to the staff, and promptly sent Howard into the field for five months. When the latter returned to Washington he began really to know Schwarz and to appreciate him. The two worked at adjoining desks. Howard did not realize fully at the time (nor for several years—it grew upon him and influenced him unconsciously) what this daily association meant, and his experience duplicates the experience of others as they came along in the ensuing years. . . . Riley, genius though he was, had no higher education (he came to America and went on a farm in Illinois at the age of seventeen); Pergande, a mechanic of no schooling; Howard, fresh from Cornell and only 23 years old; and Schwarz, a classical scholar of standing, trained with some of the best entomologists of Germany, recently from Hagen's laboratory and still more recently from the Detroit meeting of the American Association for the Advancement of Science, where LeConte praised his knowledge and his work in almost extravagant terms! What wonder that the rest of them profited greatly by this association! Howard states, "As I look back to-day, I am filled with admiration and amazement at Schwarz's bearing. He allowed us to feel no attitude of superior learning on his part. Was that not almost superhuman? It was Schwarz and his learning that kept the rest of us from rawness. Our respect for his knowledge was great, but we never, so far as I know, attempted to put ourselves in his place and consider how very crude we must have seemed to him in many ways. He was always the kindly, considerate, thorough gentleman, apparently absolutely devoid of scholastic assumption."

Schwarz's great knowledge of the literature of entomology and his card catalogues were available to all. "He was not only the most indispensable at headquarters, but was one of the best field workers in America or elsewhere and was used by Riley in all sorts of ways. Not only did he write a number of papers published without his signature, but he influenced to a large degree the character of many others. The monographic work on the Psyllidae, for example, was in the main work of Schwarz. . . .

"He played a large part for many years in many of the most important entomological investigations of the Department from the days of the cotton caterpillar work through the time of the early work on the cotton boll weevil down to the days of the investigation of the Thurberia weevil in recent years." (Howard *et al.*)

Anyone who wishes to get a deeper insight into Schwarz should read his letters in the *Journal of New York Entomological Society* (1929). One of these, to Hubbard, from Washington, January 1897, follows:

When I came to the box containing the "Cioid" and looked at the latter I came near being paralyzed and it required a superhuman effort and a swallow of whiskey to recover. Your Cioid turns out to be a most remarkable and new genus of Scolytids!! In fact it is a long time since I put my eyes upon a more odd-looking creature than this species. After recovery I mounted at once a couple of specimens, for it happened that at 5:35 P.M. I had invited [J. B.] Smith, [A. D.] Hopkins and [W. B.] Alwood to dinner at Gerstenberg's with the understanding that they should spend the evening hours in my room, all three of them to leave between 9 and 10 o'cl with the B. & O. R. R. During dinner (everything as usual fried in cockroach grease) I narrated about that Scolytid and Hopkins could hardly wait for the time to look at it. Upon returning home the specimens were at once exhibited and Hopkins became perfectly wild with excitement and cursed his miserable West Virginia Scolytids because they did not show any distinguishing characters except after a most painful scrutiny. One of your Scolytids happened to be alive and we had an opportunity to watch the movements of this wonderful species. Smith got so excited and in order to prevent further mischief I had Ida at once fetch a pitcher of lager beer. This smoothened the excitement and two subsequent pitchers were drunk to your health and it was unanimously voted that no one but yourself would have been able to unravel the secrets of the Cereus fauna.

In a postscript to a letter to Howard from Williams, Arizona, in June 1901, he quipped:

I read the other day in the Williams Calamity Howler that a party of scientists has been eaten by the natives of some of the Pacific Ocean Islands. If that party consisted of Mr. [W. H.] Ashmead and the Fishery Commission men I feel sorry for the natives.

Schwarz was one of the founders of the Entomological Society of Washington, of which he continued to be a very active member, illuminating every session with his wisdom and wit. Howard (1933) notes that he "spoke with a strong accent, complicated by badly fitting false teeth." Some of his German-Americanisms are remembered to this day. There is the story that a small boy came to Dr. Schwarz's desk and said to him, "Oh, Dr. Schwarz, here is a very beautiful beetle that I found sitting on a milkweed." "Vell," said the Doctor, "he's got to sit somevhere, ain't it?" Dr. W. Dwight Pierce, formerly in the U.S. Division of Entomology, wrote me that he was especially delighted with Schwarz's exclamations, such as, "Oh my, I schlept over myself."

Howard *et al.* say that in the Entomological Society of Washington "for many years he was the greatest single personal influence, so it was in the Department of Agriculture and in the National Museum. Younger men went to him for counsel and for help in their scientific problems and even in their personal problems, sure of his kindly interest and resourcefulness." Dr. Pierce tells me he was like a father to Herbert Barber and Raymond Shannon, two fellow entomologists.

In 1923 Schwarz was given an honorary D.S. degree by the University of Maryland. He was one of the first honorary fellows of the Entomological Society of America and received honors from a number of other entomological organizations.

Schwarz's entomological endeavors in taxonomy were largely concentrated on beetles, especially the more obscure groups, and his numerous publications were held in high regard. He became Custodian of Coleoptera in the National Museum in 1898 and introduced measures to improve the care and arrangement of the collection. Under his custodianship, the National Museum received many fine collections. Schwarz always found time to encourage young collectors to pursue the study of insects.

He himself had collected throughout much of Central Europe in his youth, and later throughout the United States, as well as in Cuba, Guatemala, and Panama. He presented the Schwarz-Hubbard collection of beetles to the National Museum, and later gave the Museum his valuable entomological library.

In 1926, according to Howard *et al.*, "while living in the home of his protégé, R. C. Shannon, Schwarz awoke one morning slightly dazed and unable to express himself coherently in speech or writing. Although he remained apparently unimpaired in intellect, his decline dated from that time. The friends closest to him were able to converse with him by formulating their questions in such a manner as to require only a nod or shake of the head in reply, and he continued to come to his office, to read the newspaper and to take interest in the entomological and personal problems of his associates. He enjoyed music as much as he always had and especially classic German music. Wagner, played on the victrola, was sure to bring happy smiles. To the last he retained his sweet graciousness and kindness to others."

Mrs. Blake (1951), a member of the staff of the Bureau of Entomology, was one of the many who loved him and she gives this description of him when he was nearly eighty. "As he sat bent over his desk he reminded me of a very shaggy little skye terrier with his long hair, his bushy eyebrows, and drooping moustache."

According to A. N. Caudell's tribute to him in 1929: "his refusal to relinquish his work until within a few days of his death, at the advanced age of eighty-four, shows to what extent he was absorbed in his chosen work. A more zealous worker, a more helpful friend, or a more worthy man never lived . . ."

Schwarz died October 15, 1928, in Washington, D.C., following a fall.

HENRY ULKE (1821–1910)

Like Schwarz, Ulke was a German entomologist who profoundly influenced the development of entomology in this country.

According to an obituary by Banks *et al.* (1910), from which the quotations in this sketch are taken, Ulke was born in Frankenstein, Silesia, January 25, 1821. As a child he studied music, painted, and collected insects. His father was a gifted, versatile man who owned a hotel in Frankenstein—a haven for artists, musicians, comedians, and naturalists. Such noted entomologists as Professor Emil Schummel and Hermann Loew visited there and taught young Henry how to use the beating net. Henry, along with his father, who knew the generic names of many insects, collected in the foothills and nearby mountains. At the age of ten he entered school, where he studied Latin and Greek. Two years later he entered art school in Breslau, and from that time on painting was his profession. He was especially interested in the natural history collections in the university there.

In 1842 he went to Berlin to continue his study of painting. Henry's youthful and radical ideas in politics brought him into conflict with the authorities, and in the revolutionary year of 1848 he was imprisoned for his radical ideas. He was soon released, but was so unhappy with the status quo in Berlin that he emigrated to the United States.

In New York City Henry held a number of jobs, including that of illustrating title pages for books. He resumed his interest in insects and

Figure 74. Henry Ulke

became acquainted with J. E. and J. L. LeConte. Then, in 1857, he moved to Washington, where he opened a photographic studio, and established himself as a portrait painter. He also began to build a collection of insects.

"There was then in Washington a number of enthusiastic young naturalists who founded the Potomac-Side Natural History Society, and Ulke, with a few congenial friends, formed an inner circle, called the Megatherium, which often made collecting excursions to the Virginia shore of the Potomac, or discussed scientific subjects at a cafe in the evening."

In 1865 Ulke married Veronica Schultze, "a lady of literary ability." She frequently accompanied him on his collecting trips in the vicinity of Washington. The Ulkes had four sons and two daughters.

> Henry Ulke was a many-sided man; a portrait painter by profession, a talented musician in his leisure hours, a most diligent collector of beetles on all occasions, and philosopher in every relation of life.
>
> As a portrait painter he will be long remembered, for on the walls of the White House and many other official buildings hang his canvases of presidents, cabinet officers, and other distinguished public men. He became known as the "Painter of Presidents" and was a friend of Lincoln and Grant.
>
> As a musician he was devoted to the German classics. During the annex meetings of our Society (Washington), whenever a piano was convenient Mr. Ulke treated the members to a masterful rendition of Richard Wagner's Pilgrim Chorus from Tannhauser; and it was amid the subdued strain of this noble composition that the dead body of Ulke was carried to its resting place at Oak Hill Cemetery.
>
> As a philosopher he took a cheerful and healthful view of life, and . . . as a companion he was unexcelled. He had the widest range of ideas, a keen humor, and an uncommon fund of anecdote. The range of his mind was broad and on the whole philosophical. He was intolerant to the last degree of sham, and his indignation when discussing false doctrines was beautiful to behold. . . . conversation never flagged when Ulke was present. Idea followed idea in quick succession, often, it is true, only half expressed, but showing the rapid-fire quality of his mind.

His collection of insects was beautifully mounted, and he set an example for others in this respect. Drs. J. L. LeConte and G. H. Horn frequently mentioned his specimens in their papers. "Often Dr. Horn, before publishing a paper, would visit Washington and invite Ulke's criticism of his manuscript and examine the Ulke collection in that group. . . . [Ulke] early began collecting the microlepidoptera, then but little known in this country, and so successfully that he discovered hundreds of new species in the vicinity of Washington. . . .

"His favorite spot was the Virginia side of the Potomac near the old Aqueduct Bridge. It was to this point that, on several occasions, he guided the famous dipterist, Baron von Osten-Sacken, who was then a resident of Washington." Ulke also collected in nearby states, as well as in Nova Scotia, Prince Edward Island, and New Brunswick, Canada. He gradually amassed one of the largest and finest mounted collections in the United States, but he published very few papers.

In 1860 Ulke revisited Germany, and on his return, at the invitation of Louis Agassiz, he went to Cambridge, where he met Dr. Hagen and also visited Dr. Asa Fitch.

"Ulke was a most diligent correspondent, and exchanged beetles with nearly every entomologist of prominence in this country from 1865 to 1890. . . .

"Mr. Ulke belonged to the first generation of American Coleopterists, having been a charter member of the Entomological Society of Philadelphia and was intimate with the men of that day such as [John] LeConte, [George] Horn, [James] Ridings, [James] Bland, [E. T.] Cresson, [H. W.] Wenzel, Wilt, and [H.] Feldman. He was a noted collector, particularly of the smaller Coleoptera, such as the Pselaphidae, Scydmaenidae and Silphidae, and also a master in technic. His collection and material were frequently used by the systematists of the time. Many references to it will be found in the writings of Cresson, LeConte and Horn. . . .

"The death of his wife in 1893, greatly depressed him; he abandoned entomology, and disposed of his magnificent collection to the Carnegie Museum of Pittsburgh." In later years his interest in insects was somewhat revived and he attended meetings of the Entomological Society of Washington.

He died of a brain concussion on February 18, 1910, at the age of 89, as the result of a fall on the stairs in his home.

THOMAS LINCOLN CASEY (1857–1925)

The names of the men associated with the construction of the Library of Congress may be seen, carved in marble, on the walls of the rotunda. Among them is Thomas Lincoln Casey, who was a Colonel in the U.S. Army, an engineer, an entomologist, a conchologist, and a student of astronomy. Casey is of interest to us because he described about ninety-four hundred species of beetles, many of them the subject of fierce controversy among coleopterists.

He was born on February 19, 1857, at West Point, New York. He came from a line of military men; his father was General Thomas Lincoln Casey, and his grandfather Major General Silas Casey. The second Thomas Casey was educated in private schools in Washington, D.C., and attended the Sheffield Scientific School at Yale University. He then

Figure 75. T. L. Casey

entered the United States Military Academy and graduated with high honors in 1879. He was commissioned in the Engineer Corps of the Army, from which he retired in 1912. As a young lieutenant, in 1882, he was a member of an astronomical expedition to the Cape of Good Hope in South Africa. His engineering work was largely devoted to river and harbor improvements and he was chairman of the Light House Board.

Casey's first papers appeared in 1884; by that time he had already gathered a considerable collection. In studying beetles, he used a binocular microscope instead of a hand lens. As Hatch (1926) notes, "He was interested in securing accuracy of measurement and eliminating the personal equation as far as possible." His measurements were painstakingly accurate and he spent "more effort on habitus and sculpture" than on color or color pattern.

Leng (1925) notes, "Of criticism there was plenty. His efforts were at first encouraged by Dr. George H. Horn, who following the death of Dr. LeConte, had become by 1884 the principal authority on American Coleoptera. Casey's insistence upon describing as new species individuals which differed in some slight respect, contrary to Horn's opinion that they did not represent 'species,' alienated the support of the older man. Horn's attitude towards Casey's descriptions was shown by Synonymical Notes in which his specific names were briefly and at times perhaps wrongfully relegated to synonomy. The industry of the younger writer and his capacity for dealing with the more minute and neglected species quickly made it impossible for Horn to cope with the situation."

J. B. Smith (1885), who sharply criticized his work, stated that Casey would later regret some of his publications, but Hatch (1926), remarking that "one of Casey's leading characteristics was that he did not regret," added, "Granting his premises about the nature of specific differences, there was nothing to regret,—for all attest to the extreme accuracy and honesty of his work." Rarely did Casey attempt to answer his critics, but in one paper he comments that it is "the veriest absurdity" to suppose that "the mere fact that every detail of an insect is brought plainly into view, must necessarily and of itself give rise to mistakes regarding specific characters." In any event many taxonomists regarded his work with mistrust, and many of his species were relegated to synonomy. W. Dwight Pierce wrote me that, in the course of a pleasant visit with Casey, Pierce had noted that Casey's descriptions were detailed and that he was called a splitter, but Casey said to him, "Pierce, I am just a generation ahead of the rest of you."

Casey described insects in many families of beetles and devoted a great deal of effort to what were then obscure families, such as the Pselaphidae, Scydmaenidae, Byrrhidae, and the like. His *Memoirs on the Coleoptera* (1910–1924) included Central and South American beetles as well as those from North America.

Many of Casey's works were privately published because of their great length and because their "radical" views were not acceptable to the editors of some entomological journals. After forty years of publication, mostly at his own expense, his studies filled a total of 8,621 pages. They were widely distributed so they could be used by others in the field.

Students of Coleoptera may criticize Casey for a number of shortcomings, and several of them are valid. He made little attempt to refer to previous literature in the field, even though he had an excellent library. His excuse was that he was too busy studying specimens to have time for the literature. Moreover, he based all his studies on the specimens in his own collection, which in later years were for the most part purchased, and made no effort to study the collections in nearby museums. In fact, while studying Brazilian Staphylinidae and though aware of the deficiencies in his own collection, he made no attempt to avail himself of the excellent collection of Staphylinidae from Brazil in the Carnegie Museum in Pittsburgh.

F. E. Blaisdell (1925), a California coleopterist, who visited Casey in Washington in 1909 wrote: "Colonel Casey was a most hospitable host. In conversation he told me that he worked but three hours a day. He explained how he arranged and studied the specimens and species before he began to write. He endeavored to be very careful in drawing up his descriptions, and he recorded the characters of the beetles as he saw and interpreted them, not always as we would have analyzed them—but then, how many of us see things as others see them? Colonel Casey has opened up many lines of discussion regarding the question of species, subspecies, and forms. Time will prove whether or not his views are correct. At present it appears that time will work in his favor, if the recent studies in *Serica* and *Donacia* are any criteria. Critics had better remain conservative until more work has been done on the genital characters. Colonel Casey's decisions have been freely criticized. He has made many more mistakes than some other writers because he has done many times more work than they have. The works of George Horn yield as many errors, if his writings on the Tenebrionidae can be taken as a criterion. Dr. Horn was too conservative and thereby retarded the science of taxonomy. Colonel Casey advanced taxonomy too rapidly."

Schwarz and Mann (1925) report that to Colonel Casey, "the classification of adult beetles was a diversion, but not his only hobby, for Conchology took up a considerable part of his time and he made notable collections of fossil shells of the Lower Mississippi and published on the family Pleurotomidae. From his pen came also papers on Astronomy as well as military engineering and other subjects connected with his official work.

"His engineering instincts and training are shown in the exactness of his systematic writings and in the exquisitely prepared specimens of his collection, as well as his method of work. For years before his retirement but two hours each afternoon could be devoted to systematic work on insects. This time he spent with mathematical regularity in the 'beetle room' at his apartment, with specimens, note pad and binocular microscope in front of him; and it is from these leisure hours that we have the greater part of his studies on the Coleoptera."

C. W. Leng (1925) says that "Physically, he was tall, well built, though somewhat overweight in his later years. His home was in Stoneleigh Court, Washington, D.C., after his retirement from the army . . . His occupations, apart from entomology, included billiards at the Cosmos Club, and music at his home, in which he and Mrs. Casey found pleasure. Many daylight hours were devoted to his beetles, studied with the help of many pipes of tobacco.

"Possessed of inherited intelligence of high order, provided with the best education schools could supply, fortunately independent of financial worries or overly onerous duties, happily married, Casey came into the study of the Coleoptera with an equipment which has seldom been equalled. His field experiences, in consequence of the various stations involved in his army life, included Long Island, Virginia, Mississippi, Texas, California, and South Africa. His generous purchases aided in building up one of the most remarkable of private collections, so extensive indeed that he was obliged to rent two apartments in Stoneleigh Court, one for his collection and books, the other for his abode.

His library was so large that he had everything, practically, in his home that the literature contained."

Colonel Casey died in Washington, D.C., on February 3, 1925, at the age of sixty-seven and was buried in the National Cemetery at Arlington with military honors. At the request of his wife, Laura Welsh Casey, the microscope that had become so much a part of the Colonel's life was placed in his casket.

According to Blackwelder (1950) his entomological collection, which numbered 19,245 species, as well as his conchological collection and his comprehensive library were all left to the National Museum. The superbly-mounted specimens were so crowded in his insect boxes, any handling of them jeopardized adjacent specimens. A special room was prepared in the National Museum to house this very important collection, and this room Mrs. Casey "generously equipped with two binocular microscopes and adorned with a portrait in oil of the famous coleopterist." In addition, Mrs. Casey endowed the Smithsonian Institution with a grant known as the Thomas Lincoln Casey Fund, "the income of which is to be used for maintenance of the Casey Collection and for the general promotion of research in Coleoptera." (Buchanan, 1935)

CHARLES WILLIAM LENG (1859–1941)

Abbott (1949), writing about "Willie" T. Davis, records how Davis and Leng first met:

Figure 76. C. W. Leng

"In March, 1879, Britton meeting Willie on the ferry boat, fished out of a pocket a letter from a young naturalist who, having found Britton's name in the Naturalists' Directory, had written to him. 'His name is Charles W. Leng,' said Britton. 'He lives on Castleton Avenue, not too far from you. You had better go and see him.'

"Willie went. Thus simply do some of the most important things in life come about.

"Leng had been born on the Island [Staten] (April 6, 1859), but the family soon moved to Brooklyn and had only now returned. He had been collecting insects since he was thirteen and had joined the Brooklyn Entomological Society when he was sixteen. He had graduated from the Brooklyn Collegiate and Polytechnic Institute when he was eighteen, with the degree of B.S. in engineering, and had immediately gone into his father's business in New York—the importing of iron and steel, with bicycles as a new and promising side line. Young Leng was a shrewd and energetic businessman; but he was also, and remained throughout his life, an exceptional entomologist."

Leng began his studies in the Coleoptera with the encouragement of members of the Brooklyn Entomological Society, of which he was one of the founders. He drew the seal of the Society, featuring *Cicindela schauppi,* supplied by a member of the Society named F. G. Schaupp.

In 1916, W. S. Blatchley and Leng published *The Rhynchophora or Weevils of Northeastern North America*. This work presents descriptions of 1,100 species and contains many useful illustrations.

At the request of an entomologist named John D. Sherman, Jr., Leng began work on his *Catalogue of the Coleoptera of America, North of Mexico*. It was published in 1920. According to Sherman (1941), "This work of nearly 500 pages is much more than a mere catalogue. Our species, more than 18,000 in number are not only listed, but their distribution is shown, and there is also a page citation of the original description of each species. At the beginning is a comprehensive essay on the various systems used in the classification of the Coleoptera: at the end is a complete bibliography of the literature. A supplement was published in 1927 and the second and third supplement, printed under one cover, appeared in 1933. In preparing these supplements after Mr. Leng assumed the directorship of the Staten Island Institute of Arts and Sciences, Andrew J. Mutchler gave valuable assistance, which was gratefully acknowledged by Mr. Leng."

Sherman describes Leng in 1894 as "a very active and energetic young man . . . with red hair and a quick temper which never lasted very long."

In 1919 Leng gave up his business interests to become director of the Staten Island Institute of Arts and Sciences. He and Davis were very much interested in the history of Staten Island, and five volumes on *Staten Island and Its People* appeared under their joint authorship, the last volume in 1933.

W. T. Davis (1941) writes of his friend that "There could hardly have been a more helpful man than the genial Mr. Leng. Whatever aid he could render was given so pleasantly and with such good will, that naturally he was often consulted. He always had time or appeared to have, which is said to be one of the attributes of the competent man. It is with pleasure that the writer looks backward over the sixty-one years of close association with one so gifted and of such happy disposition."

C. W. Leng died at his home on Staten Island on January 24, 1941. His *Catalogue* has today become as familiar to the coleopterist as Webster's *Dictionary* is to the lay public.

HENRY CLINTON FALL (1862–1939)

The following account of Fall is drawn from obituaries by Darlington (1940), Sherman (1940), and Linsley (1940).

Fall was born on December 25, 1862, in Farmington, New Hampshire. His father, Orin Tenney Fall, was a Civil War veteran, a manufacturer, and a member of the New Hampshire legislature; his mother was nee Mary Ann Hayes. Henry was educated in the public schools of Dover, New Hampshire, and Dartmouth College. There in 1884 he received a

Figure 77. H. C. Fall

B.S. and was at the head of his class. He taught physics and mathematics in Chicago schools until 1889, when his health became somehow impaired. He then moved to California, where he taught physics and chemistry at Pomona and Pasadena high schools. At the latter school he was head of the science department for over twenty years. After more than thirty years of teaching, he retired in 1917; he returned to New England with his mother and made his home with his sister and brother-in-law at Tyngsboro, Massachusetts.

Darlington notes that Fall "had many interests. As a young man in college he excelled in baseball, and all his life he was keenly interested in baseball, tennis, and other athletic sports. He collected not only Coleoptera but Lepidoptera, stamps, postmarks, and old New England railway locomotive pictures. His valuable collection of these pictures is to be placed in the Railway and Locomotive Historical Society rooms in the Baker Library of the Harvard School of Business. But it was as a Coleopterist that Mr. Fall made his scientific reputation."

When he was fifteen he had captured in Wakefield, Massachusetts, a fine specimen of the cerambycid, *Prionus pocularis,* and this was the beginning of his collection of beetles. He himself collected over a hundred thousand North American beetles, and he obtained another hundred thousand Coleoptera when he acquired the Charles Liebeck collection. All in all, he had about 250 thousand specimens, representing probably about 90 per cent of the species of North American beetles north of Mexico. According to Darlington, he "described 1,484 of the species himself, practically all unquestionably good. The types of most of them are in his own collection. He kept them in four Schmitt boxes, crowded together for a quick rescue in case of fire. His main North American series fills 292 additional Schmitts. Besides his own types, there are many co- and paratypes of other workers, and innumerable specimens which Mr. Fall compared with the types of [J. L.] LeConte, [G. H.] Horn, and [T. L.] Casey. Such specimens are carefully labeled, and there is an accurate system of cross references from the labels to Mr. Fall's notebooks, and his working set of 59 bound volumes of papers on North American Coleoptera. The whole collection is beautifully prepared and arranged."

Fall had an excellent memory. "It is stated on good authority," says Darlington, "that he knew the old Henshaw List of 11,000 numbered names so well that, with no conscious memorizing, he was able to give from memory the serial numbers of practically every species on the list, and the name which went with practically every number."

His first paper, "A List of the Coleoptera of the Southern California Islands with Notes and Descriptions of New Species," appeared when he was thirty-five. He was a thoughtful student of LeConte's and Horn's literature, and his careful work resulted in relatively little synonomy, despite the fact that his studies ranged throughout the order of Coleoptera. His last publication appeared in 1937. It was his one hundred and forty-fourth.

According to Sherman, "Early in his life he had the good fortune to make the acquaintance of Frederick Blanchard. Both of them were not only skillful and tireless collectors, but serious students as well, and they became the closest of friends. Soon after Blanchard's death (November 12, 1912) the Richmonds bought the Blanchard place at Tyngsboro, and Fall lived with them after he came back East in 1918. Their home continued to be, more and more, a shrine for students of Coleoptera to visit, as it had been during Blanchard's lifetime. It was indeed a delightful home to visit, hospitable, harmonious, inspiring. Seldom does one family, under one roof, have so much of interest to offer its guests. In this home there were three work rooms of varied sorts: one for Mr. Richmond, former patent attorney for the American Telephone and Telegraph Company; another for Mrs. Richmond, an accomplished genealogist; and the third Dr. Fall's own 'den' containing his library, his collection of both native and exotic beetles, and his other collection of butterflies, of postage stamps, and of post marks. The deepest attachment existed always between brother and sister, and he was blessed indeed to enjoy the loving devotion of this sister at all times and under all circumstances. The varied interests and hobbies of the three members of this household never precluded time and zest for other activities, and friendly discussion enjoyed by all in common."

According to Darlington, Fall received an honorary D.S. from Dartmouth in 1929. "The degree was richly deserved, but at the same time was quite unnecessary to enhance his reputation. He was already known and respected throughout the entomological world as 'Mr. Fall.'"

Among Fall's correspondents and friends were George Horn, E. A. Schwarz, A. Fenyes, Charles Liebeck, John D. Sherman, Jr., and other well-known coleopterists. He constantly helped and encouraged many younger workers, and worked and corresponded until almost the last day of his life.

Darlington observes that Fall "knew the whole Coleoptera of America north of Mexico as few persons have known them, and as few will ever know them in this age of increasing specialization."

Fall had a great influence on entomology and entomologists in California; he was the founder of the Pacific Coast Entomological Society. As Linsley says, "The early work of Dr. Fall in southern California served as a stimulus for many western coleopterists, particularly Dr. E. C. Van Dyke, Dr. F. E. Blaisdell, Dr. A. Fenyes, Mr. Ralph Hopping, and Mr. J. O. Martin. These men were assisted by his careful determinations and reciprocated by providing much of the material upon which his early monographs were based. However Dr. Fall, himself, was a good collector and field systematist."

He died at his home on November 14, 1939. His collection of insects, his correspondence, notebooks, and volumes of technical papers were left to the Museum of Comparative Zoology at Harvard.

In his tribute to Fall as friend, colleague, and teacher, Linsley predicted that future generations of entomologists would "place the name of Henry C. Fall beside that of John L. LeConte and George Horn as one of the greatest of North American Coleopterists."

EDWIN COOPER VAN DYKE (1869–1952)

Van Dyke is another in the list of medical doctors who distinguished themselves in the study of beetles. Not only was he an outstanding coleopterist, but as professor of entomology at the University of California in Berkeley, he was undoubtedly one of the best liked and most respected entomologists in the history of the university. It was my good fortune to have studied forest entomology and the history of European entomology under Dr. Van Dyke in 1933 and 1934 at Berkeley. He is one of my most pleasant memories for those years.

The following sketch is drawn chiefly from his obituary, which was written by E. O. Essig, a fellow professor at the University of California and a friend of Van Dyke's.

The latter was born in Oakland, California, on April 7, 1869. Essig

Figure 78. E. C. Van Dyke

says his father, Walter Van Dyke, "who was of direct Dutch descent, was born in western New York in 1823, came across the plains to California in 1849, and became a prominent lawyer and judge and for many years an Associate Justice of the State Supreme Court. His mother, Rowena Cooper, was born on Prince Edward Island, Canada, September 21, 1835, and came around the Horn in her father's ship to San Francisco in 1850. They were married in Uniontown (Arcata), Humboldt County, California, in 1854."

Van Dyke began to collect insects in Oakland while in high school. When, in 1885, his parents moved to Los Angeles, he continued his studies on insects as well as on the plant life of Southern California. By means of correspondence, he became acquainted with many entomologists throughout the United States, including, says Essig, D. W. Coquillet, "the tall red-haired field agent of the Division of Entomology, U.S. Department of Agriculture, who was conducting fumigation experiments in his backyard in an attempt to control the ravages of the newly-introduced cottony cushion scale in the orange orchards. Coquillet was the first real entomologist that Van Dyke met. He had a fine collection and taught Van Dyke many things about collecting and also gave him specimens and took him on collecting trips. His first insect collecting trip, in 1890, was made to Yosemite Valley, which he reached by means of donkeys." Van Dyke also became acquainted with H. C. Fall and A. Fenyes, two well-known coleopterists who lived in or near Pasadena, California.

Van Dyke published his first paper, "Butterflies of Yosemite," in 1892. He received his B.S. from the University of California in 1893 and his M.D. from Cooper Medical College in 1895 (now Stanford University School of Medicine). Until 1900 he served as an interne, surgeon, and physician in several hospitals. He did graduate work at Johns Hopkins Medical College in 1901–1902 and then undertook private and institutional practice from 1903–1912.

W. M. Mann (1948), the Director of the National Zoo in San Francisco and a specialist on ants, recalled his early association with Van Dyke: "Usually Van Dyke and I would have dinner together; when funds were plentiful we would eat either at the Poodle Dog or the Bismarck; when funds were more limited we would go to Jule's, where for seventy-five cents one could have a good French dinner, beginning with steamed mussels, and accompanied by a bottle of either Riesling or Zinfandel. After dinner we would work on beetles until time to sleep, and I was always free to use a spare cot in his office."

The San Francisco earthquake of April 18, 1906, and the ensuing fire were a catastrophe that engulfed the entire community, including the California Academy of Sciences. At the twenty-first meeting of the

Pacific Entomological Society, August 25, 1906, the first to be held after the disaster, Van Dyke reported as follows:

"The California Academy of Sciences was so much injured that it was only with the greatest difficulty that what few things were saved could be gotten out. Mr. Loomis, director; Miss Eastwood, botanist; and Miss Hyde, librarian, did what they could. There was saved simply the boxes containing the types of the Coleoptera, Hemiptera, and Hymenoptera. All else in the entomological department was lost including the Behr collection of Lepidoptera and the entire library.

"This has been a tremendous blow to us all, and rare is the person who has not lost something." The earthquake, "though causing the loss of a great many lives and much destruction of property, would not have been noticed. It was the fire that did the great damage, destroying all of the city except a fringe of residences; Dr. Blaisdell and I were fortunately in this belt. The Academy was in the midst of the conflagration."

Van Dyke gives some additional details in a 1906 issue of *Entomological News*: "As a whole our greatest loss will be our libraries, every big one in the city being burned. The Academy of Science simply saved its records, about one set of its publications, and six other volumes. Our good friends among the scientific institutions of the country could therefore be of great assistance to us if they could reserve as complete a set of their publications as possible for us until the time when one will have a proper place to deposit them and our finances in condition to begin purchasing them. For some time we will have to depend on the libraries of our two universities." He added that none of "our entomologists" was injured.

As we all know, the Academy of Sciences and the Pacific Coast Entomological Society have grown from charred ruins into two strong and flourishing institutions.

One of the mainstays of the Pacific Coast Entomological Society, Van Dyke was elected president in 1907 and served in that capacity until 1931. In 1913 he was appointed Assistant in Entomology at the University of California, Berkeley, becoming a full professor in 1927, and professor emeritus in 1939. During the school year 1917–1918 he was an exchange professor at Cornell University.

On June 7, 1915, Van Dyke married Mary Ames, a forty-three-year-old school teacher and student of natural history, who was a member of the Entomological Society. She became "a constant companion and helper to the doctor."

Van Dyke traveled and collected extensively throughout the United States and Alaska and became an outstanding authority on the Coleoptera of the Pacific Coast of North America, particularly the Carabidae,

Buprestidae, Elateridae, Cerambycidae, Cossonidae, Cleridae, Ostomidae, Scarabaeidae, Lucanidae, Circulionidae, and Scolytidae. He was especially interested in the relationship of beetles to their habitats and their distribution in North America. The adaptation by structure and color of intertidal and dry beach Coleoptera was a favorite study of his. Besides this, he was an expert on insect pests of forest trees and forest products. In 1923–1924 he collected in China, Manchuria, and Japan, and in 1933 he visited Europe and North Africa.

Essig reports in his obituary that "Dr. Van Dyke was an earnest, strenuous, untiring, careful and painstaking collector. . . . He beat trees and brush with consideration. If he turned over a stone or a log, he carefully replaced it . . . and saw to it that his students and associates followed his example. The first time I went collecting with him on Mt. Tamalpais he took great pains to see that all of us followed his example and never hesitated to reprimand those who carelessly left a trail of destruction behind them. By following his conservation methods the collecting grounds remained fruitful throughout the years and each successive class of students profited by his methods. He also kept close watch of the specimens as they were taken and whenever a rare species was discovered all hands set about to procure more. . . .

"He lectured in a loud, strong voice and used excellent diction and had a fearless demeanor. And yet he was easily approachable and had a hearty laugh for a joke or a mischievous trick, especially if it was on him."

He worked the students so hard in the mornings, they had trouble keeping up with him in the afternoon. He rarely used assistants in either the field or the classroom. His door was always open to any student.

I myself was only a beginner in entomology when, in 1932, I first met Dr. Van Dyke, and I held the strong, erect, dignified, and mustached professor in great awe. But I was soon disarmed by his friendliness.

Some years later, I paid a visit to the entomology section of the California Academy of Sciences in Golden Gate Park, and there, to my delight, was Dr. Van Dyke. I knew as I left I should never see him again, so turned to gaze back down the long corridor. He was just poking his head out of his office for a last look at me. A year later, on September 28, 1952, in San Francisco, he died.

WILLIS STANLEY BLATCHLEY (1859–1940)

An entry for Sunday, February 12, 1911, in Blatchley's *In Days Agone* throws considerable light on the man:

"On yesterday I turned down an offer of $2,500 a year to take charge

Figure 79. W. S. Blatchley

of the alfalfa weevil investigation in Utah. Ten or 15 years ago I would have accepted it with alacrity. What I want in the future is to own my own time—my only true possession here on earth; to be subject to no man's nod or beckon—free to roam as I am doing today, wherever the little god of whim may lead me. I have pondered over this offer for many days. I have made my decision and it is final. Henceforth would I have old mother earth whisper unto me her inmost secrets; her winds play unto me their sweetest music. I would see her fairest sights, taste her most delicious savors, sense her most fragrant odors. Her trees and shrubs would I have yield unto me their secrets and their fruits."

Of the several manuals Blatchley wrote—*Coleoptera of Indiana, Rhyncophora or Weevils of Northeastern America* (with C. W. Leng), *Orthoptera of Northeastern America,* and *The Heteroptera or True Bugs of Eastern North America*—he once said, "They will perhaps serve as the monuments by which I shall be best known in the years to come." And, having erected his monuments in his lifetime, he then outlined the story of his life and his accomplishments in *Blatchleyana* and *Blatchleyana II* so that no compiler could garble them at some future date.

Blatchley was born on October 6, 1859, in North Madison, Connecticut, the son of Hiram Silliman and Sarah J. (Hall) Blatchley. In 1860 his parents moved to Indiana, where his father bought a farm. Willis attended the country schools and assisted his father in market gardening. He writes that in 1877 he earned his "first money of any consequence by peddling notions on foot from house to house through the country in Putnam and adjoining counties. Attended in winter my last term of school at Bainbridge. At that time no High School (as now known) in the town, no fixed grades and no graduating exercises."

After taking a six-week summer course in a Normal School, he became a country school teacher. His first job in 1879 paid him $1.50 a day, out of which he paid $2.50 a week for board and lodging.

After several years of peddling, teaching school, and attending normal school in the summer, he married Clara A. Fordice in 1882. Their first son was born in 1883, and their second in 1885.

From 1883 to 1887 he attended Indiana University. "Worked my way through," he says, "by janitoring, collecting delinquent taxes, gathering all the plants used by the botany classes, etc., etc. During the summers sold books and maps. In spring term of 1887 taught botany in the University. Majored in Science (Botany, Zoology and Geology) under Drs. D. S. Jordan and John C. Branner. Took one term's lectures in Entomology under Dr. Branner." He wrote papers on flowers and butterflies while attending the University, and his graduating thesis in 1887 was "The Flora of Monroe County, Indiana."

From 1887 through 1894 he served as head of the science department in a high school in Terre Haute. There he taught chemistry, botany, zoology, physiology, physical geography, and physics, and pursued graduate studies whenever he could, finally receiving his M.A. from Indiana University in 1891 with his thesis, "The Butterflies of Indiana." The University presented him with an honorary LL.D. in 1921.

In the summer of 1889 he worked for the Indiana State Geological Survey, and in the summer of 1891 he accompanied an expedition to determine the height of the Orizaba volcano (18,314 feet) in Mexico. In the summers of 1892 and 1893 he served as an assistant in the U.S. Fish Commission, collecting fishes in the streams and lakes of Northern Indiana and Northwestern Ohio.

In 1894 he was elected Indiana State Geologist on the Republican ticket by a plurality of 46,313, and he served in this capacity for the next sixteen years, until the Democrats won the office. In the years prior to his election, he had published papers on birds, weeds, trees, grasshoppers, crickets, cockroaches, amphibia, reptiles, and high school teaching, among other subjects.

The sixteen years he was State Geologist were prolific years; he was required "by statute to prepare and publish an annual report on the Geology, Natural Resources and Natural History of the State. The winter months were spent in preparing these reports and in work in the State Museum, the summers in gathering data for the reports and in other ways hereafter mentioned."

As State Geologist he prepared sixteen geological reports as well as monographs on caves and cave fauna, ferns, Orthoptera, clays, mineral waters, petroleum industry, lime industry, roads, road material, and the like. He also had to edit all the other reports on geology, zoology, botany, paleontology, and other sciences.

From 1906 to 1910 he worked on the *Coleoptera of Indiana,* published as Bulletin No. 1, Indiana Department of Geology and Natural Resources, with 1,386 pages and 595 figures. Because the State refused to publish this as part of the Geological Survey, it was paid for out of other State funds. Blatchley personally paid for a private edition of 1,000 copies. This monograph has been out of print for many years and a recent (1964) catalogue lists it for fifty dollars. Serious students of Coleoptera still find this work an essential volume in their libraries.

Wade (1940) says that Blatchley was fifty-one when his official career as State Geologist of Indiana was terminated; "the remainder of his life was spent in fulfilling an ambition of a sort to make him the envy and despair of countless other students everywhere who have dreamed, albeit hopelessly, of similar fascinating and delightful achievements. During the twenty-nine years that followed, most of his time was spent in travel, collecting, and exploration, varied at convenient intervals with prolonged periods of intensive study and writing."

In *My Nature Nook,* Blatchley notes in an entry for Sunday, March 13, 1918, on the occasion of his being offered a job as taxonomic entomologist of "a great university," that he would have accepted with alacrity a quarter of a century earlier. He turned the offer down for the same reasons he had rejected the 1911 offer from Utah.

Blatchley made many long collecting trips in the unsettled parts of Florida, and traveled extensively throughout the United States and at various times in Alaska, Canada, Brazil, Uruguay, Chile, Bolivia, Peru, and Cuba. He attended scientific meetings, delivered lectures, and studied library material and museum collections. In addition to his permanent home in Indianapolis he purchased in 1913 a winter residence in Florida, where, according to Wade, he spent a considerable portion of each year thereafter.

In 1912 the *Indiana Weed Book* appeared in print, and in 1914 he began work on the *Rhynchophora or Weevils of Indiana.* After completing one-third of the manuscript, Charles W. Leng of Staten Island joined him and the work was enlarged to encompass the eastern United States and eastern Canada. As was his habit, Blatchley visited many museums to study collections. *The Rhynchophora of Northeastern America,* consisting of 682 pages with 155 illustrations, was published in 1916. The *Orthoptera of Northeastern America,* 784 pages with seven plates and 246 figures, appeared in 1920, and in 1926, the *Heteroptera or True Bugs of Eastern North America,* with 1,116 pages, twelve plates, and 215 figures.

Any one of these works alone would have been a major accomplishment for an ordinary man, and reviewers in general approved his Herculean efforts. However, there were a few, whom Blatchley (1928) calls "quit-claim" specialists, who were unhappy with his works, especially on the Orthoptera and Hemiptera. His article "Quit-Claim Specialists vs. the Making of Manuals" complains about the lack of co-operation by certain authorities:

> In 1907, realizing the great need of general works, descriptive of certain groups of insects with which I was somewhat familiar, I began the preparation for the novice or beginner of those manuals which have been put forth in the twenty years which have elapsed. In their preparation I have given free, and solely for the good of the cause, not only all my spare time but more than $12,000 of my previous earnings for their illustrating and publication. This does not include the salary of my faithful assistant, who by thirty years training and practice, has been able to put my longhand hieroglyphics of key and text into typewritten manuscript which the printer could use. The demand for such works as I have put forth is limited, and for that reason I am as yet more than $4,000 "to the bad" in their publication.
>
> In the preparation of the first two works issued, the "Coleoptera of Indiana" (1910), and in collaboration with Chas. W. Leng, the "Rhynchophora of N. E. America" (1916), I succeeded, without much trouble or delay, in getting such aid as I requested, and when completed there was little published criticism of those works.
>
> In those days there were fewer specialists than now who were imbued with the idea that the Good Lord had given them a "quit-claim deed" to all the species in this country belonging to their especial group, and that no one, not even the author of a manual who desired to include *all* species from the territory he was covering, had any right to "poach on their preserves."

In all likelihood some of the specialists resented the fact that Blatchley was raiding their preserve, and possibly they thought he might mess things up. Obstacles such as these merely spurred him on.

Besides his manuals and other publications, Blatchley wrote seven popular books on nature, *Gleanings from Nature,* 1899; *A Nature Wooing at Ormond by the Sea,* 1902; *Boulder Reveries,* 1906; *Woodland Idyls,* 1912; *My Nature Nook,* 1931; *In Days Agone,* 1932; *South America as I Saw It,* 1934. And in 1938, when he was seventy-nine, appeared his book *The Fishes of Indiana.*

Professor J. J. Davis (1941) says, "Blatchley was an individualist. He depended on no one but himself for his thoughts and ideas and was uninfluenced except by concrete information which was clearly and effectively given. . . . His memory was remarkable, and I doubt if he ever forgot observations he made during his active life. I recall a trip with him some eight or ten years ago to the Wyandotte Cave region where he collected some 15 or 20 years ago. As we would go to this or that place where he had collected before, he would often remark, 'Here is where I took such and such a species, and we should find more of the same species,' and sure enough we would."

Blatchley's wife died in 1928. Soon after, he gave up his entomological labors: "On account of failing eyesight and other infirmities I had been obliged in 1930 to give up the collecting and study of insects, but after many years of continuous activity my brain continued to demand employment of some nature. Therefore I took with me to Florida in November a large collection of cancelled postage stamps from many countries which I had accumulated and put aside since I had had to give up the hobby of philately in 1899. Many of my hours in the years since 1933 have been pleasantly passed in the study and arrangement of these stamps in suitable albums." (*Blatchleyana*-II, 1940)

According to W. T. Davis (1941), he often labored under great physical difficulties in his later years, "and it was, for instance, only his determination to finish a job once undertaken, that finally produced the manual on the 'Hemiptera of Eastern North America.' " In a letter to J. J. Davis, dated March 5, 1933, from his home in Dunedin, Florida, Blatchley writes, "I am still suffering much from neuritis. Not able to do any collecting whatever. . . . Have to sit propped up with cushions to palliate the pain. Have been that way since April 1st. During the summer I wrote, while so propped up & from my note books: 'In Days Agone.' It contains by far more notes on insects than any of the popular books I have written."

Blatchley was "a passionate lover" of books, and his valuable library and his large collection of insects with 470 holotypes were given to Purdue University.

J. J. Davis wrote me on May 15, 1961, "He was miserly and yet generous to the needs of his interests. For example, he gave me $5,000.00 in negotiable bonds, the agreement signed by his life-long secretary, me and her as Mr. X. He didn't want his two sons to know about it during his life, because, as he said, they would raise hell. I was to pay his former secretary $500.00 a year as long as she lived & if any remained on her death I was to use it for the Thomas Say Foundation. There was $2,000.00 left at the time of the secretary's death."

In this same letter, J. J. Davis says, "As I may have told you I gave the 'sermon' at his funeral. A year or two before he died he said to me, 'Davis I don't want a minister at my funeral. If they have to have a service I wish you'd give it.' So I did, with such prominent men present as Eli Lilly. I simply spoke of him as a geologist, a naturalist, an entomologist, a wit, and something else, & the only religion I brought in was at the end when I asked all to stand and repeat the Lord's Prayer.

Blatchley was not an atheist, but as he told me he had seen so many funerals which were just evangelistic sermons telling people they were going to hell, etc., that he was disgusted with some ministers. Things are different now but in days gone by I must admit he was right."

Blatchley died after a brief illness at his home in Indianapolis, Indiana, on May 28, 1940, at the age of eighty.

FRANK ELLSWORTH BLAISDELL (1862–1947)

The following account of Blaisdell is from Dr. E. C. Van Dyke's (1947) recollections of his friend.

Blaisdell was born March 13, 1862, in Pittsfield, New Hampshire, the son of Solon Greenleaf and Anna Greely Clark Blaisdell. His father was a saddler and harness maker, who served as a sergeant in the Civil War. The family, including three children, moved to San Francisco in 1870. All of the children caught scarlet fever the following year, and two of them died. Frank survived, but in such weakened condition that his parents moved to the milder climate of San Diego; there his father opened a harness and saddlery store and Frank attended the public schools. A little later his father bought a ranch close to San Diego, where he operated an orchard and an apiary and raised grain. Frank grew strong and healthy on the ranch and when older did most of the work on it, too, at the same time attending a country school.

"In 1886, he decided to study medicine, so the ranch was disposed of,

Figure 80. F. E. Blaisdell

the father starting a nursery at Coronado and selling the plants in San Diego. Frank secured a position in a store in San Diego and began reading medicine under the instruction of Dr. P. C. Remondino and Dr. C. C. Valle. He also prepared himself for the medical school examination. To a considerable extent, Dr. Blaisdell was a self-taught man. He never had a modern high school or college education, but through diligent study, whenever he could get the time, he made up for the deficiency. In 1887, he entered Cooper Medical College, San Francisco, and graduated November 12, 1889, with the degree of Doctor of Medicine. He immediately returned to San Diego and started to practice but with very poor success. Consequently, in 1892, he returned to San Francisco where he learned there was an opening in the mining town of Mokelumne Hill, Calaveras County. He left for the mountains and before long was well established there. On February 18, 1894, he married Miss Ella Katherine Peck and on March 31, 1896, his son and only child, Frank Ellsworth Blaisdell, Jr., was born."

In 1900 he became an instructor in Cooper Medical College. Just before taking over his new duties he went with his brothers-in-law on a summer vacation to Nome, Alaska, where he worked a gold claim and collected insects, birds, and plants. At Cooper Medical College, sometime after 1900, he became professor and head of the Department of Anatomy. Shortly thereafter he took a trip East to do graduate work at the Johns Hopkins University School of Medicine; he visited Washington and became friends with E. A. Schwarz and other entomologists there. When Cooper Medical College became a part of Stanford University School of Medicine, Blaisdell eventually became Professor of Surgery and was in charge of Surgical Pathology. In 1927, at the age of sixty-five, he retired as Professor Emeritus of Surgery.

During all these busy years, on his vacations and whenever else he could get the opportunity, he fervidly hunted beetles, which were his special delight. He was a very good field collector and an able naturalist with an excellent knowledge of birds and plants. He dissected birds and made fine illustrations of them. Although he was interested in all families of the Coleoptera, the Tenebrionidae and Melyridae were his special forte. His bibliography consists of ninety-nine papers.

Dr. Van Dyke thought him "a man of fine character" with an attractive personality, who was "much respected. . . . He was particularly generous in . . . giving both time and attention to beginners."

After Blaisdell retired, he pursued his interests in the Department of Entomology at the California Academy of Sciences and in 1924 deeded his collection of almost 200 thousand Coleoptera to the Academy. Failing health in 1945 forced him to move near his son in Watsonville, California. He died on July 6, 1946, at the age of eighty-four.

HENRY FREDERICK WICKHAM (1866–1933)

Wickham was born in England on October 26, 1866. The family moved to Iowa City in 1871. He graduated from the University of Iowa in 1891, and then joined its staff. He taught there for forty-two years, the last thirty as professor of zoology and entomology. He was an excellent and popular teacher, with a keen sense of humor.

He had become interested in beetles as a boy and in time acquired an outstanding collection and library. For thirty years during the summers he collected all over the United States, as well as in Alaska, Canada, the Bahamas, and Mexico. He sent many new species to T. L. Casey, G. H. Horn, H. C. Fall, and other noted coleopterists.

For a number of summers he was employed by the U.S. Bureau of Entomology on such projects as the New Mexico range caterpillars, the Mexican bean beetle, and the determination of beetle fragments in the stomachs of birds.

In later years, after building up his library, he identified many of his own beetles. He prepared about one hundred and seventy papers, including publications on local lists, color variation, structure, biology, and the distribution of beetles. With Dr. Emil Brendel he prepared *The Pselaphidae of North America* and his own *List of the Coleoptera of Iowa.*

For many years Wickham collected fossil insects at Florissant, Colorado, and he described in all about three hundred species of fossil insects. He

Figure 81. H. F. Wickham

Figure 82. H. S. Barber

also prepared the section on fossil beetles in C. W. Leng's *Catalogue of North American Coleoptera.* Wickham's collection of beetles is in the U.S. National Museum.

Wickham died at his home in Iowa City on November 17, 1933. Buchanan (1934) wrote his obituary.

HERBERT SPENCER BARBER (1882–1950)

Barber was employed by the U.S.D.A. and the U.S. National Museum for more than fifty years. He was a protégé of the famous coleopterist, E. A. Schwarz.

Barber was born at Yankton, South Dakota, on April 12, 1882. His father was an engineer who was interested in natural history and encouraged his son's interest in insects. Herbert received only a high school education although he occasionally studied languages and technical subjects at night school. He was first employed in 1898 by the National Museum as an insect preparator with the title of laborer at $20 a month. Some of his earliest efforts were devoted to the arrangement of the Hubbard and Schwarz collection of Coleoptera. In 1901 he made an extended trip with Schwarz to Arizona and New Mexico. He was employed by the U.S.D.A. in 1901, then for a few years as an aide by the National Museum, and again by the U.S.D.A. in 1908. Barber learned quickly under Schwarz's tutelage and became a skilled collector. He also profited by his association with such entomologists as L. O. Howard,

Theodore Pergande, Frederick Knab, August Busck, Nathan Banks and others. He was an active member of the Entomological Society of Washington.

Anderson *et al.* (1950) write that in 1913 "Barber published two papers on the biology of *Micromalthus debilis* Leconte, a small wood-boring beetle belonging to the obscure family of Micromalthidae. The species is remarkable because there are several distinct types of larvae and under certain conditions in the life history both eggs and living larvae are produced by larvae, in addition to the usual eggs deposited by adult females." For years many students of insects did not believe this remarkable phenomenon.

In time Barber became an authority on chrysomelid, bruchid, and lampyrid beetles. His published works, consisting of about ninety papers, were characterized "by merit, not bulk." His most important papers were concerned "with weevils infesting orchids, weevils of the genus *Trichobaris,* Chrysomelidae as a whole and particularly *Diabrotica* and related genera, flea beetles and tortoise beetles." Barber also studied fireflies and other light-producing insects as well as such primitive insects as the Protura and Zoraptera. He and Schwarz collected a great deal of coleopterous life-history.

Mrs. Blake (1951) gives the following glimpse of Barber:

"When I had been in the Bureau only a few months [1919], I had brought in some fire fly larvae and Dr. Chittenden had advised me to take them over to Mr. H. S. Barber at the Museum as he was working on fire flies and knew all about them. Mr. Barber had been Schwarz's assistant since he was a boy in his teens, or for about twenty years. He was a thin, dark-haired, intense man with a big nose and fine, nervous hands that could dissect the tiniest beetles and prepare the most beautiful specimens. He was always flaming with enthusiasm over some biological discovery. His desk, like Dr. Chittenden's, was piled up in a mountain of stuff."

Barber traveled widely, collecting in California, Texas, and Massachusetts, and other places. In 1906, together with E. A. Schwarz and O. F. Cook, he spent six weeks in Guatemala looking for enemies of the boll weevil.

Schwarz and Barber spent as much time as they could spare on Plummers Island in the Potomac River, about six miles from the District of Columbia, where they collected, reared, and studied insects.

Barber was an outdoorsman with a broad knowledge of animals, birds, snakes and flowers. He was an excellent canoeist and an expert shot with the rifle and pistol, a skill that provided many a rabbit for his and Schwarz's fare on camping trips in the Southwest. He was fond of children and enjoyed teaching them natural history, swimming, canoeing, skiing, and photography.

In later life Barber was separated from his wife. He had a daughter and a son, and the latter was killed by a kamikaze in World War II. "During the last few years of Dr. Schwarz's life, the aged scientist lived with Barber, and in spite of the latter's devotion to his old friend and teacher, the necessary care was undoubtedly something of a burden." (Anderson *et al.*)

Barber had a great influence on young entomologists and aided them in the explanation of techniques and nomenclature. His friend Snyder (1950) describes him as "earnest but friendly, helpful, and happy in his work although troubled in family life and with some feeling of lack of appreciation by his contemporaries . . ."

On June 1, 1950, he died in his sleep of heart failure at his sister's home in Washington, D.C.

10 Notable Lepidopterists

Softly—
That it may not startle
A butterfly—
The gentle wind passes
Over the young wheat.
Issa

It is not surprising that collectors of Lepidoptera exceed in number collectors of any other order of insects, for the beauty and diversity of butterflies and moths have always attracted a cult of worshipers. Here we sketch some of the distinguished men who have devoted themselves to the pursuit and study of "flying rainbows" and the more demure moths.

JOHN GOODLOVE MORRIS (1803–1895)

The Reverend John Morris, a Lutheran clergyman, was one of the earliest lepidopterists and teachers of entomology in the United States. He was the son of Dr. John Morris, a surgeon in the Revolutionary War. This lepidopterist was born in York, Pennsylvania, on November 14, 1803, and studied at the College of New Jersey (now Princeton University) and Dickinson College, graduating from the latter in 1822. In 1826 he became a Lutheran minister and served a while in Winchester, Virginia, and for thirty-three years in Baltimore.

He wrote prolifically on religious subjects and a number of entomological papers for the *Literary Record* of what was then called Pennsylvania College, at Gettysburg. In 1860, the Smithsonian Institution published his *Catalogue of the Described Lepidoptera of North America* and in 1862 his *Synopsis of the Described Lepidoptera of North America—Part I: Diurnal and Crepuscular Lepidoptera.* Graef (1914) remarks that Morris' *Synopsis* was an important work for its time and that it gave the eminent lepidopterist, A. R. Grote, "many ideas and principles pertaining to identification."

Figure 83. Rev. J. G. Morris. Courtesy of the Academy of Natural Sciences

Morris was librarian of the Peabody Institute in Baltimore from 1860 to 1865; for a number of years he taught natural history, including entomology, at Pennsylvania College.

Graef recalls how in October 1884, Morris came to his house to examine his collection. Morris was about eighty-five at the time. "After dinner he complained of extreme weakness and lay down on the sofa. For a time I was greatly alarmed that he was about to expire, but after various remedies he was relieved, as I was." He was actually ninety-two when he died—in Lutherville, Maryland, on October 10, 1895. He had remained interested in entomology to the very end.

Anon. (1895) is another source for biographical notes about him.

HANS HERMAN BEHR (1818–1904)

Essig (1931) sums up Behr's life with these words, "Physician, scientist, author, poet, humorist, savant and lepidopterist." Although Dr. Behr's contributions to knowledge of the Lepidoptera of the United States are not as great as some later workers, they are, nevertheless, important because he described a large number of Western butterflies.

Behr was born on August 18, 1818, at Cöthen in the German duchy of Anhalt. He received a classical education in the local gymnasium and the Prince's College at Zerbst and later studied medicine and natural science at the universities of Halle and Würzburg. He graduated from the University of Berlin in 1843 as a Doctor of Medicine. Always in-

Figure 84. H. H. Behr. Portrait by Kever Rasiben, 1893. Courtesy of the California Academy of Sciences

terested in every aspect of science, Behr, with the encouragement of Alexander von Humboldt and other scientists, went to Australia, where he studied the aborigines and their language as well as botany and entomology. Then he travelled to Java, the Straits Settlements, East India, and the Cape of Good Hope, always observing, collecting, and describing.

In 1847 he returned to practice medicine in Cöthen, but because of his socialistic views it became necessary for him to leave Germany a year later. He then travelled to Brazil and other countries in South America and the Philippines, and for two years practiced medicine in Manila.

Behr moved to California in 1851 and never left it except to bring back a Polish bride in 1853. She lived only a few years, but presented him with three children.

In 1854, one year after its founding, Behr joined the San Francisco Academy of Sciences. Gutzkow *et al.* (1904) describe his contribution to the Academy:

> Doctor Behr's learning and memory were truly surprising, and by no means limited to branches of natural science which he had chosen as specialties. He was an excellent linguist, speaking six or more modern languages; he had made deep study of comparative philology, and of the mythology of the East Indian cults; he had acquired a very good knowledge of Hebrew and Sanskrit, and was acquainted with the languages spoken by the Australian savages and by the Malays in Java and the Philippines as well as with the tongues of the

various branches of the great Slavonian nation. His scholarship in Greek and Latin was of much use to those who discovered new species in botany or other sciences and who did not feel safe in their Latinity. Even eastern scientists often consulted him thereon, and were sometimes astonished at the blunders he discovered in their Graeco-Latin make-ups. He frequently astounded his older friends by opening up to them a vein of knowledge of some subject with which they had never thought him to be familiar. Behr could and did say something interesting on almost every subject brought up for discussion at the meetings of our Academy. As he rarely missed a meeting during the half century of his membership (except the last few years, when the infirmities of age prevented his attendance), we seldom missed hearing something interesting or instructive from him when others remained silent. Behr was always ready to help the younger or less informed members in their researches, and assisted hundreds of farmers, fruit-raisers, and gardeners in the manifold troubles which beset the cultivation of plants and trees in a new country.

Although Behr was a kindly individual, he was the sworn enemy of medical and scientific quacks and attacked them mercilessly. As a result he made enemies. Since these could not assail him as a physician or scientist, they attacked him for being a Catholic, causing him to lose many patients at first. "He revenged himself by pointing sarcastic arrows which mutual friends never failed to wing to the target for which they were intended. For instance, discovering a particularly obnoxious louse, he named it after his enemy. Doctor Behr used to make light of the matter, but a week before his death he confessed to the writer [Gutzkow] how deeply he had suffered under those vile calumnies." (Gutzkow *et al.*, 1904) The doctor lived up to his Hippocratic Oath by helping the needy regardless of remuneration.

"But undoubtedly his happiest days were those when he escaped from his practice and made excursions into the neighboring counties in search of plants and butterflies. Experience had taught him never to go alone; so he always took along some trusty friend, a few members of the Academy, or some of his students after he became professor of botany at the Pharmaceutical College. Surely none who had the good fortune to go with him on those excursions will ever forget the genial old man, always witty, humorous, and ready to teach, and his childlike pleasure when he found a rare plant or insect. His medical friends probably envied his courage in so frequently putting aside professional cares and possible emoluments; but as those relaxations contributed so largely toward keeping him in robust health and giving him so many years of life, he proved himself in the end the wisest of them all."

Behr was a member of the famous Bohemian Club in San Francisco, where his keen, incisive wit was greatly appreciated. A raconteur of note, much of his wit was collected in a posthumous collection entitled, *The Hoot of the Owl,* which includes a story about a legendary Captain Schenck who visited a powerful king of the Cannibal Islands. "At the feast given in the captain's honor, the neighboring trees were decorated with girls bound fast and awaiting the moment they would be served at the royal table. One of the most toothsome was destined for the dinner of the distinguished guest; and when the captain was asked in what style he would have his girl served up, he astonished his cannibal friends with the words: 'Your majesty, I'll take mine raw.' "

Behr published sixteen papers, largely on butterflies, in the *Proceedings of the California Academy of Sciences,* and twelve years before his death he presented his important collection to the Academy, which remained under his care until his death. He was paid the usual curator's salary. He died in San Francisco on March 6, 1904, at the age of eighty-five. His collection was utterly destroyed in the earthquake and fire of 1906.

Anon. (1904) prepared biographical notes on Behr.

WILLIAM HENRY EDWARDS (1822–1909)

The accomplishments of amateur naturalists who were able to devote only fleeting hours to the work they loved constantly arouse the admiration of the student of entomology. Among these is William Henry Edwards, one of America's earliest and greatest lepidopterists.

Figure 85. W. H. Edwards

Edwards was born on March 15, 1822, in Hunter, New York. His mother was Helen Ann Mann Edwards. His father, William W. owned a tannery business. When William Henry was quite young, his father purchased in Green County on Schoharie Creek in the Catskills, a 1,200-acre hemlock forest for tan bark. William Henry's love for nature was nurtured in these beautiful surroundings.

According to Bethune (1909), he was sent from the village school "to Williams College, Mass., and completed his course there in 1842; he then studied law in New York, and was admitted to the bar in 1847, after which he made his home at Newburgh, N.Y. Subsequently he became interested in the coal fields of West Virginia, and removed to Coalburgh, where he was President of the Ohio and Kanawha Coal Company. He was an extensive land owner in the Virginias of the early days, a builder of railroads, and opener of coal mines, and throughout his life active in the affairs of the community. . . ."

In 1846, shortly after leaving college, he made a trip to Brazil to collect birds, butterflies, and various animals. His first publication, *Voyage up the Amazon,* appeared in 1847. This book delighted readers with its "vivid descriptions of the tropical vegetation and the strange creatures of the Amazon forests." It was one of the influences that led Alfred Russell Wallace and Henry W. Bates to undertake their 1848 expedition to the Amazon Valley.

Despite all his other activities, the study of butterflies remained a consuming passion and resulted in the publication of many papers as well as a three-volume work of the first magnitude, *The Butterflies of North America.* The first sections of this beautifully illustrated work appeared in 1868, and the first volume was completed in 1872, the last in 1897. Edwards says in the first volume: "Having, from my first study of this beautiful family, felt the want of illustrations, I long ago proposed to myself to publish a complete work on the Butterflies of North America. . . . It is a matter of regret that, in so few instances, I shall be able to say anything of the larvae. Even among our old and common species, the larvae are little more known than in the days of [John] Abbot, seventy years ago."

This shortcoming was soon overcome, for Edwards discovered in 1870 that he could "obtain eggs from the female of any species of butterfly, namely, by confining her with the growing food-plant. If the eggs mature they will be laid." And in the preface of the second volume, which appeared in 1884, he says, "The preparatory stages of North American butterflies as a whole are better known than are those of Europe." His friends then began to send him butterfly eggs from all parts of the country.

Apparently after the appearance of the second volume Edwards could no longer afford to underwrite the publication of his work, and when the third volume was ready to emerge, he had to seek the co-operation of others. Weiss (1936) says that Edwards "thought of offering his collection of North American butterflies to the trustees of the British Museum in order to obtain the money to enable him to continue his work. However, Dr. W. J. Holland, in order to preserve the types in America, offered to pay the bills for the publication of the third volume, as they became due, on condition that Edwards turn over to him his collection when he had completed his studies. This was done. Doctor Holland paid the drawing, lithographing and printing bills, and in that way the third volume was produced."

Even to the sophisticated bibliophile of this day, these are beautiful volumes. The illustrations of butterflies drawn by Mrs. Mary Peart and colored by Mrs. Lydia Brown have probably never been surpassed. Mr. Edward A. Ketterer joined Mrs. Peart in drawing the plates for the third

Figure 86. Butterflies of the genus *Asterias*. From *The Butterflies of North America,* by W. H. Edwards

volume. Skinner (1909) comments: "The monument to Wm. H. Edwards will be the three volumes he published 'The Butterflies of North America' . . . This work is one of the greatest ever published on the subject, and it has been the source of authoritative information on American diurnal Lepidoptera as a whole for nearly half a century. The author's contributions to our knowledge of life histories marks an epoch in Lepidoptera, and are of very great importance from a scientific standpoint."

Even before completion of his three volumes, Edwards was widely recognized as an authority on North American butterflies. Specimens were sent to him for identification by the Smithsonian Institution, the British Museum, the Imperial Russian Government, and by many other institutions as well as individuals.

Edwards was seventy-five when his third volume was completed, and with that his labors on butterflies ceased. An English journal (Weiss, 1936) claimed that "A labour that ended in Edwards handing over his collection under the conditions above described, must have sapped his entomological life's blood." This article also states that further encouragement might have resulted in another two volumes. This is possible, although it is also likely that a man of his age might have sought surcease from constant labor and aching eyes and found a change in pursuit refreshing. So in the last years of his life Edwards devoted himself to Shakespeare. In 1900 he entered the Shakespeare controversy by disputing the authorship as well as the spelling of the Bard's name in a book entitled *Shaksper not Shakespeare*. Nor were these his last literary efforts, for in 1903 he completed a genealogy of the Edwards family.

Bethune (1909) remarks, "To those who knew him well he was endeared by many attractive characteristics; kind, open-hearted, cheery and courteous, free from pride and ostentation, widely respected and foremost in all that pertained to the welfare of the community in which he lived . . ."

Brown (1960) published and annotated the correspondence between Edwards and S. F. Baird, the naturalist, much of which is of interest to those who study butterflies. Much of this correspondence occurred during the Civil War, but as Brown notes, no ripple of the fratricidal years disturbs the even flow of their exchange.

Skinner (1909) makes this summation: "Edwards was unquestionably the greatest Lepidopterist this country has produced, and his great work on American Butterflies is and always will be a classic one. His work on life histories has never been surpassed, and when we think that all these valuable contributions to science were carried out during the spare time of an otherwise busy man, they are all the more admirable. He published in all about two hundred papers."

Other than that he married a Miss Catherine Colt Tappan in 1851, we know nothing of his domestic life.

Edwards died on April 2, 1909, at the age of eighty-seven in his home in Coalburgh.

HENRY EDWARDS (1830–1891)

The name Henry Edwards, evokes the image of a man of divers accomplishments, as a notable actor on the stage, as a writer of geographical and theatrical articles, and as a collector of insects, especially Lepidoptera. Beutenmuller (1891), Slosson (1915), Wade (1930), and others have sketched his life.

Edwards was born in Herefordshire, England, on August 27, 1830. As a young man he studied law, but he showed little aptitude for it and entered a London counting house. Here, through the influence of fellow clerks, he became interested in amateur theatricals. Edwards showed so much talent as an actor that, despite the opposition of his parents, the stage became his chosen profession. According to Wade, "He made his first appearance as Rudolf in Byron's 'Wonder,' and in 1853, at the age of twenty-three, he resigned his position as clerk and sailed for Melbourne, Australia, on a theatrical engagement."

In time he became well known as an actor and appeared on the stage in South and Central America and Mexico, and, in 1865, San Francisco. He resided in California from 1865 to 1877, making San Francisco his headquarters, where he was associated with the famous California

Figure 87. Henry Edwards

Theatre. He appeared in Boston in 1878 and in New York City in 1879. He was connected with the stage until a short time before his death. "His last appearance was in New York in the part of Sir Oliver in 'The School for Scandal.' Most of the remaining period of his life was spent in the Catskill Mountains in a vain search for health. He died of heart disease, dropsy and other complications in New York City on June 9, 1891." (Wade)

Edwards collected Lepidoptera and other insects from boyhood and throughout his acting career, during which he obtained specimens in many lands. He devoted much of his leisure time and a great deal of money to amassing one of the finest collections of Lepidoptera extant, and in 1889 he published his *Bibliographical Catalogue of the Described Transformations of North American Lepidoptera* in the U.S. National Museum Bulletin 35. He also collected in other orders of insects as well as plants and shells for other people and was highly esteemed by his contemporaries for his generosity in presenting specimens to them.

While in California he published a number of important papers, largely on moths, especially the Aegeridae, on which he was an authority and of which he described many new species. He was an active member, and in 1875–1877 a vice-president, of the California Academy of Sciences. He was a very close friend of H. H. Behr's, the noted lepidopterist and savant, both being members of the Bohemian Club in San Francisco. When he went to New York he became affiliated with the New York Entomological Society and was one of the founders and editors of *Papilio*. He was a member of the Brooklyn Entomological Society, Torrey Botanical Club, the Boston Society of Natural History, the San Francisco Microscopical Society, the San Diego Natural History Society, and others. "It is easy to see that it was his scholarship and intense interest in the scientific questions of his day that led him into affiliations with so many of the learned societies, nor do we find these were perfunctory only, for he regularly attended the society meetings, wrote papers for them and attained to high place in their councils. He also had many entomological friends and maintained an extensive correspondence with them." (Wade)

Edwards' friend, the lepidopterist W. Schaus (1930), wrote: "Edwards loved his favorite studies quite as much as he did the stage and brought to both an ardor and freshness contagious and perennial. One of his correspondents, writing about him after his death, emphasizes his unvarying kindness and unfailing help to entomologists who were less learned than himself. 'I owe much,' said he, 'to his help and encouragement and shall miss him sorely, though I never saw his face.'"

Osborn (1937), writing of a trip to the East Coast in 1882, describes a visit he had with Edwards. "He was most cordial in his welcome and talked of entomological matters with great zest, showing his special treasures with apparently great pleasure and interest. He had a Japanese assistant, a very skillful preparator and well versed in entomology, who evidently did a large part of the mechanical labor and possibly some of the more technical separation of the species which was credited to him by Mr. Edwards in various papers."

Wade quotes from a letter to Edwards from John Muir, the noted California naturalist, written from Yosemite on June 6, 1872: "Your bundle of butterfly apparatus is received. You are now in constant remembrance, because every flying flower is branded with your name. I shall be among the high gardens in a month or two and will gather you a good big handful of your favorite painted honeysuckers & honeysuckles. I wish you all the deep far-reaching joy, you deserve in your dear sunful pursuits."

When he died in 1891, Edwards had one of the largest private collections of butterflies and beetles in the country, as the following advertisement in *Entomological News* shows:

"The collection of insects made by the late Henry Edwards, consisting of about 300,000 specimens of all orders, and well represented in large numbers of individuals and long suites of specimens from all parts of the world, is for sale. It is particularly rich in Pacific coast of North America species. A large number of Lepidoptera from this region were described by Mr. Edwards, and his types are in the collection. Institutions or private persons wishing to purchase will please address Mrs. Henry Edwards, 185 E. 116 Street, New York, N.Y."

The American Museum of Natural History acquired his collection, which was partially paid for by friends of Edwards who wished to befriend his wife. Besides about four hundred and fifty of Edwards' types, the collection contained some of A. R. Grote's types of Noctuidae and Pyralidae. The museum also acquired Edwards' voluminous correspondence and notes.

FERDINAND HEINRICH HERMAN STRECKER (1836–1901)

Strecker was born in Philadelphia on March 24, 1836, of German parentage. When he was a boy his family took up residence in Reading, Pennsylvania, where his father, who had studied sculpting in Europe, became prominent as a dealer and sculptor of marble. Herman inherited his father's talents and in time became well known for his work as an architect and a sculptor. He eventually succeeded his father in the family

Figure 88. F. H. H. Strecker

business. Because sculpturing was not always lucrative he often did tombstone lettering for a living.

As a young man, Strecker made frequent trips to the library of The Academy of Natural Sciences in Philadelphia, where he studied all branches of natural history, later concentrating on the Lepidoptera. He was a master of dead and living languages and travelled a great deal in the West Indies, Mexico, and Central America, where he studied Aztec monuments and added to his butterfly collection. In the course of forty years, and $30,000 poorer, he amassed a collection of 200,000 specimens of butterflies and moths from every part of the world, "not excepting the region close to the poles, the hearts of the wildest forests of Africa, India, Australia, South America, the smaller islands of the Indian and Pacific Oceans." At one time his collection contained 300 types and cotypes. His collection occupied an entire floor of his home in Reading.

Lepidopterists remember Strecker for his *Lepidoptera Rhopaloceres and Heteroceres, Indigenous and Exotic, with Descriptions and Colored Illustrations.* This was published from 1872 to 1878 in a limited edition of 300 copies with fifteen parts and as many plates. The illustrations were drawn on stone and colored by the author.

According to Mengel (1902), the work was published "under difficult circumstances; he was a poor man at the time. He saved sufficient money to buy a lithographic stone, and then drew the group of butterflies on the first page of the work. This was sent to Philadelphia, until all the plates were published. All the copies were sold. The demand increased, but no more were ever issued."

In 1878 he published *Butterflies and Moths of North America* with the subtitle, *With Full Instructions for Collecting, Breeding, Preparing, Classifying, Packing for Shipment, and a Complete Synonymical Catalogue of Macrolepidoptera, with a Full Bibliography, to Which Is Added a Glossary of Terms and an Alphabetical and Descriptive List of Localities.*

Strecker was not only a collector, but also a dealer in moths and butterflies, as we can observe from this letter to George A. Ehrman of Pittsburgh written from Reading on January 16, 1885:

> I have hundreds of species of Butterflies and moths to dispose of on sale, from California, Florida, Mexico, Central and all parts of South America, Europe, Asia and Australia, but I have no printed lists and it would take much time to go through all these things and write out a list and after it was done you would not probably know from the names what each species was like. Would it not be better to adopt perhaps this plan which has rendered always to my clients perfect satisfaction, which is that you send me an amount whatever you desire to spend whether a small or large one and in return I will send you Butterflies to the amt. you send from such localities as you desire—all examples sent by me are ready spread on the pins and correct name and country always given along—further add say that if you adopt this plan and are not satisfied with what I send you can return them, and if in same order as sent I will promptly refund the money. . . .
>
> N. B. The butterflies of S. America are as a rule much handsomer and finer than those of Australia or Oceanica. I can furnish all the fine Catocalus of California if desired.

Anon. (1902) noted that Strecker "had an interesting and striking personality, and was cordial, affable and full of anecdote. He was frequently visited by scientific men from this country and abroad, and his correspondence was immense. In his chosen field he ranked deservedly high." Franklin and Marshall College granted him an honorary Ph.D. in 1890.

Graef (1914) remarks that his writing is often humorous "especially in reference to A. R. Grote, with whom he was often on any but friendly terms. His language, not always Chesterfieldian, was pungent, witty and entertaining."

Strecker died from a stroke at his home in Reading on November 30, 1901, leaving a wife and two children.

ALPHEUS SPRING PACKARD, JR. (1839–1905)

Packard's great reputation rests on his studies on insects and various animals, his work in economic entomology and evolution, and his popularization of natural science.

Figure 89. A. S. Packard, Jr.

J. B. Smith (1905) describes him as "both an investigator and a teacher [who] . . . through books . . . has taught more students than any other American Entomologist. He was not a writer of numerous short papers, of hasty criticisms or of single descriptions; and yet his publications were many, some of them brief and many critical; but all had a purpose—the conveyance of knowledge or the correction of error—the correction always made without reflecting upon the honesty or capacity of him who made it necessary. Not that Dr. Packard made no errors himself; no one was more ready than he to recognize his liability in that direction, and he was not ashamed to admit it. Not infrequently he changed his conclusions, and when we compare the first edition of the *Guide to the Study of Insects,* with the *Textbook of Entomology* the full extent of the revisions of such conclusions becomes apparent."

As to his appearance, Cockerell (1943) noted that he was said to have a forehead rather like Darwin's, that he was tall, very slight, quick and nervous, with small, shapely hands and feet, and that he dressed immaculately and carried himself gracefully. "He did not," he added, "work primarily for such rewards as the world might give, and when these came he was usually surprised."

Cockerell (1920) and Dexter (1957) prepared biographical sketches of Dr. Packard. The following is largely from their accounts.

Alpha, as he was called by his family, was born in Brunswick, Maine, on February 18, 1839. His father, Alpheus Sr., who held the degree of Doctor of Divinity, was Professor of Greek and Latin at Bowdoin College and was interested in history and archaeology; his mother, Frances Elizabeth Appleton, was the daughter of President Appleton of Bowdoin

College. She died a few months after Alpha's birth, and he was reared by an aunt. In 1845 his father married again, thus providing Alpha with three stepbrothers and a stepsister, all older than himself.

According to Dexter, Alpha's daughter, Mrs. Frances Packard McClellan, preserved a copy of portions of a diary kept by her father and correspondence running from January 1, 1854, to September 22, 1899. "It is a two-volume scrapbook containing for the most part handwritten copies, made by Packard's wife, of excerpts from the original diary. Unfortunately the original documents are not extant and the portions selected by Mrs. Packard for copy may not always have been the most important ones from a scientific point of view. Nevertheless, enough has been preserved to show the development of one of the great pioneers of American entomology."

At the age of fourteen Packard was collecting shells and other natural history objects, for which he built a cabinet. He says he "devoured all the books on natural science in the library of Bowdoin College, where I was kindly allowed to browse long before entering college."

In 1855, when he was sixteen, he wrote "have had some very good presents this year. One a blank book for my Journal . . . also the *Principles of Zoology* by Agassiz and Gould. A valuable book and just the thing I wanted. . . . I have been reading a very interesting article in the North British Review on Sea-weeds, shells, and polyps, it has given me quite an idea of such things . . . I am reading now Goodriche's books on Geology. . . . Went to the library and got an article on Mollusca, with a life of [Georges] Cuvier in it, and one volume of Kee's *Cyclopedia* with a volume of engravings, and containing an interesting article on Conchology. I think I am fortunate about getting books on that subject, which is so interesting to me, but the rest of the family except Osgood [his stepbrother] laugh at me about it."

He early made himself a competent draftsman and thus was able to produce many of the drawings for the books he wrote in later years.

T. W. Harris' book, *Insects Injurious to Vegetation,* which so greatly influenced Comstock and Howard, also helped start him on the road to entomology, and in 1903, Packard wrote, "When about 16–17 I collected insects in considerable numbers. I was also aided by a maiden lady in Brunswick, Maine in naming my native plants. I formed a herbarium before entering college. From Miss Ann Jackson, when a boy, I first heard of Lamarck, and of his classification of shells, and of the Lamarckian genera of shells. With then an inborn taste for natural history, an aversion to business, and a fondness for books, my deep interest in animal life was sustained and I was impelled to devote my life to biological study. All through college I corresponded with Professor [S. F.] Baird, assistant secretary of the Smithsonian Institution, also with conchologists and entomologists, and this was a constant stimulus to the natural zeal and interest, or passion, for biology which has influenced my life. Also I was a born collector, though I have now no large collection."

Packard was born with a cleft palate, which affected his speech and made him shy. An operation in 1857, without benefit of anesthetic, corrected the defect.

Alpha was encouraged in his scientific interests by members of the Bowdoin faculty as well as by a number of entomologists elsewhere: J. L. LeConte named beetles for him; S. H. Scudder helped him with moths and butterflies, C. R. Osten Sacken with flies, and P. R. Uhler with Neuroptera. Besides insects, and shells, geology had a lifelong fascination for him, and he was especially attracted to the study of mountains and glaciers.

He was admitted to Bowdoin College in 1857, where he took the regular curriculum and studied geology and entomology on his own account. He came under the influence of P. A. Chadbourne of Williams College, who lectured on natural history at Bowdoin. In 1858 Alpha noted in his diary that he was corresponding with a young man studying with Professor Chadbourne who was very much interested in entomology, and they were planning to exchange specimens. The young man was from Boston and his name was Scudder. This was the beginning of a lifelong friendship.

One entry in his diary in 1858 says, "Been overhauling a bushel of fishes' guts; *pleasant* and profitable work." By this means he found rare shells swallowed by the fish. Shortly thereafter he began to work on geometrid moths and in 1860 to write papers on economic entomology for the *Maine Farmer;* these were the first of the publications which he turned out over forty-five years.

In the summer of that year he accompanied Chadbourne on a summer expedition to Labrador. The next year, upon graduating from Bowdoin, Packard conducted members of his class on a summer voyage in the Bay of Fundy and also worked for the Maine Geological Survey. That fall he entered the Lawrence Scientific School at Harvard to study under Louis Agassiz and on October 31 made the following entry in his diary:

"I feel well established here now, though not much advanced in study of the Museum. Professor Agassiz has agreed to remit, or not to ask, the tuition, 100 dollars for the year, upon my telling him how I was situated, and on the condition that I should do extra work, taking care of insects, etc., which would be only part of my study as an entomologist. He said he would not get me to work doing anything he would not do himself . . . I am going to have a long talk with Professor A. about staying here some years, if he will pay me enough to pay my expenses; in hopes of being one of the entomological curators of the Museum. Some of the students here he gives $300 per year, to keep them here, and I might 'tend medical lectures and get a degree meanwhile."

Packard succeeded in carrying out this plan during the ensuing three years, after which he received an M.D. from the Maine Medical School at Bowdoin.

According to Packard, Agassiz paid these salaries, to the extent of $4,000 a year, out of his own pocket. As a result Agassiz was "very hard up." When a new beautifully illustrated edition of Harris' book came out in 1862, Dr. Agassiz gave a copy to each of his entomology students.

In 1863 there was a revolt of the student assistants in the Museum of Comparative Zoology because of a desire for independence from Agassiz's discipline, the low pay, and certain new regulations which they thought unreasonable according to Dexter (1957). Packard wrote: "The situation at the Museum soon became an unhappy one because of a new ruling that assistants could not possess a private collection, that any such must be given to or deposited at the Museum, and all work done at the Museum became property of the institution. This led to a withdrawal of nearly all of Agassiz's student-assistants [Verrill, Putnam, Morse, Hyatt], including Packard." They left Cambridge and found employment elsewhere; there were, however, no ill feelings between Packard and Agassiz, and they remained friends for many years.

The year 1864 was an eventful one for Packard. Not only did he receive an M.D. from the Maine Medical School but a B.S. from the Lawrence Scientific School; he also joined another expedition to Labrador and received a commission as assistant surgeon of the First Maine Veteran Volunteers. Packard served with them in some stiff battles in Virginia and won from General Hyde the commendation, "he was a dare devil and knew no fear."

In 1865, after being mustered out of the army, Packard published a *Synopsis of the Bombycidae* (native silkworms) of the United States. For a year he served as acting librarian and custodian of the Boston Society of Natural History; then in 1867 he was made Curator of the Peabody Academy of Science in Salem. The same year he married Elizabeth Derby Walcott, by whom he subsequently had a son and three daughters, helped found the *American Naturalist,* and began to write scientific articles in a popular style. For the next few years he lectured on economic entomology at Maine State Agricultural College (University of Maine) and Massachusetts Agricultural College at Amherst (University of Massachusetts), and on entomology and comparative anatomy at Bowdoin. During these years he also studied marine life at Key West, in the Tortugas islands, and off the coasts of North and South Carolina. His *Guide to the Study of Insects,* over seven hundred pages with many illustrations, appeared in 1869. This book, which Agassiz thought highly of, went through eight editions.

In 1872 Packard travelled through Europe and met many outstanding naturalists. The following summer he assisted Agassiz in his natural history school on the Island of Penikese off Massachusetts.

Among his publications designed to popularize entomology were *Our Common Insects* (1873) and *Half-Hours with Insects* (1877).

Packard had a temporary association with the Kentucky Geological Survey in 1874 to investigate animal life in Mammoth and other caves; his publication, *The Cave Fauna of North America* (1888), resulted from these studies. He was also connected with the U.S. Geological Survey in 1875–76, when he wrote papers on the insects of the Great Salt Lake and on the Mallophaga, myriapods, and moths of Colorado. His *Monograph of the Geometrid Moths,* with 600 pages and thirteen plates, appeared in 1876.

In that year, Congress appropriated money for the U.S. Entomological Commission to study the Rocky Mountain locust and other injurious insects. Three outstanding entomologists were appointed to this Commission, C. V. Riley as chairman, Packard as secretary, and Cyrus Thomas as treasurer. Riley was thirty-four, Packard thirty-eight, and Thomas fifty-two. The country was divided among the three of them for study. Packard travelled by stage and horse to study the locust and many other insect pests. Five famous reports were published by the Commission; the fifth and last report in 1890, on insects injurious to trees, was largely Packard's work.

Howard (1930) complains that Packard had certain peculiarities in writing his manuscripts "in the way of economizing paper. He would write his manuscript on any scraps of envelopes, wrapping paper and so on that were available. These were numbered and constituted his manuscript. Considering this peculiarity together with Packard's not too legible handwriting, one can imagine the difficulty I had in editing the big report on 'Insects Affecting Forest and Shade Trees' with its 957 pages."

In 1878 Packard was appointed Professor of Zoology and Geology at Brown University. There he wrote *First Lessons in Zoology* in 1886, *Entomology for Beginners* in 1889, the previously mentioned *Fifth Report of the U.S. Entomological Commission* in 1890, *The Labrador Coast* in 1891, and a *Textbook of Entomology* in 1895. The latter was quite different from his earlier *Guide to the Study of Insects* and dealt with the anatomy, physiology, embryology, and metamorphosis of insects. Many an entomologist got his start from Packard's books.

He apparently liked teaching, but at times found the path of the pedagogue a rocky one, as Walton and Osborn (1937) report that the students occasionally hired an organ grinder to play under the windows of the lecture hall, which sometimes irritated him so that he would dismiss the class. Packard must have been a good teacher, for, according to Kingsley (1888), he was "a very pleasant and entertaining companion, and not the least among his good qualities is the interest he takes in all who show any predilection toward scientific work. There he is always ready to assist and encourage to the extent of his ability."

According to Cockerell he had a genial disposition; he was fair, modest, retiring, and dignified, with a keen sense of wit; he was kind and courteous. He delighted in music, concerts and the theater; he loved art and had some artistic ability. "In his dealings with others, he was entirely honorable, and never forgot to give credit for assistance of any kind to younger men, in a day when this was by no means invariably the custom. . . . Although unassuming and of a quiet disposition, he had a tremendous spirit when roused to just anger."

Cockerell says that he was "a very keen collector of insects"; that his "enthusiasm never waned, and he never became a purely indoor and laboratory naturalist." According to Cockerell his work on geometrid and bombycid moths was good, although many of his species were synonymized because of poor descriptions by some of the earlier workers. Upon mastering a group and writing a monograph on it, he rarely returned to it. He was one of the first entomologists to perceive that there were more insect orders than those described by Linnaeus, and he added to the list.

Besides Lepidoptera, Packard described insects in the Aptera, Diptera, Homoptera, Mallophaga, Hymenoptera, and Mecoptera orders. He also described the Arachnida, Diplopoda, Pauropoda, Crustacea, Mollusca, Bryozoa, and others for a total of fifty genera and 580 species of insects and other animals.

In his later years he studied evolution and the philosophy of zoology. He devoted less and less time to the description and identification of specimens and more to their structure, growth, and the principles of their distribution. Even as a boy, he had been fascinated by Lamarck, the French naturalist and student of evolution, and in 1901, following a visit to France, he wrote a biographical study, *Lamarck, the Founder of Evolution*.

Packard was a member of the National Academy of Sciences, and an honorary member of a number of international scientific societies.

After a year of illness, he died in Providence on February 14, 1905.

GEORGE DURYEA HULST (1846–1900)

One of the early authorities on moths, especially the Geometridae and Pyralidae, Hulst was born in Brooklyn, New York, on March 9, 1846. He was interested in plants and insects from boyhood on. He graduated from Rutgers in 1866, receiving a medal for his knowledge of the classics, and in 1869 graduated from the New Brunswick Theological Seminary. He thereupon became pastor of the South Bushwick Reformed Church in Brooklyn, where he remained until his death, highly esteemed by his congregation, according to Weeks (1900).

Despite the demands of his congregation, which he did not neglect, he continued his studies in natural history and entomology. He lectured at Rutgers College on entomology, served for a time as State Entomologist of New Jersey, and was editor of *Entomologica Americana* from 1887 to 1889. He studied both macro- and micro-lepidoptera and became a recognized authority on geometrid moths.

J. B. Smith (1900) speaks of his long friendship with Hulst, which began in 1880: "I owe to him encouragement and assistance at the outset of my career, and the ready liberality with which he allowed me to take specimens from his cabinet was characteristic of the man."

Hulst became an active member of the Brooklyn Entomological Society shortly after it was organized in 1872. "At its inception the society was a meeting of collectors for informal discussion, and these discussions were not less attractive because of a social glass with which the members

Figure 90. Rev. G. D. Hulst

modified any dry problems. When Dr. Hulst joined, the glass was eliminated as part of the regular program, in deference to the cloth; but he was not a total abstainer nor a severe judge of those who made temperate use of alcoholic refreshments." Smith notes that he was tolerant of others whose religious beliefs were not as strong as his own.

Hulst was a member of numerous other scientific organizations, including the Brooklyn Institute of Arts and Sciences, where he was President of the Department of Botany and an officer of the Department of Entomology. Rutgers University granted him an honorary Ph.D. in 1891. He died suddenly at his home in Brooklyn on November 5, 1900. Many of his types are in the Brooklyn Museum and his collection is at Rutgers.

AUGUSTUS RADCLIFFE GROTE (1841–1903)

Grote, an outstanding specialist on the noctuids and sphingids, was as attracted to moths as Scudder was to butterflies. He was born in Liverpool, England, on February 7, 1841; his father was German, his mother Welsh. He moved to the United States at the age of seven and spent his youth on a large farm purchased by his father on Staten Island, where, as Bethune (1903) notes, he delighted in roaming the woods and upland meadows. In his "Synonymical Catalogue of North American Sphingidae with Notes and Descriptions," which Bethune refers to as "Hawk Moths of North America" Grote describes the joys of these rambles: "the early dawn is a profitable time for the collector of

Figure 91. A. R. Grote

lepidoptera, who may then surprise the moths on their first resting places after the fatigues of the night. On Staten Island my early rising was awarded by many captures at the hour when the cat bird sings and betrays to none but chosen ears her relationship to the many-tuned mocking birds of the South."

E. L. Graef (1914), in his recollections of some early Brooklyn entomologists, says that he, Grote, and Frederick Tepper were schoolmates in the 1850's in Brooklyn and in 1854 began to collect insects. They were at that time unable to find a good reference collection or anyone to name the insects for them. They used J. W. Meigen's *Handbuch fur Schmetterlingsliebhaber* to help them with their butterflies. Says Graef:

> One may readily appreciate our delight about this time in *discovering* a man of great value to us, from whom we learned much about rearing specimens, etc., besides being able to buy from him good insect pins, nets, setting boards, and other entomological supplies. This important personage was the late John Ackhurst, who then lived on Prospect Street where now stands the anchorage of Brooklyn Bridge. I shall never forget his genial, good-natured greeting, "Good morning lads," as the trio of us stepped inside his door. He appeared to me then about the same as he looked thirty years later, with long hair, great goggle spectacles, smooth shaven, the head surmounted by a square paper cap.* He took a fatherly interest in us and gave us much valuable information about collecting. For more than fifty years he took pride in breeding Lepidoptera, particularly rare Sphingidae and Bombicidae and his success in this respect induced the United States Agricultural Department to send him 100 pupae of *Platysamia cynthia* with a view to testing their availability as a substitute for the better known silkworm, *Bombyx mori*. He bred them in captivity for three years, then liberated a number on the ailanthus trees, from which undoubtedly all specimens found in this region are descended.†

The works of the Reverend John G. Morris of Baltimore, particularly his *Synopsis of North America Lepidoptera,* acquainted Grote with the principles of taxonomy.

* Mr. Ackhurst had long flowing hair, which made him look very conspicuous. He explained to me that this was because he used great quantities of arsenic in his work as a taxidermist and his long hair kept the poison from entering his system. E. L. G.

† The ailanthus, the tree from China on which the caterpillars which provide pongee silk of commerce feed, was planted in Brooklyn about 1866 as a substitute to shade trees infested with span worm pest, the *Ennomos subsignaria*. H. A. Graef was the pioneer in asking for spraying the trees with poison. An appropriation of $25,000 for this purpose was defeated in the Board of Aldermen. A physician of Newark advocated importing the English sparrow for the same purpose. This was done, and, whatsoever may now be said to the detriment of the bird, it did its work. The shade tree span worm has never since been a nuisance in the cities, although three years ago the whole city was white with them one September night, brought by the wind.

Unfortunately, his father invested his money in the Staten Island Railway, which fell into William H. Vanderbilt's hands in the panic of 1857, resulting in financial disaster for Mr. Grote. "Meanwhile young Grote had been preparing for Harvard University, but was obliged by the straitened circumstances of the family to abandon his prospective career; later on he was enabled to go to Europe and completed his education on the Continent; after his return he received the degree of A.M. from Lafayette College, Pennsylvania." (Bethune)

Grote married in 1870 and lived somewhere in Alabama until his wife died in 1873. According to Bethune, "During his residence in Alabama, Prof. Grote studied the cotton worm, and brought the subject before the public in a lecture; he then went to Washington and tried to interest the Government in the matter, but without success. Subsequently 'The Entomological Commission,' consisting of Messrs. Riley, [A. S.] Packard and [Cyrus] Thomas, was appointed by act of Congress, and Prof. Grote was keenly disappointed at his failure to obtain a place upon it, this failure, he rightly or wrongly, attributed to the adverse influence of Dr. Riley, and for many years he took every opportunity of criticizing in vehement language the work of this distinguished Entomologist. He was, however, employed by the Commission in 1878 to visit Florida, Georgia and Southern Alabama for the purpose of investigating the insects injurious to the cotton plants and especially to make observations upon the supposed migrations of the moth. His brief report is incorporated in the large volume published by the Commission in 1885."

Anyone who carefully reads the old numbers of American entomological journals will become aware of Grote's antipathy for C. V. Riley. In fact, Grote's article (1889) on Dr. T. W. Harris appears to accentuate the virtues of Dr. Harris at the expense of the unnamed Professor Riley.

Grote wrote five articles from 1871 to 1875 on his studies of the southern cotton worm. Upon his wife's death, he moved to New York State to become associated with the Buffalo Society of Natural History. While with the Society he began to publish its bulletin, to which he contributed many articles. Another enterprise of his, *The North American Entomologist* collapsed after the publication of the first volume. Howland (1907) states that many of his lectures on a variety of subjects were subsequently printed in the *Scientific American* and in *Popular Science Monthly.*

Milburn (1913), recalling Grote's years in Buffalo, says: "I am sorry that I cannot tell you much about Grote. He was the director of the museum on a small salary and in the front rank of entomologists. He was, I should say, in the middle of the '70's, somewhat over thirty years of age, dark, slight in build, and of a nervous mercurial temperament. He was a man of general culture, quite a poet, and devoted to music. . . . He was a delightful companion and a good talker. Those were the days when Darwinism was spreading fast, and of what used to be called the conflict between science and religion. Grote was rather the leader of the group of young men to which I have referred, and for them every problem in the universe was in the melting pot.

"When Grote was not occupied in identifying or describing some new species, he was writing a lyric, composing an opera, or recasting theology. He was a very vivid and interesting personality."

Grote had begun to publish on moths at the age of twenty-one, his first papers appearing in the *Proceedings of the Academy of Natural Science* under the aegis of the American Entomological Society of Philadelphia. While in Buffalo he became closely associated with Coleman T. Robinson, a wealthy New Yorker, and the two of them described new species of moths in such papers as *A Synonymical Catalogue of North American Sphingidae, with Notes and Descriptions* and a *List of the Lepidoptera of North America.*

Grote also wrote *The New Infidelity,* which was both praised and damned by the critics and largely ignored by the public. "The versatility of his talents was further shown by his devotion to music; while in Buffalo he was organist of one of the principal Episcopal Churches, and subsequently composed many pieces of music, one of which only was published, and even attempted the composition of two operas, which, however, he found too great a task and never completed." (Bethune)

When his father died in 1880, Grote left the Society in Buffalo to return to his old home on Staten Island. At his request the Society turned over to him his famous collection of moths. On Staten Island he had a profound influence on the young W. T. Davis, who was to become a famous naturalist, cicada specialist, and historian.

According to Abbott (1949), "Grote was one of the founders of the New York Entomological Club (not to be confused with the New York Entomological Society, a later organization). Davis was only a little over eighteen when he attended one of the early meetings of the Club as a visitor, and was on the same occasion elected an active member—undoubtedly proposed by Grote. He did not attend regularly, but he was there occasionally."

In 1884 he moved to Bremen, Germany, where he lived for ten or eleven years and remarried. He then moved to Hildesheim; there for the last nine years of his life he was an honorary assistant of the Roemer Museum. His notable collection was sold to the British Museum. For a long time it was offered for sale in the United States for $5,000, to satisfy debts Grote had incurred; but there were no takers.

Grote wrote from Germany that he "was desperately homesick for

America and wanted to find a position in this country so that he could come back. Davis was unable to help him, in this respect, but wrote cheerfully about things he knew would be of interest. . . . Davis kept his letters and his old leather-covered collecting box all his life." (Abbott)

While abroad Grote still contributed to periodicals in North America. He visited London twice and helped in choosing a collection for the Centennial Exhibition in Philadelphia. His failure to obtain a position on the staff of the British Museum was a keen disappointment to him.

Although he made a few bitter enemies, most people found him genial and kindly, and his friends, warmly affectionate. He aided collectors of every caliber in identifying moths, always with the hope of being repaid for his efforts by discovering a rare or new species.

He died at Hildesheim on September 12, 1903.

WILLIAM JACOB HOLLAND (1848–1932) *

Holland's *The Butterfly Book* and *The Moth Book* introduced many a youngster—and oldster—to the collection and study of butterflies and moths. He concentrated especially on the study of butterflies from exotic places made famous by early explorers and missionaries. His talents had many facets, and he often served concurrently as a minister of the Gospel,

Figure 92. Rev. W. J. Holland

* This sketch, with some variation in wording, first appeared in *Entomological News* 79 (5): 125–135, 1968.

a university chancellor, a professor of ancient languages, a museum director, a zoologist, an entomologist, and a paleontologist.

Much that is known about Holland is from answers, given when he was eighty-one, to a questionnaire from H. D. Gunder. Holland was very unhappy with the 1929 article Gunder prepared from this questionnaire, especially with the latter's interpretation of his attitude towards his religious activities. And we can see in Holland's own copy of *Entomological News,* in the old man's shaky handwriting, this notation to Gunder's article: "N. B. I do not recognize this sketch as my work. It is a transmogrification of what I wrote as replies to Gunder's questionnaire." Nevertheless, Gunder's profile is a very valuable source of information.

Besides articles by Gunder (1929) and Avinoff (1933), some biographical material about Dr. Holland comes from men who knew him. Another source of valuable information was a very dusty box containing a small part of Holland's correspondence (boxes of his papers were thrown out by a former secretary), found by Dr. George E. Wallace in the catacombs beneath Carnegie Museum.

Holland's father, the Reverend Francis Raymond Holland, was a Moravian missionary in Bethany, Jamaica, where William was born on August 16, 1848. His mother was from Bethlehem, Pennsylvania. Both parents were greatly interested in natural history, his father being an amateur conchologist, entomologist, and botanist. According to Gunder (1929), Holland wrote: "Father's home in Jamaica seemed to be headquarters for naturalists and sooner or later lovers of nature found their way there. C. B. Adams, Professor of Zoology at Amherst College, lived for a long time with us while pursuing his studies on the Island."

The Holland family returned to the United States in 1851, where the father was made pastor of the Moravian Church at Dover, Ohio, and later at Tuscarawas, Ohio. The Reverend Francis Holland brought his collection of shells, plants, and insects from Jamaica. "As a child I was permitted the examination of these collections and on rainy or snowy days I delighted to look them over and gradually came to know some of their Latin names. My mother taught me to draw and to paint and I still have among my papers a number of sketches of those Jamaican shells and butterflies which I drew from life before I entered my teens." (Gunder) His father encouraged him to collect plants and told him the Latin names whenever he could, and the two of them explored the woods and fields.

In 1858 the Reverend Francis Holland was transferred to Salem, North Carolina, and for a time William was educated by private tutors in a parochial school for boys. "All the spare time I could command was devoted to collecting birds, birds' nests and eggs, and to fishing or shooting. My father taught me how to prepare bird-skins and to rear and mount insects. The back verandah of our house was covered in the summer and fall with breeding boxes in which I reared many lepidoptera." He also studied books on natural history in the library of the Salem College for Women, including Thomas Say's *American Entomology,* and he made drawings and water colors of the illustrations. "Being provided with a carpenter's bench and tools I learned to make my own insect boxes, a few of which I still possess."

He had reason to remember his return from the South. "In the fall of 1863 the family came north by the 'Underground Route,' a trip full of adventure for a boy, as well as for the adults." They arrived in Bethlehem, Pennsylvania, to live in the home of Mrs. Holland's father, who was no longer living. William entered Moravian College there. He already had a mastery of elementary Latin, Greek, and mathematics, and "German and French were in a measure 'mother-tongues' to me, for from my earliest childhood I had been taught to use these languages."

While in Bethlehem, he studied drawing and oil and water-color painting under a German artist. In 1867 he completed his courses at Moravian College and entered Amherst in 1869, where he had classes in chemistry, physics, geology, astronomy, paleontology, and Kant, Hegel, and Fichte. "My room-mate during my senior year was Neesima, the first Japanese educated in America. He taught me Japanese in return for assistance given him in the study of Greek." At the end of his senior year Holland became principal of Amherst High School (1869–70) in Massachusetts and then of Westborough High School, Massachusetts (1870–71).

In 1871, at his father's request Holland entered the Princeton Theological Seminary to prepare for the ministry. He studied there for three years, devoting himself especially to Hebrew, "Chaldee," and Arabic. "Writing in my eightieth year, I may say that my Arabic has largely evaporated, but my Hebrew still abides with me in some strength."

Then in 1874, he went to Pittsburgh to become pastor of the Bellefield Presbyterian Church, a position he held until 1891. He was almost immediately made a Trustee of the Pennsylvania College for Women (now Chatham College), where he also held the position of Professor of Ancient Languages. From 1891 to 1901 he was Chancellor of the Western University of Pennsylvania (now the University of Pittsburgh); there he taught anatomy and zoology. The University grew rapidly under his direction.

In 1887 Dr. Holland went to Japan as a member of the United States Eclipse Expedition and in 1889 departed for Africa on a similar expedition, pursuing a number of biological investigations of special interest to him. He visited Europe in 1892 to consult technical literature in various libraries and to study collections of African Lepidoptera. On his trips to the museums of England, France, and Germany he met many scientists with whom he carried on a lively correspondence for most of his life.

In 1899, while in the Rocky Mountains of Wyoming Dr. Holland suffered an appendicitis attack which almost resulted in his death.

According to Gunder, Andrew Carnegie often confidentially discussed with him his "plans for the development of the cultural institutions which he wished to establish in Pittsburgh and elsewhere," and in 1898, though still Chancellor of the University, he was elected Director of the new Carnegie Museum in Pittsburgh. This position he held with "signal success" until 1922, when he became Director Emeritus.

In 1901 he relinquished the chancellorship at the University in order to devote himself entirely to the guidance of the Museum, which was greatly enlarged in 1907. Nevertheless, he was still on the Board of Trustees of the University and deeply involved in many civic activities in Pittsburgh, including those of the Filtration Commission, which kept the city practically free of typhoid. "If I had done nothing else for Pittsburgh than this, I should be happy." He was Vice President of the Carnegie Hero Fund from 1904 to 1922, a member of the Carnegie Corporation, and Belgian Consul in Pittsburgh for a number of years after World War I. He also founded the American Association of Museums at the Smithsonian Institution in 1907.

The Butterfly Book: A Popular and Scientific Manual Depicting All the Butterflies of the United States and Canada, with 48 plates in color photography, was published in 1898. Henry Skinner (1899) wrote in a review of it: "The plates represent the highest type of what is known as the three-color process and are successful to a remarkable degree. Where they are not quite satisfactory it is owing to the fact that the forms figured are so closely related as to make any process insufficient. The majority of figures leave nothing to be desired, as they are close to perfection. The work is excellently done, and the author is to be sincerely congratulated. This book will do more to stimulate an interest in these insects than anything heretofore printed."

In his preface to the first edition, Holland says: "Its aim is to guide the amateur collector in the right paths and to prepare him by the intelligent accomplishment of his labors for the enjoyment of still wider and more difficult researches in this and allied fields of human knowledge. The work is confined to the fauna of the continent of North America north of the Rio Grande of Texas. It is essentially popular in its character. Those who seek a more technical treatment must resort to the writings of others."

The preface notes that *The Butterflies of North America,* by W. H. Edwards, in three volumes, was selling for $150, "even at this price below the cost of manufacture," and that *The Butterflies of New England,* by S. H. Scudder, in three volumes, sold for $75, which "likewise represents at this price only a partial return to the learned author for the money, labor, and time expended upon it." The initial price of *The Butterfly Book* was $5. The preface to the Revised Edition (which appeared on Holland's eighty-second birthday) states that 65,000 copies had been sold of the original edition.

The Moth Book: A Popular Guide to a Knowledge of the Moths of North America appeared in 1903, with 48 plates in color and numerous black and white illustrations. It did for the study of moths what *The Butterfly Book* did for the study of butterflies, at the same original price. Both books were written during the author's spare time.

Avinoff (1933) remarks that Holland's collection was rich in the Lepidoptera of North America, Africa, and Asia. "Especially valuable portions of the collection of Dr. Holland were the Pyralids collected by Pryer in Japan and the well-known collection of North American Rhopalocera assembled by [W. H.] Edwards with all the types described by these noted entomologists. Dr. Holland's collection included types of many hundreds of new species which were described by him during his life time." Holland concentrated largely on the butterflies of equatorial Africa and Central and South America. He left his collection of four to five hundred thousand mounted specimens and his library, rich in books and periodicals, to the Carnegie Museum.

Much of Holland's collection was acquired through the efforts of professional collectors and amateurs, who were paid for their efforts. Among the latter was Maria E. Fernald, the wife of Professor C. H. Fernald, and the Reverend A. C. Good, a missionary in Africa. Holland purchased the famous William H. Edwards collection of North American butterflies just as the latter was negotiating to sell it to European collectors. "I agreed in exchange for the collection to pay all the expenses of producing the third volume of his celebrated work, *The Butterflies of North America.*" (Gunder)

Extremely conservative in the classification of butterflies, Holland heartily belabored those of his colleagues in correspondence and print who took liberties in "splitting" genus and species. Thus he had little use for taxonomists who used Jacob Hubner's *Tentamen* as a basis for changing long-established generic names, and he was very unhappy about specialists who used high powered microscopes to find "invisible distinctions." In *The Moth Book* he views "splitters" and "lumpers" as follows: "When a man cultivates the habit of discrimination to excess, he is apt to become, so far as his labors as a systematist are concerned, 'a splitter.' A 'splitter' magnifies the importance of trivial details; he regards minute differences with interest; he searches with more than microscopic zeal after the little things and leaves out of sight the lines of general resemblance. "The 'lumper' on the other hand, is a man who detects no differences. Questions of structure do not trouble him. General resemblances are the only things with which he deals. The 'lumper' is the horror of the 'splitter,' the 'splitter' is anathema to the 'lumper'; both are the source of genuine grief and much hardship to conscientious men who are the possessors of normally constituted minds and truly scientific habits."

Holland was also greatly devoted to the study of paleontology, especially dinosaurs; under his direction the Carnegie Museum acquired a famous collection of these reptiles from fossil beds in the West. Among the discoveries was *Diplodocus carnegiei,* whose huge skeleton in the Carnegie Museum is viewed by thousands each year. Nine plaster of Paris replicas were made for museums in Europe and both Americas; and for this discovery, Holland received many foreign decorations, which he loved to wear for special occasions on his frequent trips to Europe.

In response to Gunder's questionnaire, Holland wrote:

"Personal hobbies? Well, in my youth I loved all outdoor sports and was a good swimmer. I was fond of horses and rode a great deal. In the last few years I liked a good game of golf, but rarely get the chance to swing the clubs! I always enjoy a game of whist, but have grown rusty as a chess player. Years ago in Japan I achieved quite a reputation over the chess board. They were fine players over there, too.

"You ask me about my outstanding mental traits. It seems that nature has always endowed me with a good memory, not quite so good today as it used to be, but I still retain a reading knowledge of a number of the ancient languages and can remember where many lepidoptera species for example, were figured and described. I am naturally of an active and industrious turn and the only thing that troubles me is lack of time in which to do the things I would like to do."

Holland ran the Museum like the skipper of a tight ship. He had little patience for fools, and, at times, even ordinary mortals. He occasionally treated the scientific personnel in the Museum as so many hired hands, and although many of them had great respect for his abilities, a number had little affection for him. He was a somewhat large, overbearing individual, whose humor, with age, took on a sarcastic edge. There can be no doubt that he occasionally rubbed some of his scientific colleagues the wrong way: one paleontologist who did not appreciate his practical jokes repaid him by naming a small extinct animal

"Dinohyus hollandi" after him. At first this would appear to be a great honor: however, when the name is translated it means "Terrible pig of a Holland."

Dr. O. E. Jennings, the famous botanist, once told me this anecdote about Holland and his staff archaeologist. The latter argued long and hard with Holland about some illustrations for an article. When he found that he was making no impression on the obdurate Dr. Holland, he left in disgust, and headed for a local tavern to drown his sorrows. Several hours later the tavern owner called Holland to tell him he had forgotten his hat. Holland indignantly replied, "I have not been down there." "Well, sir, your hat is here with your name in it." The staff archaeologist made a point of not appearing in Holland's office for some time. Another anecdote recalls that a salesman once asked Holland if he wanted to buy a new encyclopedia. Holland said, "No." Salesman, "Would it be all right to see the other members of the museum?" Holland, "No, if there is anything they want to know, let them come and ask me!"

Once during World War I, when one of his collectors in South America inundated him with a series of gloomy epistles, the doctor finally could no longer bear his burden of grief and erupted as follows:

My Mondays are meatless
My Tuesdays are wheatless
I grow more eatless each day
My house it is heatless
My bed it is sheetless
They've been sent to the Y.M.C.A.

The bars are all treatless
My coffee is sweetless
Each day I am poorer and wiser
My stockings are feetless
My trousers are seatless
My God, how I do hate the Kaiser!

Yet if at times Dr. Holland showed a lack of humility and drove a hard bargain to get what he wanted, he had a sympathetic ear for the members of his staff in the matter of their personal problems and despite the demands on his time was gracious to many who needed his help.

To a woman who thought a fortune could be realized from collecting Lepidoptera he wrote, "the return which a collector of lepidoptera makes is found in the pleasures which is derived from the study of these beautiful creatures, and not to any great extent in dollars and cents." He ended a reply to an inquiry about the name of a butterfly from an old gentleman as follows: "You say you wrote me about fifty years ago from Keytesville, Missouri, by which sign both you and I are getting to be antiques and therefore far more valuable than we used to be."

The Fourth International Congress of Entomologists at Ithaca, New York, honored Holland by acclamation on his eightieth birthday, which occurred during the course of the Congress. He was, among other things, in large part responsible for obtaining the financial aid that enabled many of the foreign entomologists to attend the meeting.

Holland continued his many activities at the Carnegie Museum as Director Emeritus long after he retired in 1922 and continued to write articles on a great variety of subjects. In all, he completed some five hundred papers during his lifetime.

Despite a stroke which partially paralyzed him in 1932, he continued his correspondence and dictation up to the time when another stroke completely disabled him. He died on December 13, 1932.

JOHN BERNARD SMITH (1858–1912)

In placing this accomplished taxonomist, outstanding economic entomologist, teacher and pioneer in the practical control of mosquitoes among the lepidopterists, we have no intention of eclipsing his competence in other fields of entomology—he shone in all.

Torre-Bueno (1948) provides this portrait: "John B. Smith (nee Schmidt) was one of our greatest economic entomologists, and had the distinction of having cleared the Jersey marshes of mosquitoes (pro tem.). His father, an old time German, was a cabinet maker and collector of insects, and he devised and made the justly famous Schmidt insect boxes. . . . he became a great entomologist, his specialty being the night-flying moths. He had a great sense of humor, and had a truly Teutonic fondness for beer. He was short and rotund; his face was of the shape and color of the sun in effulgence, and was surrounded with rather thin whiskers, his hair rather thin on top."

Smith was born in New York City on November 21, 1858, and educated in the public schools there. He was admitted to the bar in 1879, but law was not his dish, for he found "a fly on the wall was more interesting than the case in hand." And after four years as a lawyer he abandoned law to accept appointment as Special Agent to the United States Department of Agriculture under C. V. Riley.

In all probability the Brooklyn Entomological Society and its enthusiastic members were largely responsible for switching Smith from the "legal local" to the "entomological express." There, with such

Figure 93(A).

Figure 93(B). (A) J. B. Smith. (B) Caricature of Smith done in 1910. From *Report from Rutgers,* February 10, 1959

members as Charles Fuchs and Frank Schaupp, he quaffed beer and talked insects.

In 1886 he became Assistant Curator of Entomology of the U.S. National Museum. In the three years he was there he worked on cranberry and hop pests and published papers on the Sphingidae, Arctiidae, Saturniidae, Noctuidae, as well as on the genus *Lachnosterna* of the Coleoptera. During the years he was in Washington, 1884 through 1889, he became very well known among entomologists and other students of natural history and was editor of *Entomologica Americana* and the *Bulletin of the Brooklyn Entomological Society.*

In 1889, under the influence of the Reverend George D. Hulst, Smith resigned his position in the National Museum to become Professor of Entomology at Rutgers University, Entomologist in the New Jersey Agricultural Experiment Station, and, in 1894, State Entomologist of New Jersey. Grossbeck (1912) says that "In these three capacities he brought honor and renown to the institution he served. His annual reports, which all told, form several bulky volumes, are mines of information, and rank with the best ever produced by any experiment station."

According to Howard, Webster, and Hopkins (1912) he was one of the first state workers to enter an agricultural experiment station, "and from the very start his energy and capacity made him one of the foremost workers in this line of research. He was interested in and active in every new development of economic entomology and was a prominent figure at every meeting of the Association of Economic Entomologists." He worked like a Trojan in entomological societies and associations. It was only natural that he should be elected president of the American Association of Economic Entomologists in 1895 and of the Entomological Society of America in 1910.

Despite the pressure of many meetings, voluminous correspondence, and experimenting with insecticides, he still found time for his systematic studies, especially on the Noctuidae, on which he was a world authority.

Grossbeck (1912) says that, "On more than one occasion I have heard Professor Smith say: "I *was* a coleopterist, I am *now* a lepidopterist," and "My collection of Coleoptera was sold to the National Museum." Because of the pressure of his many other duties, he finally had to limit himself to the family Noctuidae; he named and described over nine hundred species of moths, mostly noctuids.

"In the years that I spent at the Experiment Station," says Grossbeck, "the collection was kept in a small basement room in the fire-proof college library, and here it was that I frequently spent the Sunday with him [Smith]; working on the Hulst collection of Geometridae. This was the one day of the seven in the week when Professor Smith found time to discuss things lepidopterological, and in this room I learned much concerning questions of nomenclature, and of old entomologists who left this world before I stepped into the field of entomology."

In reference after reference Smith is described as a jovial, good-natured man, ever willing to help others. He made excellent illustrations of insects for many papers by other entomologists.

Hunt (1921) describes him as "Stout, thick set, [with] a beard which covered his face, legs which scarcely raised him above the meadow grass, a little blue soldier hat with a vizor like that of a pullman conductor, walking through meadows with a net, always smiling, always optimistic."

And Barber (1912) remembers when "I had the great pleasure of being out in the country one full day with Dr. Smith and Mr. Schwarz, and the way they seemed to enjoy each other, after a long separation, was one of the prettiest things I ever saw in my life. Both of them were just overjoyed all the time."

According to Osborn (1912), "No small part of his service to entomological science may be found in the numerous collections which he has identified for various students, and the specimens freely loaned for investigation in other hands."

Taxonomy was Smith's first interest, and, enmeshed as he was in all aspects of applied entomology, he occasionally yearned for the days when he could spend more time on moths. He ended a letter to Dr. W. J. Holland at the Carnegie Museum in May 1898 about finding an entomologist for the Museum with the proposal: "Why don't you create a chair of Entomology in your Institution, combine it with the curatorship of the Museum, with a good salary, and offer it to me?"

According to Hansens and Weiss (1958) and Dickerson (1911), Smith's activities in economic entomology were overwhelming. As soon as he assumed office at Rutgers in 1889, he issued a bulletin requesting the farmers to report their pest problems. The response was massive: Smith was inundated with problems. During his twenty-three years with Rutgers he was in almost perpetual motion. He established a system of nursery and bee inspection. He studied hundreds of insect pests and recommended remedies for their control. At first he concentrated on the horn fly, elm leaf beetle, clover leaf beetle, asparagus beetle, yellow-necked apple tree caterpillar, plum curculio, fall webworm and the periodical cicada. He conducted tests with London purple on the codling moth, hellebore and tobacco dust on the grapevine sawfly, tobacco dust against the white cabbage butterfly, Paris green against the cutworm, carbon bisulfide against melon lice, arsenate of lead against a host of pests, and cyanide for fumigating nursery stock. Probably every kind of injurious insect in New Jersey came under Smith's expert scrutiny: fruit insects, vegetable insects, nursery insects, forest insects, and livestock insects.

Smith's work on the San Jose scale in the nineties brought him national recognition as an outstanding economic entomologist, and in 1891 Rutgers awarded him an honorary D.Sc. In 1896 when New Jersey sent him to California to study biological and other control methods to curb the San Jose scale, he recommended spraying with fish-oil soap. Later he worked with crude oil as a dormant spray, with miscible oils, and lime sulfur. Hansens and Weiss (1958) say that "Smith pointed out the desirability of finding some method of treating the petroleum by which it could be made directly soluble, or miscible in water. The result was the material known commercially as 'Scalecide,' which was soon widely adopted for use against this and other pests."

The cranberry industry in New Jersey was important even at that time, and he recommended a number of insecticides as well as the "hopper dozer," a machine for collecting grasshoppers, and the flooding of bogs. There can be no doubt that the cranberry grower profited by Smith's exertions. But of all Smith's accomplishments in economic entomology, the most outstanding was his success in the control of pest mosquitoes. It can be said that he prepared the blueprints for pest mosquito control used by later workers. But he did something more than point the way, he actually influenced the State Legislature to appropriate $10,000 for work on the biology and control of mosquitoes—and in 1902 this was little short of a miracle. He once said that "mosquitoes were not a necessary affliction" and proved it.

In 1912 L. O. Howard commented, "when we consider the condition that exists in the State of New Jersey, and the indefatigible and successful work of Doctor Smith in the handling of the most difficult problem of the species that breed in the salt marshes, and of his persistent and finally successful efforts to induce the State legislature of that wealthy but extremely economical State to appropriate a large sum of money to relieve New Jersey from its characteristically traditional pest—we must hold up our hands in admiration." And in 1920, when Howard addressed a meeting of mosquito workers, he told them, "The work you are doing now in New Jersey is based almost entirely on the early scientific investigations of the late John B. Smith whose discoveries as to the exact biology of the salt marsh mosquito were revolutionary and of supreme importance." (Leslie, 1963)

The routine ditching and draining of marshlands to control pest mosquitoes was J. B. Smith's idea. Herman H. Brehme, John Grossbeck, and Henry L. Viereck were some of the early mosquito workers who helped Smith prove that it was possible to blunt the mosquito's pesky probe. Hansens and Weiss (1958) write: "In his special report, *The Mosquitoes of New Jersey,* Smith showed that a general mosquito problem anywhere within forty miles of the salt marsh required the elimination of both salt-marsh and fresh-water breeding species. He was the first entomologist to prove that the Atlantic Coast *Aedes sollicitans* breed exclusively in salt marshes but rise and fly many miles inland, nullifying completely the efforts of an inland municipality to free itself from the pest. Thus the failure of some local attempts at mosquito control was explained, and thus in the necessary consideration of the size of an area, elimination of salt-marsh mosquitoes became essentially a state rather than a local problem."

Smith had to participate in the no-holds-barred politics of his day to encourage the flow of funds from the State Legislature. He could hold his own in rough political infighting, and when an outside mosquito specialist was brought in to challenge his position, he vigorously resented it in an exchange of letters to the editor of *Entomological News* for 1906:

"Just a word to acknowledge a few minutes of real enjoyment in reading the letters of Mr. Henry Clay Weeks in the January (1906) News. Its very violence makes it unnecessary for me to reply, but I do wish to disclaim any feeling of jealousy. So far as I am aware, Mr. Weeks has never done anything that any one need be jealous of. That he has done and is doing work in New Jersey may be true, because even in New Jersey there are men with more money than brains. As for the rest, my original letter stated facts which are easy of verification. I cannot say the same for the answer."

Although Smith was good-natured, he was stubborn and somewhat opinionated. Thus, when he and the sarcastic H. G. Dyar locked horns on a number of occasions the reverberations were felt throughtout the entomological world. Such journals as the *Entomological News* record a few encounters between the two behemoths. In fact, according to several of their contemporaries, the feud reached the stage where Dyar named an insect *corpulentis* in honor of the rotund Smith, and Smith reciprocated the honor by naming one *dyaria!*

Besides *The Mosquitoes of New Jersey,* Smith published three great catalogues of the insects found in that state, an *Explanation of Terms Used in Entomology,* a systematic work on the Noctuidae, two lists—*Lepidoptera of Boreal America* and *A Contribution toward a Knowledge of the Mouth Parts of the Diptera*—a popular book, *Our Insect Friends and Enemies,* and *Economic Entomology,* the text for his course at Rutgers.

He was in constant demand as a speaker by farm associations, scientific institutions, and public schools.

This man of "indomitable energy," who worked even when confined in bed, died of Bright's disease after several years' illness on March 12, 1912. He was survived by a wife and two grown children.

RICHARD HARPER STRETCH (1837–1926)

Stretch was born on November 27, 1837, at Nantwich, England. His father died when he was eight years old. He attended school in England,

Figure 94. R. H. Stretch

worked in a draper's shop and as a cashier and bookkeeper for a manufacturer. In 1861 he decided to visit some relatives in Illinois. On arrival in New York, instead of heading straight for Illinois, he took a steamer to Panama and gradually worked north to Illinois, collecting insects along the way. A short time after he returned to England, he moved to the United States.

In 1863 he joined an emigrant party on its way to California and literally kept his scalp because of a fortunate incident. According to Coolidge and Newcomb (1920): "Mr. Stretch's party consisted of but five, a very insufficient number in view of the numerous hostile Indian bands. But while other and larger emigrant groups were attacked, and in some cases, wiped out, this small party passed unmolested. At the outset of the journey one of Mr. Stretch's collecting bottles, containing a various assortment of insects in alcohol, had rolled from one of the wagons unobserved. A friendly Indian had found and returned it. Mr. Stretch displayed his entomological wares to this red skin (sic), who viewed them with intense interest, but not in a scientific way. And all along the line of the emigrant trail word was passed among the Indians that a Big Medicine Chief was coming, and apparently orders were given that no harm should befall him."

A keen, perceptive man, with an eye possibly sharpened by his entomological pursuits, he soon became expert in observing minute differences in mining ore. While in Virginia City, to his astonishment, he was elected State Mineralogist of Nevada. From this time on he became a mining authority and expert map-maker of mines. In 1868 he returned to Virginia City to make maps of the famous Comstock lode for the U.S. Geological Survey, which greatly enhanced his reputation.

On his travels throughout the West (he even crossed Death Valley) and Mexico, he constantly collected insects, especially moths and butterflies, and undoubtedly made some available to such friends and fellow lepidopterists as A. R. Grote, H. H. Behr, Henry Edwards, S. H. Scudder, and William H. Edwards (no relation to Henry). Henry Edwards was one of his closest friends.

Stretch is especially well known in entomological circles for his monograph, *Illustrations of the Zygaenidae and Bombycidae of North America,* illustrated with many fine drawings by him. He was one of the first to realize and proclaim the potential hazards to Californians of the introduction of the cottony cushion scale.

In 1885 Stretch gave his entomological library to the Mechanics Institute in San Francisco and his collection of Lepidoptera to the University of California. Upon Henry Edward's death in 1891, Stretch lost his interest in insects and spent the rest of his life as an engineer in the West and Alaska. He published a widely-used book on mining in 1899.

After living many years in Seattle, Stretch died at the home of his daughter on March 22, 1926.

HENRY SKINNER (1861–1926)

Well known in the Philadelphia Academy of Sciences as a lepidopterist and one of the founders of *Entomological News,* Skinner won equal fame for editing this magazine from 1890 to 1910, when it was one of the liveliest and most interesting journals in the annals of entomology. "His humor and his incisiveness, even caustic at times, gave it a characteristic tone which was appreciated far and wide." He was also one of the founders of what was then called the Entomological Society of America, of which he became president in 1908.

Born in Philadelphia on March 27, 1861, Skinner attended public schools there and graduated from Rugby Academy. In 1881 he received his B.S. and in 1884, his M.D., from the University of Pennsylvania. He specialized in gynecology. For a time he was an assistant in gynecology to a Dr. William Goodall. Then, in 1900, he abandoned medicine to study the Lepidoptera and to devote himself to his other interests in the Academy, where, from 1884 until his death, he served as a curator or custodian of various insect collections.

Skinner first collected the Lepidoptera in Philadelphia and later netted specimens from many localities throughout the United States and

Figure 95. Henry Skinner

Canada. He was a specialist in the Hesperiidae, or skipper butterflies, and published many papers on them. All in all he described over one hundred species and subspecies in the western hemisphere.

Skinner's policy as editor of the *Entomological News* was to produce a periodical that had broad appeal, both to the professional and to the amateur. In this he succeeded, largely through a sense of humor that provoked a chuckle in the reader, thereby bridging the gap between the learned and the learners. For example, after receiving some specimens in poor condition, he wrote: "We may also say in passing that we have coined a new word, 'Sloppydoptera,' which has reference to specimens captured with a baseball bat or temporarily loaned to the new baby as playthings before being sent out." He also spiced the pages of his journal with poems and stories to overcome the dullness of taxonomic descriptions.

Skinner lived with his family in a beautiful home in Narberth, Pennsylvania, until he died on May 29, 1926, in Polyclinic Hospital, Philadelphia, after a brief illness.

Calvert (1926) and Holland (1926–1927) wrote obituaries of him.

HARRISON GRAY DYAR (1866–1929)

This sketch could as readily appear in the chapter on "Notable Dipterists," for L. O. Howard has said that "Dyar was probably the best posted man on the classification of mosquitoes of his time, with the

Figure 96. H. G. Dyar

possible exception of F. W. Edwards, of the British Museum of Natural History."

Forbes and Aldrich (1929) wrote of him that "The world has produced many entomologists with a good eye for species—a number who have been able to comprehend the major groups of insects—several who have carefully and intensively studied the biology and early stages of one or another group. There have been hardly any who could do all these three things, and see a group of insects as a whole," as could Dyar.

Dyar was born in New York City on February 14, 1866. His father made a fortune from inventions relating to dyes and left Harrison financially independent. Harrison attended the Roxbury Latin School in Massachusetts, received a B.S. in chemistry from the Massachusetts Institute of Technology in 1889, and, from Columbia University, an A.M. in 1894 and a Ph.D. in 1895.

In 1894 he published an important paper entitled "A Classification of Lepidopterous Larvae," in *Annals of the New York Academy of Science*. His Ph.D. thesis, "On Certain Bacteria from the Area of New York City," was concerned with specificity and variability in bacteria. In 1897, because of his work on the larvae of the Lepidoptera, L. O. Howard invited him to become custodian of the Lepidoptera in the U.S. National Museum, a position he held until his death on January 21, 1929. In 1902 the National Museum published his list of North American Lepidoptera, and thereafter he produced a constant flow of papers on the Lepidoptera of North America.

According to Howard (1929), Dyar was also interested in the larvae of mosquitoes. When the Carnegie Institution made a grant to Howard in 1903 for a monograph on the mosquitoes of North and Central America, Dyar and F. Knab collaborated with him. Dyar and Knab were responsible for the taxonomic part of the four-volume work, *The Mosquitoes of North and Central America.* Many important papers on mosquitoes appeared under Dyar's authorship. After Knab's death Dyar began to work on adult mosquitoes as well as the larvae and became an authority on the Culicidae. Through his studies of the male genitalia he helped to stabilize the classification of these insects. He also studied flies such as the Simuliidae, Psychodidae, and Chaoboridae.

Dyar's Law, which he propounded, arose from his study of the larvae of caterpillars. This law states that the "width of the head capsule of the larva follows a regular geometrical progression in the successive instars." This enables a student to determine the various stages by measuring the head.

Dyar owned and edited the entomological journal *Insecutor Inscitiae Menstruus,* which ran to fourteen volumes from 1913 to 1927.

The name Dyar figures very prominently in entomological journals as a source of constant controversy, largely because of Dyar's inclination to rip people apart in print. He was an extremely critical man, and his victims replied in kind. His verbal encounters added a good deal of spice to the entomological journals of that day, something missing from today's sedate scientific periodicals. Among Dyar's antagonists were D. W. Coquillet, J. B. Smith, and H. Skinner.

Dyar's review (1905) of Henry Skinner's *A Synonymic Catalogue of North American Rhopalocera,* Supplement No. 1, is typical of the remarks that ruffled the feathers of a number of authors.

"Dr. Skinner has given us a very useful little supplement to his catalogue of butterflies. It is somewhat bristling with typographical errors and blunders, but we are used to that sort of thing from Philadelphia. There are some comments, indicating a new synonymy and one new name (varietal) is proposed. The number of species added to our list is not large. The generic names have not been brought up to date, the author expressly stating he is 'not interested' in the subject, which he is pleased to designate as 'generic fantasies.' This is, we think a fault. It is easy to stigmatize what one will not take the trouble to understand; but a good opportunity of correcting the antiquated nomenclature of the North American butterflies has here been lost."

An entomologist who knew Dyar wrote to me that "Dyar was most able, even brilliant, and a terrific producer in the taxonomy of both the Lepidoptera and the mosquitoes; but I remember him as a cantankerous person (I don't recall ever seeing him smile), and with one exception his associates at the Museum had little to do with him. The exception, of course, was Knab, only a year or so younger than Dyar, and also a brilliant worker. (Unfortunately he died as a comparatively young man—in his early fifties, I think.) They seemed to understand each other well; and, as you know, they collaborated in the development of some excellent work in the taxonomy of mosquitoes. They had numerous publications under joint authorship."

During the thirty-one years he was in Washington, Dyar received no pay for his work, although for a few years he was on the rolls of the Bureau of Entomology—he was not even recompensed for his many field trips throughout the United States and various foreign lands.

AUGUST BUSCK (1870–1944)

Busck described over six hundred American species of microlepidoptera, especially in the groups formerly known as the Tineina and the Tortricina. Heinrich and Loftin (1944) note: "If he had so chosen, he could have added a thousand more names to our lists, but he was

Figure 97. August Busck

loath to describe a new species unless there was some real scientific or economic need for a designation. He was no species monger, nor was he fixed in any convenient rut of taxonomic procedure. Thoroughly abreast of every new departure in technique and taxonomy, he was ready to try any one that promised to advance a more natural classification however much it might disturb traditional practices or concepts, and was one of the first to apply genitalic characters consistently to the classification of the various categories in the Microlepidoptera."

Busck was born in Randers, Denmark, on February 18, 1870, and educated at Ordrup College and Royal University, Copenhagen, receiving an M.A. and a Ph.B. in 1893. He was an instructor in zoology and botany at Ordrup College and Copenhagen High School from 1889 to 1893.

In 1893 he visited the World's Columbian Exposition in Chicago and then entered the wholesale flower business in Charleston, West Virginia, where he became a citizen of the United States.

Busck was appointed in 1896 as an assistant to Theodore Pergande in the Department of Agriculture's Division of Entomology. At some point he also became associated with the U.S. National Museum, where he served as a specialist on microlepidoptera until his retirement in 1940 at the age of seventy; through his efforts, the National Museum became noted for its collection of microlepidoptera.

In the early 1900's he did a great deal of traveling through Cuba, the West Indies, and Panama to help in a survey of mosquitoes for several scientific organizations. Later he studied the pink bollworm as a cotton pest in Mexico, British Guiana, and the West Indies. In 1908 he visited England and helped his friend Lord Walsingham prepare the volume on microlepidoptera in the *Biologia Centrali-Americana.*

Busck, or "A. B." to his intimates, used to go out of his way to assist friends. "His convivial nature made him many friends wherever he travelled, and his robust enthusiasm for entomology converted many younger people into amateur collectors." (Heinrich and Loftin, 1944)

Upon his retirement, Busck went to Hawaii on a Yale fellowship and helped arrange and identify the microlepidoptera for the Bishop Museum there. When he died on March 7, 1944, he left a wife, two sons and a daughter, and a legacy of over one hundred and fifty papers to his fellow entomologists.

CARL HEINRICH (1880–1955)

Heinrich was born in Newark, New York, on April 7, 1880; his father was a talented musician, mathematician, and lawyer. When Carl was six months old, his family moved to Omaha, Nebraska. Wade and Capps (1955) report that many of the famous musicians of that period, while passing through Omaha on tour, would stop over for visits with the family. The musicians, accompanied by his father, would often sing and play most of the night, and Carl developed a considerable interest in music.

He won a scholarship to the University of Chicago, where he majored in Greek and Greek drama from 1898 to 1901.

Figure 98. Carl Heinrich

In 1902 Heinrich moved to Washington and worked for business organizations for a number of years. In 1908 he went to New York to study music with the composer Edward McDowell, but McDowell died before Carl could begin. He joined the United States Department of Agriculture in 1913, where he remained until his retirement at the age of sixty-nine. He served briefly in the field of applied entomology; later he worked in the U.S. National Museum on the classification of Lepidoptera. According to Wade and Capps, he helped to "establish characters by which many of the economic pests such as the European corn borer and the pink bollworm could be identified with certainty." He wrote numerous technical papers on Lepidoptera, some of which were of great economic importance.

Heinrich was a gregarious, congenial, kindly person, with a wide circle of friends, many made on his numerous field trips out of the country. Among his colleagues during his years of museum work were J. M. Aldrich, Nathan Banks, H. G. Barber, August Busck, A. N. Caudell, J. C. Crawford, R. A. Cushman, A. H. Clark, Harrison Dyar, H. E. Ewing, W. S. Fisher, A. B. Gahan, A. D. Hopkins, L. O. Howard, F. Knab, Theodore Pergande, E. A. Schwarz, and H. L. Viereck.

Heinrich published a book of poetry and various newspaper articles and even wrote fiction. After his retirement he continued to work on the taxonomy of moths. His monograph *American Moths of the Subfamily Phycitinae* was published posthumously in 1956.

He died at the age of seventy-five in the Garfield Memorial Hospital, Washington, on May 31, 1955, of a heart attack.

WILLIAM SCHAUS (1858–1942)

Schaus put forty years of work into making the collection of tropical Lepidoptera in the National Collection in Washington one of the finest in the world. Heinrich and Chapin (1942) report that he contributed generously to other institutions, notably the British Museum of Natural History, the Carnegie Museum in Pittsburgh, and the American Museum of Natural History; but that "the bulk of his collection and his valuable library were given to the U.S. National Museum . . . He described over five thousand new species, mostly from tropical America. With few exceptions the types of these are deposited in the National Collection."

William Schaus, Jr., was born in New York City on January 11, 1858. His father, who was born in Germany, had become an American citizen and occupied himself as an art collector and dealer. William was educated at Exeter Academy and in France and Germany. Although his training was primarily in art, music, and languages, he became interested in the Lepidoptera through the influence of Henry Edwards and despite parental objections devoted himself to the study of these insects.

Schaus made frequent trips through Mexico, Costa Rica, Guatemala, Panama, Cuba, Jamaica, Dominica, St. Kitts, the Guianas, Colombia, and Brazil. He lived in England from 1901 to 1905 and later made several trips to the Continent. All in all, he collected over two hundred

Figure 99. William Schaus

thousand Lepidoptera, and he published 121 papers, most of them on moths.

He joined the staff of the U.S. Bureau of Entomology in 1919, where he remained until his retirement in 1938, first as a specialist and later as an entomologist. In 1921 he was appointed honorary assistant curator of insects in the U.S. National Museum. In 1925 he received an honorary doctorate from the University of Pittsburgh.

Described by Heinrich and Chapin as "an accomplished linguist, a lover of art and music, a charming host and the most generous of friends," William Schaus died on June 20, 1942.

WILLIAM BARNES (1860–1930)

The parents of William Barnes were early settlers in Decatur and had some of the finest farm land in the area. The father was a physician. William was born in Decatur, Illinois, on September 30, 1860. He graduated from Decatur High School in 1877, from Harvard in 1883, and from Harvard Medical School in 1886. He did graduate work abroad in Germany and Paris and was proficient in both French and German. In time he became one of the foremost surgeons in Illinois and was one of the founders and backers of the Decatur and Macon County Hospital. He married in 1891 and had a son and daughter.

For many years Barnes collected Lepidoptera in the Rockies during the summer. Eventually his collection of North American butterflies and

Figure 100. William Barnes

moths became one of the largest ever privately owned. He employed successively J. H. McDunnough, A. W. Lindsey, and F. H. Benjamin as assistants.

Engelhardt (1930) writes: "Only those who have been privileged to visit Dr. Barnes at Decatur, Ill., can have an adequate idea of the size, composition and scientific importance of his collection. Housed in a separate building of fire-proof construction, the main collection is placed in oak cabinets of some 1,200 drawers, while reserve and exchange material is contained in 2,000 or more so-called 'Schmidt' boxes arranged on shelves on the wall. Type specimens, including Holotypes, Paratypes, Homotypes, etc., are represented to the number of nearly 7,000. Five hundred thousand would seem a conservative estimate as to the total number of specimens in the Barnes collection. To assemble such a collection has been the work of a life time of indefatigible labor and unstinted expense. Himself an enthusiastic worker in the field, Barnes also supported most liberally experienced collectors and dealers in Lepidoptera in all parts of the country and usually on terms which gave him the first choice on the season's captures or receipts. Nearly every species in the Barnes collection has been finally determined by comparison with the original type wherever located, over here or abroad."

Barnes also purchased private collections, including the North American species in the famous Oberthur collection, except for certain families. He made several trips to Europe to study types, where he became acquainted with European collectors.

Barnes apparently left it to his curators for the most part to undertake scientific studies of his collection. According to Ferguson (1962), McDunnough wrote, 'Barnes, in my time, never did any lengthy studies on the Lepidoptera; he used to drop in [in] the morning and usually for a short time in the evening to discuss work, etc., and when time was available, spent it studying Phalaenids, particularly members of the genus *Euxoa* in which he was greatly interested." According to Dr. A. W. Lindsey, "Dr. Barnes' work in entomology was chiefly the lavish support of collectors and the provision of a superb opportunity for the research done by his curators. He described some species, but did no actual taxonomic research as far as I know." (Ferguson, 1962)

Gunder (1929) says that Barnes "is a stanch Republican and does not care much about the blue-law church people. He is broad-shouldered and athletic despite his years. Collected since a child. He knew Oliver Wendell Holmes and Louis Agassiz while a student at Harvard." He told Gunder he had never had any narrow escapes from Indians but that the Apaches "got two of my collectors in an early day in Arizona." He added that he had personally known W. H. Edwards, Henry Edwards, C. F.

McGlashan, E. L. Graef, G. D. Hulst, A. S. Packard, S. H. Scudder, F. H. Strecker and most of the other older collectors. Strecker, Barnes said, "was a very peculiar man when visiting you and it was better to stay close by his side as a protection to valuable specimens! He had a most peculiar habit of crawling in between the sheets with his boots and clothes on."

In 1920 Barnes made arrangements for the sale of his collection, the proceeds to go to the Decatur and Mason Hospital. A bill was introduced in Congress in 1922 for the sale of the collection to the Federal Government for $310,000—$300,000 for the collection and $10,000 for moving it from Decatur. The U.S. National Museum acquired the collection.

Dr. Barnes died May 1, 1930, in Decatur, after a protracted illness.

JAMES HALLIDAY McDUNNOUGH (1877–1962)

Ferguson (1962) states that McDunnough "was one of the most prominent figures in systematic entomology for half a century and for much of this period was perhaps the leading authority on this continent in two diverse orders of insects—the Lepidoptera and Ephemerida."

McDunnough was born on May 10, 1877, in Toronto, Ontario, and received his early education at private schools and at Jarvis Street Collegiate. In 1897 he went to Berlin to continue his education and to prepare for a career in music. There he studied under the celebrated

Figure 101. J. H. McDunnough

violinist, Josef Joachim. McDunnough played for one season for the Glasgow Orchestra.

In 1904, abandoning professional music though not music itself, he returned to Berlin to study zoology at the Kaiser Wilhelm Institute and at the same time enrolled at Queen's University, Kingston, Ontario, for an extramural course. He received his Ph.D. from the Berlin institution and an M.A. from Queen's University in 1909. In this same year he married Margaret Bertels of Berlin.

His first job, after getting his doctorate, was with the Marine Biological Laboratory at Woods Hole, Massachusetts. He had collected Lepidoptera in his youth and was a member of a nature group which included Arthur Gibson, E. M. Walker, and R. S. Lillie. His Ph.D. thesis was on the anatomy and histology of the lacewing, *Chrysopa perla* L. Then in 1910 he became a curator in William Barnes' private museum in Decatur, Illinois. He worked with Barnes until 1919, and sixty-seven papers were published under their joint authorship. Although Barnes provided the collection, facilities, and salary for McDunnough, the latter was almost entirely responsible for the contents of the papers. Many of their joint publications are in Barnes's journal, *Contributions to the Natural History of the Lepidoptera of North America.* In 1917 McDunnough and Barnes published their *Check List of Lepidoptera of Boreal America,* and their beautifully illustrated paper, "Illustrations of the North American Species of the Genus *Catocala,*" appeared in 1918. While at Decatur, McDunnough continued playing the violin and the viola and was a member of a local string quartet.

In 1918 he arranged the collection of macrolepidoptera in the Canadian National Collection in Ottawa and in 1919 left Barnes to become chief of the newly created Division of Systematic Entomology at Ottawa, where he served for twenty-eight years. There he organized insects in all orders and helped accumulate one of the finest insect collections in North America. In 1921 he began to publish papers on insects other than the Lepidoptera. His association with the National Collection in Canada resulted in the publication of one hundred and fifty-three papers on the Lepidoptera, thirty-eight on the Ephemeroptera, five on the Odonata, two on Diptera, and one on the Hemiptera. Some of his most important studies were on the geometrid, agrotid, and tortricid moths, and the mayflies, of which he described over two hundred species. His *Check List of the Lepidoptera of Canada and the United States of America* appeared in 1938 and 1939.

When McDunnough retired in 1946, it was as chief of the Canadian National Collection. He was then appointed a Research Associate in the Department of Insects and Spiders at the American Museum of Natural History in New York City. He lived in New York for three years and produced three important monographs, including *The Saturniidae of North America.* Although he enjoyed his work in the Museum, he apparently did not like living in New York. After three years of it, he moved to Halifax, Nova Scotia, where he spent the remainder of his life. There, as a research associate for both museums, he worked in the Nova Scotia Museum of Science, while the American Museum published his papers.

Ferguson (1962) describes him thus: "The attitude of his earlier years must have been somewhat different, but during the eighteen years that I knew McDunnough he showed little patience with new ideas and tended at times to be critical of anything he did not understand. He shunned meetings and social gatherings, and perhaps for this reason had few scientific friends apart from immediate associates. He enjoyed association with others but seemed unwilling to admit it. His steadfast independence was a most pronounced characteristic, yet whenever he lived or worked too long in isolation his disposition deteriorated. Those who made the effort to gain his acquaintance found McDunnough a meticulous person of superior manners, humour and generosity, combined with a manner of unusual frankness which, on occasion, probably led to misunderstandings of his personality."

McDunnough had little interest in other fields of natural science than his own. He had "a good eye for species" and proposed about fifteen hundred new names with apparently little synonymy. According to Ferguson (1962), "He usually worked rapidly on just one problem at a time, and it seemed that he could not rest until it was finished." He was austere, orderly, an agnostic, and not mechanically inclined. Ferguson adds that "Apart from music, he enjoyed golf, bridge, crossword puzzles, and was an avid reader of mystery and crime fiction." He was a Fellow of the Entomological Society of Canada and received many honors in his lifetime.

McDunnough died in Halifax on February 23, 1962, at the age of eighty-four.

FOSTER HENDRICKSON BENJAMIN (1895–1936)

When Benjamin was a public school student of the age of twelve he became acquainted with George P. Engelhardt and Jacob Doll of the Brooklyn Museum, who took an interest in him and introduced him to the study of insects, especially butterflies and moths. With their aid and encouragement and that of two other entomologists, John Grossbeck and John B. Smith, he eventually became an expert lepidopterist.

Born in Brooklyn on September 17, 1895, Benjamin attended high

Figure 102. F. H. Benjamin

school in New York City and, later, Cornell University, where he received a B.S. in 1920 and an M.S. in 1921. His master's thesis was "A Revision of the Noctuid Moths of the Genus *Lampra*."

He served in the Navy in World War I in the lighter-than-air aviation unit and then worked for a short time as an entomologist with the Mississippi State Plant Board. From 1922 through 1927 he was curator of the Barnes Collection in Decatur, Illinois. It was there that he established himself as an authority on the Lepidoptera, writing over a hundred papers on this order. He then joined the Bureau of Entomology of the U.S.D.A. and worked on a variety of problems, including the Mexican fruit fly in Texas and the Mediterranean fruit fly in Florida. In 1931 he transferred to the taxonomic staff of the U.S. National Museum, where he specialized in noctuids and fruit flies.

Benjamin died on January 24, 1936, in Washington, D.C., at the early age of forty, thus cutting short a very promising future. He left behind a wife and two young children. Muesebeck and Heinrich (1936) wrote his obituary, from which this sketch is drawn.

GEORGE PAUL ENGELHARDT (1871–1942)

Engelhardt was for many years Curator of Natural History at the Brooklyn Museum and an authority on the clear-wing moths.

He was born in Hanover, Germany, on November 23, 1871, and was educated in a gymnasium in Baden-Baden. He moved to the United States in 1889, where he became a citizen. After working as a wrapping-

Figure 103. G. P. Engelhardt

paper salesman in New Orleans, he entered the entomology department of a New York biological supply house in 1900. Later he helped develop the natural history collection in the Children's Museum in Brooklyn. Around 1912 he became Curator of Natural History in the Brooklyn Museum, a position he held until 1930, when he retired as Curator Emeritus.

Engelhardt journeyed to all parts of the country on collecting trips, often accompanied by his entomological friends. A kindly man, he liked to encourage young naturalists. Teale (1942) says of him: "In collecting and studying *Aegeriidae,* Mr. Engelhardt was far more than a classifier of dead insects. He well recognized the truth of the saying of Benjamin Franklin's Poor Richard: 'What signifies knowing the names if you know not the nature of things?"

Engelhardt was an active member of the Brooklyn Entomological Society; he was business manager for its publications and was treasurer for twenty-five years. His monograph, *The North American Clear-Wing Moths of the Family Aegeriidae,* was published posthumously, after being completed by his friend August Busck. Engelhardt died on May 24, 1942.

AUSTIN HOBART CLARK (1880–1954)

Clark was a world authority on echinoderms. Although his study of butterflies was an avocation, he was highly regarded as a lepidopterist.

He was born on December 17, 1880, at Wellesley, Massachusetts. He

Figure 104. A. H. Clark

graduated from Harvard University in 1903 and in 1901 began the first of several expeditions to Margarita Island, Venezuela. In 1906 and 1907 he was the acting chief of the scientific staff of the *S.S. Albatross,* a scientific vessel belonging to the U.S. Bureau of Fisheries. Clark joined the Smithsonian Institution in 1908, and in 1920 became curator of echinoderms, until he retired in 1950.

Clark wrote on ornithology, entomology, and echinoderms as well as other aspects of marine biology, in English, French, Italian, German, and Russian. He was a member of the Cosmos Club in Washington where, besides being an excellent bridge player, he was famous for his omniscience. When one member mentioned the need for an up-to-date encyclopedia, another said, "Why bother, Austin Clark is a member of the club."

In 1932 Clark wrote *The Butterflies of the District of Columbia and Vicinity* as Bulletin 157 of the U.S. National Museum, and in 1951 in cooperation with his second wife, Leila Gay Forbes, published *The Butterflies of Virginia.* Among Clark's books are his *Monograph of the Existing Crinoids, Animals of Land and Sea, Nature Narratives, The New Evolution,* and *Animals Alive.* He published 650 papers.

He was president of the Entomological Society of Washington and a director of the Press Service for the American Association for the Advancement of Science. "In this position he was eminently successful in stimulating and encouraging accurate and understandable writing to make science popular and known to the layman."

He died on October 28, 1954. Snyder, Shepard, and Clarke (1955) wrote his obituary, from which this sketch is drawn.

Figure 105. R. C. Williams, Jr.

ROSWELL CARTER WILLIAMS, JR. (1869–1946)

Williams was born in Brooklyn, New York, on August 21, 1869, and died in Philadelphia on March 7, 1946. Bell (1946) wrote his obituary.

He received a B.S. from Adelphi College in Brooklyn and an M.E. from Cornell in 1892. For a time he was associated with Charles P. Steinmetz and later with an electrical engineering and contracting firm in Philadelphia. During World War I he was a captain in the Ordnance Department.

Williams was a student of natural history, especially the Lepidoptera. For many years he was active in the Academy of Natural Sciences of Philadelphia. He was president of the American Entomological Society from 1926 to 1935. His chief interest was the Hesperiidae; he published numerous papers and described many new tropical and North American species in this family. He collaborated with Dr. Henry Skinner on a study of the male genitalia of North American Hesperiidae, and with E. L. Bell and A. W. Lindsey on *The Hesperioidea of North America.* At various times he collected in tropical America, Ecuador, Argentina, and other countries.

WILLIAM PHILLIPS COMSTOCK (1880–1956)

Anyone who has seen the *Butterflies of the American Tropics,* a monograph on the genus *Anaea* of the Nymphalidae, and observed the extremely beautiful illustrations, can understand Comstock's enthusiasm

Figure 106. W. P. Comstock

for these butterflies. This work, in whose preparation Frank Johnson assisted Comstock, appeared posthumously.

Comstock was born in New York City on March 1, 1880, and received his education there, completing it with a B.A. in 1903 from Columbia University. That same year he entered his father's publishing business, the William T. Comstock Company. In 1910 he became a consulting engineer in construction work.

When construction and engineering were at a low ebb during the depression in 1932, he began to devote more and more time to his hobby, butterflies. As a youth he had collected Rhopalocera in Van Cortlandt Park and in nearby fields. For a time he was associated with the Newark Museum as a research assistant. In 1944 he became a research associate with the American Museum of Natural History. He wrote numerous papers on the Lepidoptera; his main interest was the Lycaenids.

Dos Passos (1956) says that when he was at the American Museum of Natural History "he usually wore a blue smock and invariably a blue bow tie and a striped shirt. Blue was his favorite color. With hair that was grey, he looked more like an artist than an entomologist."

Comstock died in Neptune, New Jersey, on September 23, 1956.

ARTHUR WARD LINDSEY (1894–1963)

Lindsey was born January 11, 1894, in Council Bluffs, Iowa, and attended high school and Morningside College in Sioux City, obtaining

Figure 107. A. W. Lindsey

his A.B. in 1916 and an honorary Sc.D. in 1946. Having collected butterflies as a boy, his first publication, "The Butterflies of Woodbury County, appeared in 1914. In 1919 he received his doctorate from the State University of Iowa, his thesis being *The Hesperioidea of America, North of Mexico*. This monograph was revised in 1931 with the collaboration of R. C. Williams, Jr., and E. L. Bell and entitled *Hesperiidae of North America.*

While working on his doctorate he visited Dr. William Barnes at Decatur, Illinois, and studied his famous collection. There he became acquainted with J. H. McDunnough, who had been curator of this private collection since 1910. When, in 1919, McDunnough moved to Ottawa, Lindsey succeeded him as curator. Lindsey was with Barnes until 1921, and as a result of this association a revision of the moth family Pterophoridae appeared in that year under their joint authorship, although the work was really Lindsey's.

In 1921 Lindsey taught at Morningside College and in 1922 became professor and head of the zoology department at Denison University in Granville, Ohio, a position he held until he retired from teaching in 1960. During the summers of 1916 through 1918 he held various positions with the U.S.D.A. and later was associated with several biological field stations throughout the United States.

During his first twenty years at Denison he published about a dozen papers on the Hesperioidea; later additional duties and his interest in experimental evolution left little time for butterfly studies.

Voss (1963) writes that his students in evolution, genetics, embryology, and other subjects profited from "his remarkable clarity and precision of expression," and that he "inspired many with excitement of research." His premedical students did well in medical colleges and reflected the competence of their teacher. As Voss remarks, "to many, he appeared a bit remote and aloof; yet in fact he was ever approachable and willing to counsel with students."

Lindsey was the author of five textbooks on general zoology, evolution, and genetics, and from 1945 through 1948 he was managing editor of the *Annals of the Entomological Society of America*.

His collection of 6,000 specimens with twenty-eight types was obtained by the Carnegie Museum through the efforts of W. J. Holland. After retiring, Lindsey collaborated with R. M. Fox, H. K. Clench, and Lee D. Miller on a manuscript entitled *Butterflies of Liberia*.

On March 8, 1963, he died of a heart attack in Lancaster, Ohio, and his former students recall with "affection and appreciation" this lover of Bach, butterflies, and daffodils. Voss (1963) and Fox (1963) wrote his obituaries.

ERNEST LAYTON BELL (1876–1964)

Bell was born on November 21, 1876, in Flushing, Long Island. Ruckes and Dos Passos (1965) write that he "lived his entire life in the town of his birth and was greatly beloved by his neighbors, particularly the children whom he befriended and who usually paid him daily visits in his home." He worked in a local bank and eventually became head of one of its departments.

Figure 108. E. L. Bell

Bell collected Indian artifacts in his locality, on sites now obliterated by row after row of monotonous apartment houses. He was also interested in snakes, stamps, and coins. In 1919 he began to study butterflies, especially the Hesperioidea, and in time became an authority on this superfamily, describing a number of genera and over two-hundred species and subspecies. He prepared joint papers with such authorities as A. W. Lindsey, R. C. Williams, Jr., William P. Comstock, K. J. Hayward, and W. H. Evans. With Lindsey and Williams, he was co-author of *The Hesperioidea of North America,* which was published by Denison University. In 1934 he was appointed a research associate at the American Museum of Natural History. He collected throughout the United States, Central America, and the Caribbean Islands, especially Jamaica.

Bell died in Flushing on December 12, 1964, at the age of 89.

11 Notable Hymenopterists

Obituary of an Entomologist *

O gentle reader drop a tear
For one beneath this stone
In life he named 7,000 bugs
To science, all unknown

But now, alack! he is condemned
In a place I dare not name
With his own books, through endless years,
To identify the same.

A. Victim
Entomological News, 13(9):297, 1902

Ants, bees, and wasps have intrigued man from the earliest days of recorded history, and the writings of the ancient Chinese, Egyptians, and Hebrews contain numerous references to them. In recent times there has been a revival of interest in these insects. Men of great competence have studied their taxonomy and behavior and have given us solid facts instead of the hearsay so common in earlier accounts.

EZRA TOWNSEND CRESSON (1838–1926)

American entomologists remember Cresson as one of the earliest hymenopterists in this country and, as a founder of the American Entomological Society. P. P. Calvert (1928), who knew Cresson well, is the source of much that follows.

According to Calvert, Ezra's father was Warder Cresson, "a deeply religious man, successively a member of various sects, U.S. consul at Jerusalem, Palestine, in 1844 and finally a convert to Judaism." His mother was nee Elizabeth Townsend. Ezra was born on June 18, 1838, in Byberry, Bucks County, Pennsylvania, the seventh of eight children.

* T. D. A. Cockerell wrote, "You think the ponderous idiocy in the middle of the page 297 is funny." Apparently, Dr. Henry Skinner, the editor, thought it was.

Figure 109. E. T. Cresson

Ezra's father lived alone in Palestine from 1844 to 1848 and from 1852 to 1860, the rest of the family remaining in the United States.

Ezra attended public schools in Philadelphia through the eighth grade, when he had to quit to help support the family. The only information available as to his youthful activities or interests is that he participated in a balloon ascension from Lemon Hill, Philadelphia, in 1857 under the direction of a certain M. Goddard. And that Cresson's future father-in-law, James Ridings, a lover of nature and a collector of butterflies and beetles, kindled Ezra's interest in insects.

According to Calvert, the Entomological Society of Pennsylvania was organized in 1842 with F. E. Melsheimer, S. S. Haldeman, D. Ziegler and J. G. Morris as members. They met in their various homes in the Susquehanna Valley several times a year. In February 1859, Cresson, James Ridings, and George Newman gathered at Cresson's house to found an entomological society more durable than the one begun in 1842. Cresson, Calvert says, "was the youngest of the three founders, at this time not yet twenty-one years of age. He was between five feet five inches and five feet six inches in height, slender, with light brown hair and blue eyes." The society they founded in 1859 was the Entomological Society of Philadelphia, whose name was changed in 1867 to the American Entomological Society. This is the oldest existing entomological society in the United States.

At the time of the founding of the new entomological society, Cresson was working as a clerk in the treasurer's office of the Pennsylvania Railroad Company. He was secretary to the Society, a position he resigned on May 8, 1859, in order to go with members of his family to Texas, where they hoped to start a ranch in New Braunfels. This project failed and Ezra was back in Philadelphia on October 24, 1859, as indicated in the records of the Society.

The first paper to appear in the *Proceedings of the Entomological Society of Philadelphia,* Vol. 1, No. 1, 1863, was Cresson's "Catalogue of the Cicindelidae of North America." This was his first and last paper on beetles, for from that time on he worked on the Hymenoptera exclusively. It was appropriate that Cresson's paper should appear first in the *Proceedings* because he actually printed the magazine. Thus Baron Osten Sacken quotes a letter written to him by Cresson on September 4, 1961: "We do our own printing, as you already know; I am the compositor and also assist in the press work, and although I have had little or no experience in setting type (I have set the type for *all* the pages of the *Proceedings* thus far), yet be assured that I will do my best to have your paper got up in as neat and scientific a style as possible." (Calvert, 1926)

At a later date, on the death of Charles A. Blake, Cresson wrote, "I remember the many nights Mr. Blake toiled with me in the publication of the Proceedings and Transactions, and he was ever ready and willing to help me when no others volunteered; we worked together side by side at the case, and while I rolled on the ink, he pulled the press—being the stronger."

According to Weiss (1936), since there were already existing catalogues on the Coleoptera by F. V. Melsheimer, the Diptera by C. R. Osten Sacken, and the Lepidoptera by J. G. Morris, Cresson "believed that students needed a catalogue of the described species of North American Hymenoptera and he proposed to publish a series of catalogues of American species, starting with the sawflies and continuing with other families as time permitted, or until the appearance of M. [H. L. F.] Saussure's expected work on the Hymenoptera of North America." Thus his second paper in the *Proceedings* was a "Catalogue of the Described Species of Tenthredinidae and Uroceridae Inhabiting North America."

During the early years of his association with the Entomological Society of Philadelphia, Cresson was employed as a secretary to Dr. Thomas B. Wilson, a great patron of both the Society and the Academy. Cresson escaped the Civil War because according to Calvert (1928), "When the draft came, he was prevented from entering the army by his employer, Dr. Wilson, who paid for a substitute, and whose 'sympathies during the war were distinctly Southern.'" Dr. Wilson's death in 1865 was a severe blow to the fortunes of both Cresson and the Society.

The next year Cresson became curator of the Society. However, as it could only raise $752.50 of his $1,250 salary, he had to take a job with the Franklin Fire Insurance Company of Philadelphia, of which he became General Superintendent in 1874, a job he held until ill health forced him to resign in 1910.

Until 1887, Cresson actively pursued his studies of the Hymenoptera, though, as he wrote G. W. Belfrage of Texas, a collector, on January 11, 1872, "The labor of determining and describing species is great, and when I think that I must do all in the evenings, having to work all through the day-time I am sometimes almost persuaded to give up Entomology after I finish my memoir on your Hymenoptera Texana. Should you have more undetermined species, please send them as soon as possible, so that I may include them all in my paper & make one job of it. If I could work at insects in the day-time I could accomplish a great deal in a month.—With gas-light one works under difficulties, as some colors appear to change more or less, & besides this it is very hard on the eyes." In a later letter he hoped that Belfrage would present his types to him because his finances did not enable him to purchase insect specimens; nevertheless he was prepared to pay fifteen cents, if necessary, for each type.

All in all, Cresson published sixty-six papers on the Hymenoptera between 1861 and 1882. Among Cresson's most notable contributions to the study of Hymenoptera are the following, all published by the Society he founded: the "Catalogue of the Described Species of North American Hymenoptera," in *Proceedings,* (Vol. 1, 1863); in *Transactions,* "A List of the Ichneumonidae of North America," (Vols. 1, 2, 6, 1867, 1868–1869, 1877), "Catalogue of North American Apidae," (Vol. 7, 1878–1879), "Synopses of the Families and Genera of the Hymenoptera, North of Mexico," (Supplement, 1887), and "Catalogue of the Tenthredinidae and Uroceridae of North America," (Vol. 8, 1880); and "The Cresson Types of Hymenoptera," in the *Memoirs.*

In 1865 Cresson undertook the publication of the *Practical Entomologist,* the first journal devoted to economic entomology in the United States. The original editors were Cresson, A. R. Grote, and J. W. McAllister. B. D. Walsh, an economic entomologist from Illinois, took over the editorship of Volume II in October 1866. This publication perished the next year of a common ailment, monetary malnutrition.

After he reached fifty, there was a notable cessation of his taxonomic studies, and it is believed that this was the result of trouble with his eyesight, very likely caused, as may be inferred from his letter to Belfrage, by studying insects under dim gaslights.

Nevertheless, Cresson continued to be a vital factor in the growth of the American Entomological Society and American entomology for many years. He served the Society as recording secretary in 1859, as corresponding secretary from 1859 through 1874, as curator from 1866 through 1874, as editor of the *Transactions* from 1871 through 1912, and as treasurer from 1874 through 1924.

Cresson had married, in 1859, Mary Ann Ridings, the daughter of James Ridings, one of the founders of the Society. The Cressons had four sons and one daughter. Two of the sons, George Binghurst Cresson and Ezra T. Cresson, Jr., inherited their father's love for entomology, becoming entomologists in their own right.

In 1883 the Cresson family moved to Swarthmore, Pennsylvania, a short distance from Philadelphia. Ezra wrote L. O. Howard on August 18, 1887, "I live some 10 miles out of the city, in plain informal country fashion, & if you can put up with such you will be welcome. I left the city some five years ago to try a home in the country, principally on account of my children, of whom I have five. Their health has been so much benefited by the change, we expect to remain country-folk until the little ones grow up." Cresson lived in Swarthmore for the rest of his life; after his wife died in 1909, he lived with his married children by turns.

Of the two sons interested in entomology, George was curator of the American Entomological Society from 1879 to 1881 and 1888 to 1889 and librarian from 1892 to 1896. He specialized in ants. Ezra Jr., entomologically, was the more famous of the two brothers. He edited the *Transactions* and *Memoirs* of the American Entomological Society from 1926 to 1945 and published 144 papers on such families of the Diptera as the Ephyridae, Ortalidae, Trypetidae, and the like. He was also an assistant curator at the Academy for many years.

Ezra Sr. wrote W. J. Holland Sept. 8, 1888, in reference to his son George's curatorship: "George is now at the Academy, working at the collections and library of the Soc. half of each week day. For the past year we have been out of funds to pay for the services of the curator and have been obliged to dispense with them, much to the injury of the collections. We have now collected together some funds for the purpose and so George is able to give part of his time to caring for our treasures. Wish we had a well invested fund for the support of a curator, so that we would not be obliged to go from hand to mouth as we do now."

When, in 1901, Cresson *père* presented his collection of the Hymenoptera to the American Entomological Society, it consisted of 2,367 types and 3,511 species. In his memoir of 1916, *The Cresson Types of Hymenoptera,* he listed a total of 2,737 types as well as his entomological papers.

In Calvert's words, "In summing up Cresson's work it is evident that by his kindliness, persistence, methodical habits and attention to detail he furnished the nucleus around whom those of his associates interested in insects could and would gather and so form an active working unit—

the entomological society which he helped to found—and this may really constitute his chief service to science. Beyond this he was a student of the taxonomy of the adult Hymenoptera."

Cresson died at Ezra Jr.'s home on April 19, 1926, at the age of eighty-eight.

GEORGE WILLIAM PECKHAM (1845–1914)

According to Muttkowski (1914): "In dealing with the work of Dr. Peckham, we cannot separate therefrom the work of his wife and collaborator. From the time of their marriage these two are inseparably linked in all phases of their work, in their researches, in their travels,

Figure 110. G. W. Peckham

in their very thoughts. Scientifically, their researches followed two definite lines—each in a way, logically the outcome of the other, that of psychology of spiders and wasps, and that of taxonomy of spiders." So the Peckhams could be classified as properly under the arachnologists as under the hymenopterists.

A search of the literature reveals surprisingly little about this husband-and-wife team; it is fortunate their work speaks for them. Muttkowski provides a little information about George Peckham but very little about Peckham's wife, nee Elizabeth Gifford, whom he married in 1880. It is believed that their three children shared some of their parents' enthusiasm for arthropods, for occasionally the children are mentioned in their writings.

Peckham was born in Albany, New York, on March 23, 1845. He went to Milwaukee in 1853, joined the Union ranks in 1863, and for his valor was made a lieutenant in the field at the age of nineteen, where he was in charge of a battery. When the war was over he went to Antioch College in Ohio and later to a law school in Albany, where he was admitted to the bar. In 1870 he began a medical course at Ann Arbor, Michigan, and received his medical degree in 1872. Instead of practicing medicine, he began to teach biology at the East Division High School in Milwaukee. Eventually he became principal. He was later made Superintendent of Public Instruction and Director of the Milwaukee Public Library.

Their book, *Wasps Social and Solitary,* provides a glimpse of the Peckhams in pursuit of one of their entomological interests:

> We were at our summer home near Milwaukee, where meadow and garden, with the wooden island in the lake close by, offered themselves as hunting grounds, while wasps of every kind, the socialistic tribes as well as the extreme individualists of the solitary species, were waiting to be studied. . . .
>
> We wanted to estimate the amount of labor done by a worker in a day, and so, rising one morning at the first bird call, we went out in the freshness of dawn, and for an hour had the world to ourselves; but a little before five a few straggling wasps that had stayed out all night began to bring in loads, and by half past seven they were fairly under way. From half past four until twelve we counted all that passed, 4534 going out and 3362 coming home; and with all this activity there seemed to be no pleasure excursions, for each one carried food when returning, and took out a pellet of earth when leaving. We once raised a little garden from the pellets that were dropped on our porch table where we kept a bowl of water. Wasps are great drinkers, and when they find such a provision they come frequently to refresh themselves, dropping their loads as they alight. This habit of holding on to their loads until they settle down may perhaps make them a factor in extending the boundaries of plant distribution, both under ordinary conditions and when, as must often happen with little creatures flying so high, they are blown to long distances from home.

In this same work the Peckhams devote a chapter to the activities of the singular wasp *Ammophila,* which uses a pebble to tamp down the earth with which it covers the opening to its burrow. In this burrow the wasp had placed a paralyzed caterpillar, on which it had laid its egg.

The Peckhams made their first scientific contributions with their studies on the mental capacity of spiders, and later, in 1889 and 1890 published their observations on sexual selection in jumping spiders. In 1909 their monograph "Revision of the Attidae of North America" appeared in *Transactions of the Wisconsin Academy of Science.*

William Morton Wheeler (1927) reports from an entry made in his

Figure 111. A drawing of an *Ammophila urnaria* using a pebble to pound earth over its nest, by James H. Emerton. From *Wasps, Social and Solitary,* by George H. and Elizabeth G. Peckham. Courtesy of Constable Publishers, London

diary in 1885: "Soon after my return to Milwaukee my old friend, Dr. George W. Peckham, who had long been making important contributions to arachnology and was beginning his well-known studies on the behavior of the solitary and social wasps, persuaded me to take a position as teacher of German and physiology in the school of which he was principal. Peckham was a very learned and charming man, deeply steeped in the evolutionary literature of the time and keenly alive to the possibilities of the new morphology that had been inaugurated by Huxley in England and a host of remarkable investigators in the laboratories of the German universities. Every year he most conscientiously read, as a devout priest might read his breviary, Darwin's *Origin and Animals and Plants under Domestication.* We became very intimate and I find from my diaries that for some years I regularly spent my Sunday mornings in his home drawing the palpi and epigyna of spiders to illustrate papers which he wrote in collaboration with his equally gifted and charming wife. I was privileged to collaborate with them in one paper (on the Lyssomanae) and to help them during the summers in their field work on the wasps at Pine Lake, Wisconsin. Under Peckham's management the biological work of the Milwaukee high school was carried far beyond that of any similar institution in the country."

The Peckhams clinched their claim to fame with the publication in

1898 of *On the Instincts and Habits of Solitary Wasps.* This was later revised and enlarged and published under the title *Wasps Social and Solitary,* with illustrations by James H. Emerton, who was an arachnologist as well as an artist.

Muttkowski says their work "bore at once the impress of scientist, scholar and poet." He then continues, "Dr. Peckham, as the writer knew him, was a small man, somewhat bent with age, rheumatism and the close application necessitated by his myopia. The scholarly stoop, the silvery white hair, and the moderate gait impressed everyone as attributes of a man who has made his mark on the world. On public or semi-public occasions the thoroughness and breadth of Dr. Peckham's information was surprising, even as the modesty and moderation with which it was put forth won him innumerable friends. Amiable, moderate, modest, kindly and scholarly—in these words his personality is best described."

Dr. Peckham died in Milwaukee on January 10, 1914, after which there seems to be no further mention of Mrs. Peckham.

ALBERT KOEBELE (1852–1924)

The story of the discovery of the Vedalia, the ladybird beetle that sparked great interest in biological control, has dramatic overtones. There are the featured actors, C. V. Riley, of whom we have already written, and Koebele, the supporting cast, D. W. Coquillet and F. S.

Figure 112. Albert Koebele

Crawford, and the minor characters, who play their subordinate but essential roles.

Koebele was born in Waldkirch, Germany, in 1852. There is no information on his activities in Germany as a youth, nor is it known exactly when he came to the United States, but he became a naturalized citizen in New York in 1880. There he was a member of the Brooklyn Entomological Society, and met C. V. Riley, who was so impressed with Koebele's beautifully mounted insects that he invited him to move to Washington and enter the Federal service. Koebele apparently was so overwhelmed by the offer that he immediately resigned the job he was holding. Unfortunately, it was months before Riley could fulfill his promise. He finally got Koebele a job in the Department of Agriculture in 1881.

Howard (1925) reports that Koebele found himself in congenial company in Washington. He encountered first E. A. Schwarz, T. Pergande and G. Marx and then O. Lugger and John B. Smith, "all of whom were of German origin and all of whom were ardent collectors." Early in 1882 Koebele was sent to the South to work on the cotton worm, *Alabama argillacea.* During the winter of 1882–1883 he went on an expedition to Brazil to study cotton pests and naturally took advantage of the trip to collect a large number of insects.

After an unhappy love affair, Koebele asked to be transferred from Washington to another locality and in 1885 he went to the West Coast. In California he made Alameda, across the bay from San Francisco, his home and headquarters. For the next three years he studied the life histories of economic insects and worked on insecticides. In 1886 he recommended a resin wash for the cottony cushion scale, *Icerya purchasi* Maskell, a pernicious pest of citrus and other plants in California.

In February of that year Koebele was transferred to Los Angeles as an associate of D. W. Coquillet's, who was also working on the cottony cushion scale. Apparently Koebele and Coquillet did not get along; as a result of Koebele's complaints Riley furloughed Coquillet for a year. This was a move that Riley lived to regret, for it was during this year that Coquillet developed HCN fumigation for citrus, the most widely used method for the control of citrus pests until the end of World War II. Coquillet was rehired in 1887.

Doutt (1958), in an absorbing article, assembled an array of material that portrays the dramatic story of Koebele, Riley, and the Vedalia. According to Doutt, Waldemar G. Klee, formerly of Denmark, as State Inspector of Fruit Pests in California corresponded with W. M. Maskell and Frazer S. Crawford in Australia about the cottony cushion scale. Riley was also making inquiries about the scale in Australia. For a short time Riley believed that the Island of Mauritius in the Indian Ocean rather than Australia was the original home of the cottony cushion scale. The California growers, in dire need of immediate relief from the ravages of the cottony cushion scale, put pressure on the U.S. Department of Agriculture for instant relief. Riley recommended that an entomologist be sent to Australia. The California growers suggested Coquillet, but Riley sent Koebele instead.

In 1888 the appropriation bill for the Department of Agriculture had a rider that prohibited foreign travel for Department employees (aimed at Riley and his foreign junkets). The resourceful Riley got around this by having Koebele travel as a representative of the State Department to the International Exposition in Melbourne. F. M. Webster went later as Commissioner to report on agriculture at the Exposition. Koebele sailed to Australia from San Francisco on August 25, 1888. Upon his arrival he went to Adelaide, where Crawford introduced him to the parasitic fly, *Cryptochaetum iceryae*. Koebele immediately began to send this parasite and thousands of parasitized cottony cushion scales to California. While collecting the parasitic fly he also gathered some predacious ladybird beetles, including the Vedalia, *Rodolia cardinalis* (Mulsant). Coquillet received three small shipments of the Vedalia between November 30, 1888, and January 24, 1889. He placed them on an infested orange tree enclosed in a tent on a ranch owned by J. W. Wolfskill of Los Angeles. By the beginning of April, the ladybirds, both larvae and adults, had destroyed the cottony cushion scales on the tree. Then one side of the tent was opened, and the ladybirds were permitted to spread to adjacent trees. Within a few months it was apparent that the miracle sought by the desperate growers had been wrought by the Vedalia, with its voracious appetite for the cottony cushion scale.

People came in hordes from many localities in the state with small receptacles, such as pill boxes and similar containers, to gather the larvae and adults from the orchards. Thus it was Riley, Koebele, Coquillet, and indirectly Crawford as well as others who were responsible for the introduction of the Vedalia that ushered in the modern method of controlling insects with insects.

In appreciation for the success of the beetle, the California growers gave Koebele a gold watch and his wife a pair of diamond earrings. In the first flush of success, the roles of Riley, Coquillet, and Crawford were largely overlooked, and Riley complained out loud. In 1891, under pressure of the California growers, Riley was forced to send Koebele on another expedition to New Zealand, Australia, and other Pacific islands. As a result of this trip Koebele introduced to California several valuable predators of scale insects, mealybugs, and other insects.

By 1893, the friction between Riley and the Californians had reached the kindling point, and he recalled Coquillet and Koebele to Washington. Coquillet in time attained great distinction as a dipterist in the National Museum. Koebele foresaw only a stormy future, so he resigned from the Federal service to work for the Hawaiian provisional government—at a greatly increased salary.

Koebele did outstanding work in biological control of injurious insects, first for the Hawaiian government and later as a member of the staff of the Experiment Station of the Hawaiian Sugar Planters' Association. For years he travelled throughout Australia, Ceylon, China, Japan, Mexico, the Fiji Islands, much of the United States, and elsewhere and introduced many beneficial predators and parasites into Hawaii.

According to Perkins (1925), who worked with him in Hawaii:

> Koebele was *par excellence* a field worker in entomology and his knowledge of living insects was of a most extensive character, at one time or another he paid special attention to all orders, but chiefly to Coleoptera and Lepidoptera, to some of the minute Hymenoptera and to scale insects. At one period he did much rearing of micro-Lepidoptera for Professor Riley. As may be judged from the nature of his field work, the Coccinellidae or ladybirds were his especial favorites, and he collected great numbers of species in the various countries he visited. He was not a great reader of the entomological literature, but certain systematic works he used continually, e.g. Maskell's and Green's Coccidae, and especially Crotch's book on the Coccinellidae, which accompanied him on all his travels. Of the classification and specific character structures of these groups he had an extensive knowledge, though he published no notes of a systematic nature on others excepting some official reports and even these were to him an uncongenial task.
>
> His success in the field was due to his acute perception of the habits of insects, and unsurpassed perseverance, and he was naturally a very quick worker, so that with insects that are rare and difficult to obtain he could collect a greater number in a given time than most of the best field workers we have known. Under any circumstances he was a most pleasant companion on a trip, for even when the hardest and most uncomfortable conditions were added to ill success he remained cheerful and good humored, hoping to the last to achieve something by which a failure might be converted into a triumph. He met with many adventures in his varied traveling, and in unhealthy countries contracted many fevers, which failed to lessen his enthusiasm for his work, but he rarely spoke of his adventures.

According to Essig (1931), Koebele asked to be relieved of active duty in 1910 because of ill health, including eye trouble. He returned to Germany for the treatment of his ailments. Nevertheless he continued to serve as consultant with the Hawaiian Sugar Planters' Association and to collect beneficial insects. Although a naturalized American citizen, he was not permitted to leave Germany during World War One and was in dire circumstances during the war years. After the war arrangements were made to have him return to his home in Alameda, California. However, he had become too feeble to make the trip and died in Waldkirch, Germany on December 28, 1924.

WILLIAM HARRIS ASHMEAD (1855–1908)

Ashmead published more than two hundred and fifty papers, largely on the systematics of parasitic wasps. Surprisingly, there is a dearth of published information about him. This sketch is largely based on his obituary by Howard, Schwarz, and Banks (1908).

W. Dwight Pierce worked with Ashmead in the National Museum and recalls him as a large man with little hair. Pierce, who was then working on the Strepsiptera parasites of bees, wrote me that he found Ashmead very helpful; that he was like August Busck in that he seemed to require only three or four hours of sleep a night, while Pierce needed ten. "As I understand he went to the Cosmos Club early in the evening and played cards with the first installment—probably bridge or whist. Then when they went home the game was poker and the liquor stronger. He outlasted all, then took his short sleep." He added that in those days "every entomologist was a character," and that Ashmead used to use checks and even money as bookmarks.

"Like Chittenden, Ashmead didn't file things but he could lay his

Figure 113. W. H. Ashmead

hands on what he wanted in the piled littered desk. Every paper had its place but only he knew where to find it."

Ashmead was born in Philadelphia on September 19, 1855, the son of Elizabeth and Captain Albert Ashmead, both of colonial ancestry. After receiving an education at private and public schools in Philadelphia, he entered the publishing house of J. B. Lippincott Company. Some years later he and his brother went to Jacksonville, Florida, and established a printing firm which published agricultural as well as other books. There they founded an agricultural weekly and a daily newspaper. Ashmead edited the scientific part of the weekly and became interested in studying and collecting insects. In 1880 he published his *Orange Insects: A Treatise on the Injurious and Beneficial Insects Found on Orange Trees in Florida.*

C. V. Riley recognized Ashmead's ability and in 1887 appointed him a special field entomologist in the Division of Entomology, with injurious insects of Florida as his special province. The following year Ashmead was appointed entomologist in the State Agricultural College and Experiment Station at Lake City, Florida, where he wrote one of the first Florida publications on insects. Ashmead returned to the Division of Entomology in 1889 as an assistant entomologist and investigator. He then took a leave of absence and studied in Berlin from 1890 to 1891. In 1895 he was appointed Assistant Curator of the Division of Insects in the National Museum, a position he held for the rest of his life.

Howard *et al.* (1908), who also remark on his sleeping habits, say that "As a worker Doctor Ashmead was possessed of an enthusiasm and industry that has rarely been equaled. . . . The amount of work accomplished was thus enormous."

When Ashmead first became interested in insects he wrote about insects injurious to the orange; he was at that time also occupied with the Hemiptera and ladybird beetles. Later he collected galls and studied the Cynipidae or gall wasps. As a result of rearing the cynipids from the galls, he obtained many parasitic chalcids, and began to study them as well as other parasitic Hymenoptera. All in all, he described 3,100 new species and 607 new genera. He wrote important monographs and papers on the classification of the wasps of the superfamilies Proctotrypoidea, Sphecoidea, Vespoidea, Cynipoidea, as well as the Chalcidoidea.

Howard *et al.* note that when he went to Washington "he was a man of large property, which, however, was greatly reduced by the disastrous Jacksonville fire. This, however, did not appear to prey on his mind."

Professor J. J. Davis turned over to me a letter from Ashmead, dated December 10, 1891, to T. J. Monell. Since Riley was the editor of *Insect Life* at the time, the following paragraphs are worth quoting:

Fortunately, I have still a few separata left of my "Synopsis of the Coccidae," a copy of which I mail to you today.

Insect Life gives me "faint praise" for this brochure which took me three years of hard work to finish, although by those competent to judge it has been pronounced most excellent.

I account for it from the fact that in publishing it I stepped on somebody's preserves.

According to Howard *et al.* (1908), "Few were more interested in our [Entomological] Society than [Ashmead]. Rarely did he miss a meeting, and when present always had something to say. He contributed a great number of papers, and his ardent and exultant manner in presenting 'two new and remarkable genera' will never be forgotten. The details of their structure, their relation to other forms, their effect on the classification, were all explained to admiring listeners." Like so many famous entomologists he constantly helped and encouraged younger workers.

Ashmead received an M.S. from Florida State Agricultural College (now University of Florida) and a Ph.D. in 1904 from the Western University of Pennsylvania (University of Pittsburgh). His thesis, a monograph of the chalcid flies collected in South America by H. H. Smith, was published as a memoir by the Carnegie Museum.

Ashmead died after a long illness in St. Elizabeth's Hospital in Washington, D.C., on October 17, 1908.

THEODORE DRU ALISON COCKERELL (1866–1948)

Cockerell, with a long flowing beard like some ancient prophet's, had the broad view of natural history so characteristic of the Victorian scholar. Because of his many interests, his kindly nature, his encouragement of young people, and the fact that he lived to a ripe old age, T. D. A. had a great influence on American biologists and especially on entomologists. Fortunately, some who knew him well such as Essig (1948), Calvert (1948), Rohwer (1948), Linsley (1948), Michener (1948), Schwarz (1948), and Ewan (1950), recorded their recollections of him.

The eldest son of Sydney and Alice Cockerell, T. D. A. was born on August 22, 1866, in Norwood, a suburb of London, England. He attributed his early interest in natural history to his father, to a kindly teacher, to excellent museums, and to books on the natural history of England. His father, who was a student of nature and an avid reader of Darwin's works, encouraged his interests, but unfortunately died when T. D. A. was eleven. The teacher who so greatly influenced him and permitted

Figure 114. T. D. A. Cockerell

him and his brother Sydney to share her fascinating library was Mrs. Sarah Marshall of Bickenham School. Cockerell, who had native ability as an artist, was torn between his love for natural history and the arts, but science ultimately emerged as victor.

His interest (Cockerell, 1935) in natural history began as early as he could recall. "As I was a weakly child (I remember some one saying, be kind to the little boy, he will never grow up) it was considered expedient to send me, as often as possible, to the country. For the same reason, I had comparatively little schooling. Down in Sussex, near Robertsbridge, I used to wander about the fields and woods, admiring the birds and butterflies and the wild flowers, watching the men at work, and 'helping' in such small ways as were permitted. It was a wholesome life, and although the country folk knew nothing of science, as we understand it, they did know a lot about the events of the country side."

Upon the death of his father he was taken by a friend to the island of Madeira; there he made "what I think of as my first scientific discovery. I was apparently the first to find and report on the caterpillar of the finest Madeiran butterflies, *Pyraemis indica occidentalis.*" Also as a youngster he developed an interest in snails and slugs which he maintained until the end of his days.

While in London, upon completion of his schooling, Cockerell worked for flour agents, and in the course of his employment developed tuberculosis. On June 27, 1887, at the age of twenty, he sailed to the United States in search of a climate that would cure him of his illness. The small town of Westcliff, Colorado, provided this climate, and after three years there, he recovered his health. During these years Cockerell cultivated his taste for natural history and worked on a catalogue of the flora and fauna of Colorado. In the process of preparing this work, he became acquainted with entomological literature and corresponded with entomologists as well as naturalists interested in flowering plants, molluscs, and various animals. Already Samuel Scudder was referring to him as the "industrious entomologist of Colorado."

In 1890 Cockerell returned to England to work in the British Museum. During this period in England he met and was profoundly influenced by no less a biologist than Alfred Russell Wallace. Cockerell had corresponded with Wallace and had impressed him with some of his observations. In time, Wallace invited Cockerell to help him prepare a new edition of his book *Island Life*. Cockerell thus became well acquainted not only with Wallace, but through him with many other well-known naturalists. In his conversation in later years, Cockerell constantly referred to Wallace and always kept a fine picture of him in a prominent place in his office.

It is likely that Wallace recommended Cockerell for his next job as Curator of the Public Museum in Kingston, Jamaica, which he undertook in 1891. It was in Jamaica that he first became interested in scale insects. After two years in this humid climate, his tuberculosis reappeared, and his frail wife, Annie Fenn, gave birth to a son who died in infancy. Luckily for Cockerell, he was able to switch jobs with his friend, Professor C. H. T. Townsend, of the New Mexico College of Agriculture at Las Cruces. There, in 1893, he became Entomologist of the Experiment Station and Professor of Entomology and Zoology of the College. Shortly after his arrival in New Mexico, his wife died as she was giving birth to another son. This son died of diphtheria at the age of eight.

In the opinion of other entomologists, Cockerell's greatest contribution to the study of insects was on wild bees, his special interest. Cockerell described in his own words (Calvert, 1948) how this interest began: "On the flowers in great abundance, I found an entirely yellow species of bee which seemed to play a game of hide and seek with the similarly yellow predatory bug *Phymata,* also sitting on the flowers. It seemed strange that such an interesting and (in that region) common insect had not been described, but [William J.] Fox at once confirmed my opinion that it was new. Starting thus, I continued to investigate the genus *Perdita*. From this time onward I have never ceased to work on bees and have published [to March, 1938] 5,480 new names for species, subspecies and varieties and 146 names for genera and subgenera."

Linsley (1948) adds that "New Mexico was momentous in Cockerell's life in another way. It was here that he met his [second] wife Willmatte

Porter. Through the years Mrs. Cockerell was a constant companion and strong support, accompanying him on expeditions and field trips, sharing his interests and collaborating in many of his projects. There is little doubt that she had a most profound influence in increasing the effectiveness of his scientific career." Cockerell met Miss Porter, a biology teacher in the local high school, while out collecting bees. Thereafter she often came to Cockerell for information about the local plants and animals. "The Professor occasionally fondly referred to the incident with a chuckle saying, 'And I found the young Stanford graduate so ignorant that I took her on as a permanent pupil.' "

Rohwer (1948) describes his recollection of the Cockerells, who moved to Boulder, Colorado, in 1904,

> the year I entered the State Preparatory School. Mrs. Cockerell was my teacher of high school biology. All of us were enthused, stimulated and charmed with the interesting and perhaps unusual way the subject was presented. It was not the customary text and laboratory manual course. It was a humanized presentation and we soon found we were learning fundamentals without quite knowing how. They seemed to become a part of our knowledge and thus had lasting influence on our development. Professor Cockerell was associated with the University and often frequented the laboratory and lecture room of the "Prep" school. More than that, occasionally he would talk to us. The word "talk" is used rather than "lecture" since it describes more accurately the delightful way these treats were presented. In fact this was true of all the many "lectures" that I heard in succeeding years—either in the University or on more formal occasions. One never thought of Professor Cockerell as a lecturer, yet there are few who could hold an audience as well. With informal natural presentation, almost as if it was directed personally to you, he would tell his story, logically and effectively in simple words, with appropriate wit and humor to emphasize significant points. Clear and direct as his writings are, the talks are even better.

Mrs. Cockerell not only accompanied her husband on his collecting trips but also was co-author of some of his papers and often shared the lecture platform with him. Schwarz (1948) notes, "The number of species that have been named *wilmattae* further testify to this partnership of joint interest, which extended through a married life that nearly attained its Golden Anniversary."

In 1903 Cockerell acceped a position as Curator of the Museum at Colorado College, Colorado Springs. Shortly thereafter he taught zoology, including entomology, at the University of Colorado at Boulder until his retirement in 1934, though he had no academic degree, as he once carefully pointed out on a postcard to the Carnegie Museum:

> I noticed that on the envelopes of the Carnegie Museum publications sent to me I was designated "Dr." Kindly note that I have no doctor's degree, * I have a friend who was similarly designated in error, and the thing got into print, and now it looks as if it could not be stopped. So I wish to nip the error in the bud . . .

According to Linsley (1948), "he not only continued his prolific work on bees, but the nearness of the Florissant and Green River Shales turned his interest to Paleoentomology and Paleobotany. He also found time to study fish scales, color variation in sunflowers, anatomy of rodents and various other subjects." Cockerell was a competent botanist and an authority on the genera *Rosa* and *Hymenoxys*. He was the author of two widely used books, *Zoology, A Textbook for Colleges and Universities* and the *Zoology of Colorado*.

Speaking of his early efforts in entomology, Cockerell reports (Calvert, 1948) that he "found the Coccidae extremely interesting, not only on account of their importance for economic entomology, but as illustrating the process of evolution and diversification through the reduction and suppression, as well as the modification of parts." Schwarz (1948) notes that some of his discoveries, "like the cochineal insect *Dactylopices opuntiae,* have proved of great economic importance. In Madras this insect, it is reported, has cleared 40,000 square miles of cacti and it has been used successfully in other areas as well where the prickly-pear is an agricultural obstacle."

According to Linsley (1948), "It is doubtful if any man of his generation made so many original contributions to the taxonomy of so many different fields as did T. D. A. Cockerell. . . . The number of species of plants and animals, living and fossil, which he has named must be in the vicinity of seven or eight thousand!" He wrote over three thousand articles and notes on bees, scale insects, fossil plants, fossil insects, biography, geology, and a great variety of other subjects. Naturally any man who spreads himself so thin, may at times be lacking in depth and accuracy and be criticized for some of his conclusions based on a limited knowledge of certain subjects; yet there were few biologists who knew so much about so many different fields.

Art and poetry were other avenues for his vast energy, and he even published a book of poems. He was also concerned with the economic and labor problems of his time and was a sympathizer of the British Labor movement even presenting a course at Colorado College entitled "Modern English Reformers."

Anyone interested in Cockerell should read Ewan's (1950) account

* Nor, indeed, any degree of any sort.

of him. Ewan describes the professor in his narrow room at the University of Colorado with its high ceiling: "It was plainly a naturalist's work room, and he made it peculiarly his own. Even the dry unventilated smell of the insect boxes ranged along the wall, mingling with the reminiscent use of insecticides down through the years, was of a different intensity here from that of most museums. In his seventy-fourth year Professor Cockerell was slender, as he had always been, of average height, soft-voiced and even a little sibilant of speech yet steady of finger and with little dimming of the wide blue eyes that his students remember so well. He usually dressed in a light grey suit and wore his coat and waistcoat in even the warmest weather invariably closed at the throat by a more unconventional ample dark green cravat of the sort commonly associated in this country with artists. He was crisp but unhurried in conversation; in movements purposeful."

Cockerell collected in Europe, Japan, Siam, India, Siberia, Australia, Morocco, Africa, Canada, and in Honduras in 1947 when he was eighty-one years old. Many of his expeditions were to collect wild bees. The terrible earthquake of 1923 hit the Japanese mainland just as Cockerell and his wife boarded ship, fortunately with no injury to either of them.

After retiring from teaching at the University of Colorado, he shared his time and interests between Colorado and Southern California, where he spent his winters. In 1942 until the winter of 1946–1947 he was in charge of the Desert Museum in Palm Springs. He was noted for the time and effort he spent helping other researchers with information, references, constructive criticism, and painstaking letters written in minute calligraphy. As an example of his encouragement of young workers, Linsley (1948) recalls the aid Cockerell gave him during his Palm Spring days. "It was during this period that the writer had most of his personal contact with him. Ever ready to lend encouragement to a newcomer in the study of bees, he sent specimens and literature, loaned types and unpublished manuscript notes, placed me in contact with other workers in the field, shared his home when I went to study his collection and, in short, did everything possible to assist me. The same encouragement was offered throughout his career to anyone who would accept it."

Cockerell died on January 26, 1948, in San Diego as the result of several strokes.

WILLIAM MORTON WHEELER (1865–1937)

No entomologist can be long ignorant of Wheeler's contributions to our knowledge of ants and other social insects.

Figure 115. W. M. Wheeler

Wheeler was born on March 19, 1865, in Milwaukee, Wisconsin, to Julius M. and Caroline Georgiana Wheeler. In his diary (Parker, 1938), Wheeler described his early schooling in Milwaukee: "Owing to my persistently bad behavior soon after I entered the public school my father transferred me to a German academy founded by Peter Engelmann, an able pedagogue who had immigrated to the Middle West in 1848. The school had a deserved reputation for extreme severity of discipline. To have annoyed one of the burly Ph.D.'s who acted as my instructors, as I had annoyed the demure little schoolmarms in the ward school, would probably have meant maiming for life at his hands . . ."

Wheeler also attended the German-American College, a normal school associated with this institution, from which he graduated in 1884. He always attributed his broad education and his interest and competence in the classics to his training at these institutions in Milwaukee. He writes (1927) that his life and that of his friend, Carl Akeley, were influenced by an event that occurred in 1883:

> Professor H. A. Ward, proprietor of Ward's Natural Science Establishment in Rochester, New York, which was not so much a museum as a museum factory, learned that there was to be an exposition in Milwaukee in the fall of 1883 and that the local German academy, which I had attended, possessed a small museum. He decided therefore, to bring a collection of stuffed and skeletonized mammals, birds, and reptiles, and an attractive series of marine invertebrates to the exposition, and to persuade the city fathers to purchase the lot, combine it with the academy's collection, and thus lay the foundation of a free municipal museum of natural history. I had haunted the old academy museum since childhood and knew every specimen in it. Indeed, Dr. H. Dorner, my instructor

in natural science, had often permitted me to act as his assistant. Of course, I was on hand when Professor Ward's boxes arrived, and I still remember the delightful thrill with which I gazed on the entrancing specimens that seemed to have come from some other planet. I at once volunteered to spend my nights in helping Professor Ward unpack and install the specimens, and I worked only as an enthusiastic youth can work. He seems to have been duly impressed by my industry, because he offered me a job in his establishment. I was quite carried away with the prospect of passing my days among the wonderful beasts in Rochester. Not the least of Professor Ward's attainments were his uncanny insight into human nature and his grim business and scientific acumen. He offered me the princely salary of nine dollars a week, six of which were to be deducted for board and lodging in his own house.

I entered Ward's Establishment February 7, 1884. My duties consisted in identifying, with the aid of a fair library, and listing birds and mammals. Later I was made a foreman and devoted most of my time to identifying and arranging the collection of shells, echinoderms, and sponges, and preparing catalogues and price lists of them for publication. Such is the present state of conchology that my shell-catalogue is still used. At this time Akeley entered the establishment as a budding taxidermist, and for once Professor Ward's estimate of human nature appears to have been at fault, for as Akeley informs us *In Brightest Africa* he was given a salary of $3.50 a week, without board and lodging. He attached himself to William Critchley, a young and enthusiastic artisan, with the voice and physique of an Italian opera tenor, who had attained the highest proficiency in the taxidermic methods of the time, but did not seem to give promise of advancing the art. In the course of a year Akeley had more than mastered all Critchley could teach him, and was longing for wider opportunities than could be offered by an establishment, which, after all, was neither an art school nor a scientific laboratory, but a business venture. But even so there was reason to believe that its standards of workmanship were higher than in any of the museums that had grown up in various parts of the country.

The relations between Akeley and myself soon ripened into a warm friendship. We were nearly of the same physical age, but I was the younger and the more unsettled mentally, for he had been reared by sturdy parents on a quiet farm and I had been brought up in a bustling city with a superheated atmosphere of German Kultur.

In the course of time our relations settled into those of affectionate older and younger brothers. I cannot recall that we were even on the verge of a quarrel, and this must have been due to Akeley's self restraint and sympathetic tolerance, because I was often irritable and unwell in those days. Owing to the fact that we did not work in the same building, our companionship was largely limited to evenings and Sundays. As I read the diaries of 1884 and 1885 I marvel at the multiplicity of our youthful interests and occupations.

Wheeler and Akeley, having derived all they could from the Ward Establishment, left it in 1885, Wheeler to return to Milwaukee. There he studied spiders under the able guidance of G. W. and E. Peckham,

taught in the high school for two years, and then became custodian of the new Milwaukee Public Museum. While he was with the Museum, he studied embryology under Dr. William Patton of the newly established Allis Lake Laboratory biological station, which was under the directorship of Professor C. O. Whitman, and became very competent in embryology and cytology as well as a master of microscopical technique.
Wheeler (1927) writes:

In the meantime the Milwaukee Public Museum had been established according to the plan suggested by Professor Ward, and I saw an opening for Akeley as taxidermist. I persuaded him to come to Milwaukee and live with me. He arrived November 8, 1886, and although he was not officially appointed to the institution until November 20, 1888, he was given a certain amount of its work. We converted a barn on my mother's place into a shop and here he worked at least during the evenings for several years. I was made custodian of the Museum, September 19, 1887, and held the position until August 29, 1890. By that time my association with Peckham, Whitman and Patton had converted me into a hard-boiled morphologist, and I was induced by Whitman to accept a fellowship at Clark University, where he had become professor of zoology a year earlier. Till October 1, 1890, when I left Milwaukee for good, Akeley and I had spent so many happy hours together that the parting was painful. After leaving the high school I had fitted up a laboratory in the house and when my eyes grew weary with the microscope I repaired to his shop and read to him while he worked or more rarely he read to me. My diary mentions the volumes we read and I wonder at Akeley's patience and apparent pleasure in listening to Bryce's *American Commonwealth,* translations of Aeschylus, Max Nordau, and similar highbrow stuff. I patiently read a whole small library for at that time I had serious conscientious objections to beginning a book without reading its every word. Perhaps Akeley only heard occasional important fragments and had found that he could carry on his own train of inventive thought better when we were together and I was making a continual but not too disturbing noise.

In this period, he published "The Embryology of *Blatta germanica* and *Doryphora decemlineata*" in the *Journal of Morphology* (1899). At Clark University in Worcester, Massachusetts, Wheeler served as Assistant in Morphology and continued to publish on insects. He received his Ph.D. in 1892. His thesis was on insect embryology. The same year he followed Dr. Whitman from Clark to the University of Chicago, where he became Instructor in Embryology. During 1893–1894, Wheeler went to Europe and studied at the University of Wurzberg in Germany, at the Naples Zoological Station in Italy, and at the Institut Zoologique at Liège, Belgium. Then from 1894 through 1899 he resumed his connections with the University of Chicago as an assistant professor of embryology. His entomological interests began to encompass fields other than embryology.

In 1898 he married Miss Dora Bay Emerson of Rockford, Illinois.

Wheeler was an omnivorous reader, with a "phenomenally retentive memory," according to Melander and Carpenter (1937), not only in the field of biology, but also of sociology, psychology, philosophy, and metaphysics, as well as Greek and Latin classics. He read fluently in Greek, Latin, German, French, Italian, and Spanish, and his mastery of the classics "enabled him to coin readily the multitude of euphonious terms that constantly embellished his writings." This coining of new terms to describe a multitude of biological phenomena has brought on his head both blessings and abuse from his colleagues, some of whom objected to the prolific generation of such verbal concoctions.

In 1899 Wheeler was made Professor of Zoology at the University of Texas. There with the aid of a single instructor and several laboratory assistants he gave all the zoological instruction in the University. This included courses in general biology, comparative anatomy, embryology, histology, and special work in entomology to a group of several more advanced students. During his four-year stay in Texas he began to study ants and write about them. C. T. Brues and A. L. Melander, who later became entomologists themselves, went to the University of Chicago to study under him, but finding him transferred to the University of Texas followed him to Austin and studied in his laboratory for several years. Under his direction, the reputation of the laboratory was greatly enhanced.

Although Wheeler regretted giving up teaching, he welcomed the opportunity in 1903 to become Curator of Invertebrate Zoology at the American Museum of Natural History and the chance to devote more time to the study of ants. There he organized and arranged the Hall of Invertebrates and worked on the taxonomy, structure, functions, distribution, habits, ecology, and social relations of ants. For five years he published numerous important papers on ants in the *Bulletin of the American Museum of Natural History.* During this period he also worked on his book *Ants,* which was published by the Columbia University Press in 1910.

In 1908 he decided to return to teaching again, and became Professor of Economic Entomology at Harvard University, where he had laboratories in the Bussey Institution, a graduate school for research. He became dean of the Institution a few years later. During his tenure he fought for a first-class staff with appropriate salaries and attracted a body of graduate students who in time became prominent entomologists themselves. In 1929 he resigned the deanship to move to new laboratories in Cambridge, where he taught until his retirement in 1934.

Brues (1937) says that Wheeler was an inspiration to his students and young associates and stimulated them "to accomplishments quite beyond their own expectations. He was always enthusiastically interested in his own work and however deeply immersed in it, was always ready to welcome the student who wandered into his laboratory at any time."

Wheeler could never be accused of having been a "closet-entomologist"; he collected in Mexico, the Bahamas, Puerto Rico, Panama, Costa Rica, Guatemala, Cuba, British Guiana, Galapagos, Australia, Morocco, and the Canary Islands as well as throughout the United States, returning more than once to a favorite area. He obtained and studied fossil insects from Florissant, Colorado, where T. D. A. Cockerell too had worked. He made numerous trips to Europe, serving as an exchange professor at the University of Paris in 1925.

Besides his book on *Ants,* a wonderfully absorbing source of information on "Their Structure, Development and Behavior," as it is subtitled, he produced a number of other important works: *Social Life Among the Insects* (1923); a translation (1926) of an unpublished manuscript by R. A. F. de Réaumur, *The Natural History of Ants; Foibles of Insects and Men,* and *The Social Insects, Their Origin and Evolution* (both 1928); *Demons of the Dust, A Study in Insect Behavior* (1930), an inquiry into the biology of ant lions; and, with T. Barbour, *The Lamarck Manuscripts at Harvard* (1933); also many shorter works.

Parker (1938) describes Wheeler at work in Cambridge: "In his new Harvard surroundings he settled down with great complacency having two private laboratories, one in the Museum of Comparative Zoology among the insect collection and the other in the Biological Laboratories. That he spent more time in the latter than in the former resulted from his habit of smoking while at work. Smoking because of fire risk was prohibited in the Museum, but was allowed in the Biological Laboratories.

"Wheeler always arrived early in Cambridge for his day there, being usually driven in a car from Boston by his daughter. By nine o'clock he was to be found, as a rule, at his laboratory table in the biological building. Here he commonly worked until noon, when he repaired to the Museum, where in the quarters of its Director, he took lunch. This mid-day rendezvous called by its frequenters 'the eateria' was a center to which were invited many of the biological notables temporarily in Cambridge. It was therefore an interesting and stimulating gathering to which Wheeler added much and in which he took great delight. In the afternoon he usually worked either in his quarters in the Museum or in those in the Biological Laboratories. In the late afternoon he was driven back to his home . . ."

Wheeler's friends remember the keen sense of humor which constantly flavored his conversation. He was said to be extremely outspoken in voicing either approbation or disapproval.

"To those who knew Wheeler personally," says Parker (1938), "he was a quiet, modest, unassuming man, the last in the world to reach for distinction and yet happy in its reception. Nevertheless he could be roused to passion, even to strong passion, particularly when the situation seemed to him to carry with it injustice, covered deceit, or insincerity. To none of these indirections would he yield a point and friend or foe must answer to him in the open. Yet this passionate side of his nature was not shown to all."

Henderson *et al.* (1937) wrote that Wheeler was "extraordinarily good company. He laughed with one and inoffensively, at one; and he was one of the very rare individuals whose idiomatic knowledge of three or four languages was such that he could laugh with real gusto in all of them. During his later years he spent most of his evenings in his study in West Cedar Street where one would find him sitting at a deskful of books—with more books on chairs and on the floor and with sheets of manuscript under and over them. The casual visitor was installed in an armchair and the maid was sent down for a bottle and the cigars."

Wheeler wrote almost five hundred publications, most of them on systematics but many on a wide variety of subjects that revealed his encyclopedic knowledge. His writing expressed his sensitivity in a style noted both for its lucidity and workmanship. At times he wrote satirical pieces such as "The Dry-Rot of Academic Biology." Thus Klopsteg (1963) notes Wheeler's opposition to the National Research Council as " 'that superorganization of superorganizers,' and to its parent, the National Academy, as 'a distinguished organization of distinguished scientists whose principal occupation is writing each others' obituaries, and it's a pity they don't have more to do.' "

Wheeler died on the night of April 19, 1937, on the platform of the Harvard Subway Station, probably on his way to his laboratory. He was survived by his wife, a son, and a daughter.

Wheeler received numerous honors from institutions in the United States and foreign lands, but none more appropriate than the remarks of men who knew him. Professor Alfred North Whitehead is reported to have said that he was "the only man who would have been worthy and able to sustain a conversation with Aristotle," and former President Lowell of Harvard expressed the thoughts of Wheeler's friends and associates when he said upon Wheeler's death, "A great light has gone out of the intellectual firmament."

HENRY LORENZ VIERECK (1881–1931)

J. A. G. Rehn (1932) wrote a sensitive account of Viereck, a specialist in wild bees and ichneumonids. The following is from his memoir.

Figure 116. H. L. Viereck. Courtesy of J. J. Davis

The youngest of five children, Viereck was born in Philadelphia on March 28, 1881. His father, John A., emigrated to the United States in 1856 and was wounded at the second Battle of Bull Run.

Both young Viereck and Rehn were known as Charles W. Johnson's "boys." Johnson was Museum Curator of the Wagner Free Institute of Science in Philadelphia and an excellent teacher of natural history. The boys collected not only insects, but also birds, molluscs, and reptiles; one of their favorite collecting spots was on the slopes of the Blue Ridge on the farm of a Pennsylvania Dutch family at Lehigh Gap.

Viereck was educated in Philadelphia public schools and later in Brown's Preparatory School, where he did not finish. He enlisted under age in the Spanish-American War, almost dying of typhoid in a Georgia army camp. His mother died while he was convalescing and a few years later "his kindly, white-bearded father, beloved by all the son's boyhood friends, was killed by an express train at Lehigh Gap."

According to Rehn, "The crusader spirit was always strong in him, in his scientific studies and in his personal attitude toward social and political problems of the day. Compromise with wrong or injustice was unthinkable to him, and . . . his strong convictions and demand for what he considered fair dealing more than once caused him to shift the scene of his activities. While a blithesomeness which his old associates will never forget was one of his chief possessions, tragedy stalked through his life, loved ones were taken from him with startling suddenness . . ."

When in 1900 he became a Jessup student of the Academy of Natural Sciences in Philadelphia he spent many hours studying the Cresson types of Hymenoptera. He accompanied Rehn on a several months' visit to New Mexico and proved to be a capable and diligent collector. It was there that he began to make plans for "a monograph of the bee genus *Andrena,* which remained through the rest of his life the one contribution he wished to complete, and toward which a number of his published papers were preliminary, but the final study never appeared."

From 1903 to 1905 he studied medicine at the Jefferson Medical College in Philadelphia but never completed that course. In the summer of 1903 Viereck was associated with John B. Smith in mosquito control. In 1904 and 1905 he was connected with the Connecticut Agricultural Station, where he worked with Dr. W. E. Britton and studied mutillid wasps and bees in co-operation with T. D. A. Cockerell and other hymenopterists. He specialized in the Hymenoptera, especially during his twenties, averaging six or less hours' sleep a night. His association with the Connecticut Agricultural Station eventually resulted in the manual, *Hymenoptera of Connecticut,* in whose preparation he assisted.

Rehn remarks that Viereck's nervous temperament and idealism became restive "under official requirements or regulations" and he frequently severed connections. From 1905 to 1907 he was an assistant in the pathology laboratory of a medical college which Rehn does not name. In 1907 and 1908 he worked for the Pennsylvania Department of Zoology; from 1909 through 1913 for the Bureau of Entomology in the National Museum, where he studied the ichneumonids; and in 1914 he was an entomological explorer for the California State Horticultural Commission. In the same year, Viereck established himself in Sicily and discovered the Sicilian citrus mealy-bug parasite, *Leptomastidea abnormis* (Girault). From 1916 through 1923 he was on the staff of the U.S.D.A. Biological Survey, and sometime between 1923 and 1926 he was Assistant Entomologist with the Entomological Branch of the Canada Department of Agriculture.

Viereck married Ida Adele Davis, a widow, in 1918, and never recovered from her death the next year of pneumonia. He travelled with friends to Colombia in 1922 as their guest, and there he became acquainted with tropical insect life. In 1926 he resumed work at the Academy in Philadelphia but lived with a sister in Irvington, New Jersey.

By 1928 Viereck had written ninety-two papers on the Hymenoptera. In addition, he was largely responsible for the section on the Hymenoptera in J. B. Smith's 1910 edition of the *Insects of New Jersey.* Although his main specialty in bees was the Andrenidae, he also published papers on the ichneumon flies.

Viereck's life, replete with tragedies, ended with one: on October 8, 1931, while collecting along a road near Loudenville, Ohio, for that state, he was killed by a hit-and-run motorist.

Figure 117. L. H. Weld

LEWIS HART WELD (1875–1964)

Weld was an authority on cynipid gall wasps. He was born on December 30, 1875, on a farm near Medina, New York. In 1896 he entered the University of Rochester and graduated with an A.B. in 1900. He then undertook graduate work at the University of Michigan, the University of Syracuse and Cornell. In 1904 he became an instructor in biology at the Evanston Academy in Illinois and remained there until 1917. During the summers he traveled widely, collecting and studying cynipid galls. He was a member of the U.S. National Museum from 1919 until 1924, when, being of independent means, he gave up his regular employment but remained a collaborator with the U.S.D.A. on cynipid wasps. He published many papers, monographs, and handbooks on cynipid galls, some of these privately printed. In 1952 he summed up all his work on cynipids in a large volume entitled *Cynipoidea (Hymenoptera), 1905–1950*. This was privately printed, as were the *Cynipid Galls of the Pacific Slope* in 1957, the *Cynipid Galls of the Eastern United States* in 1959, and the *Cynipid Galls of the Southwest* in 1960.

Weld died at his home in Arlington, Virginia, on April 22, 1964, at the age of eighty-eight. His obituary was written by Burks (1965).

JAMES CHAMBERLIN CRAWFORD (1880–1950)

Crawford, a specialist on bees, chalcids, and thrips, was born in West Point, Nebraska, on August 24, 1880, and died on December 20, 1950.

Figure 118. J. C. Crawford

He attended the University of Nebraska and for a short time was head of the Biology Department of the University. Later he obtained an M.S. from George Washington University. In 1904 he served as a Special Field Agent on cotton insects for the U.S. Bureau of Entomology and published articles on the Hymenoptera. In 1908 he succeeded W. H. Ashmead of the U.S. National Museum becoming Assistant Curator and later Associate Curator. There he worked primarily on bees and chalcids.

In 1919 he left the National Museum to engage in business. From 1923 to 1929 he was employed by the North Carolina Department of Agriculture at Raleigh and from 1930 to 1940 did quarantine work and insect identification for the U.S. Bureau of Entomology and Plant Quarantine in the Port of New York. There he studied the Hymenoptera and Thysanoptera and in 1940 became a specialist on thrips in the Bureau's Division of Insect Identification, where he remained until his retirement a few months before his death. He wrote about seventy-five papers on bees and chalcids, and over twenty-five on thrips. Gahan *et al.* (1951) wrote his obituary.

ROBERT ASA CUSHMAN (1880–1957)

Cushman was born in Taunton, Massachusetts, on November 6, 1880. He studied at the University of New Hampshire and Cornell University. In 1906 he was appointed a field agent in the U.S. Bureau of Entomology, and his first job was concerned with the cotton boll weevil and its

Figure 119. R. A. Cushman

parasites. In 1911 he worked on fruit insects in Virginia and became associated on a part-time basis with taxonomists in the U.S. National Museum, where he studied ichneumon flies and chalcids. After 1911 he was shifted to North East, Pennsylvania, to work on grape pests and their parasites. In 1927 he was sent to the Philippines to arrange and pack the C. F. Baker collection for shipment to the Museum. Cushman was very active in the Entomological Society of Washington and was president in 1925. Upon his retirement from the Museum in 1944 because of ill health, he moved to California. He died at his home in Altadena on March 28, 1957. Muesebeck (1957) wrote his obituary.

ARTHUR BURTON GAHAN (1880–1960)

Gahan was born on December 9, 1880, on his parents' farm near Manhattan, Kansas. His mother died when he was seven, leaving eight other children.

Gahan attended the local schools and Kansas State College and worked during the summers harvesting wheat. In 1903 he graduated from Kansas State College; in 1904 he became an assistant in the Department of Entomology in Maryland Agricultural College (University of Maryland). He received an M.S. from the College in 1906 and remained there as Assistant Entomologist until 1913. While there he became interested in parasitic Hymenoptera and began the study of the braconid parasites of aphids. In 1913 he was appointed Assistant Entomologist in the Division of Cereal and Forage Insect Investigations of

Figure 120. A. B. Gahan

the U.S. Bureau of Entomology and was assigned to the National Museum. As a member of the Museum he worked on the Ichneumonidae, Braconidae, Proctotrupoidea, and Chalcidoidea. Later he concentrated on the chalcids and became an authority on these parasites.

Gahan married in 1908 and made his home in College Park, Maryland. Cory and Muesebeck (1960) report that "One of his hobbies was gardening, especially the growing of dahlias, gladioli and Amaryllis; and he was fond of sports. Although a quiet man and slow of speech, he had a dry wit and a surprising sense of humor that made him always a delightful companion."

Gahan was active in civic affairs and in the Entomological Society of Washington, and was president in 1922.

He died on May 23, 1960, and was survived by his widow, a son and a daughter. Five pages of publications listed in his obituary are a tribute to his accomplishments as an entomologist.

HERBERT FERLANDO SCHWARZ (1883–1960)

Schwarz was born on Fire Island, near Long Island, New York, on September 7, 1883. He received his B.A. from Harvard in 1904, where he studied literature, writing, and languages. In 1905 he obtained an M.A. in philosophy and in 1907 an M.A. in Elizabethan literature, both from Columbia.

Keenly interested in natural history and Indian lore, Schwarz travelled

Figure 121. H. F. Schwarz

to the Southwest in 1904 and 1905 to study the lives and myths of the Navajo and Pueblo tribes.

From 1909–1917 he was connected with a publishing firm in New York and helped F. E. Lutz to edit and publish the latter's *Field Book of Insects*. Gertsch (1961) writes, "Lutz and the gentle, reserved Schwarz were about the same age and they quickly became close personal friends. They were opposites in many ways, with Lutz a man of penetrating mind who loved nothing more than to shock friend and foe with piercing barbs. Lutz kindled in Schwarz his first interest in insects and, because of his own liking for biology and physiology of the bees, directed Schwarz's attention to the study of these captivating social insects. On many occasions Herbert Schwarz expressed his great admiration for Frank E. Lutz and regarded him as his teacher and mentor."

In 1919 Schwarz accompanied Lutz on a field trip to Colorado, where he met Professor T. D. A. Cockerell, who further encouraged him in his studies of bees. In 1921 Schwarz was appointed a Research Associate in the Department of Entomology in the American Museum of Natural History. From 1921–1925 he was the editor of *Natural History Magazine*. When Lutz died in 1943, Schwarz was acting chairman of the Department of Entomology and an authority on the taxonomy, biology, and natural history of the stingless bees (Meliponidae). He wrote several papers on the Meliponidae which were printed in several journals published by the American Museum of Natural History. He collected widely throughout the United States as well as in Colombia, Mexico, and Central America.

Fluent in several languages, he was, according to Gertsch (1961), regarded as "one of those rare individuals whose high code of honor and genuine sincerity charmed all who met him."

He died October 2, 1960.

ALEXANDRE ARSENE GIRAULT (1884–1941)

An economic entomologist and taxonomist, Girault made a name for himself both in the United States and Australia. He was born on January 9, 1884, in Annapolis, Maryland. The year after he received a B.S. from the Virginia Polytechnic Institute in 1903, he became a field assistant in the U.S. Bureau of Entomology. For four years he conducted biological studies on the plum curculio, the Colorado potato beetle, and the lesser peach borer. Then, in 1908 he became an assistant in the laboratory of Dr. S. A. Forbes, the State Entomologist of Illinois, and from 1909 to 1911 he served as assistant in entomology at the University of Illinois. There he conducted studies on insects infesting stored products, the Colorado potato beetle, and the bedbug. His investigations on the biology of the bedbug were extremely thorough.

In 1911 he was employed as an entomologist by the Bureau of Sugar Experiment Stations in Queensland, Australia, where he became an expert on the classification of parasitic Hymenoptera. Returning to the United States in 1914, he renewed his association with the Bureau of Entomology to work in Washington on the classification of chalcids. In 1917 he was appointed assistant entomologist at the Queensland Department of Agriculture and Stock, thus resuming residence in Australia. There he prepared his monograph of over nine hundred pages, *Australian Hymenoptera Chalcidoidea,* published in the *Memoirs of the Queensland Museum* (*1912–1915*). He also specialized in thrips.

Girault wrote in all about 325 papers, most of which, according to Muesebeck (1942), were merely brief notes and descriptions of new genera and species, some of them "badly printed on a small press of his own. These privately printed papers were distributed to comparatively few institutions and workers and, therefore, have not been readily available to all students of the chalcid flies. Brilliant, industrious, and a keen observer, Girault might have contributed vastly more than he did to sound progress in the study of the Chalcidoidea. Unfortunately he was erratic and not suited temperamentally for participation in undertakings that require co-operative effort. He worked alone, largely ignoring other investigators in the same field and apparently unmindful of difficulties he might be creating for future workers. This was, indeed, carried to the point of irreparably damaging type specimens of many species of chalcid flies in efforts to see certain specific structures, to which he happened at the moment to attach special importance. Furthermore, his descriptions are mostly inadequate for the recognition of the genera or species to which they apply. Undoubtedly, however, these shortcomings in large measure are ascribable to continuing ill health, and it must be acknowledged that in spite of them Girault added appreciably to our knowledge of a large and exceedingly complex group of Hymenoptera."

Muesebeck concludes his remarks by noting that "his comparatively few papers on the biology and control of various economic insects reflect an amazing capacity for observing and recording minute biological details."

Girault died in Brisbane, Australia, on May 2, 1941.

WILLIAM M. MANN (1886–1960)

Mann was born in Helena, Montana, on July 1, 1886. He attended Staunton Military Academy in Virginia and studied entomology in Washington State College and Stanford University under A. L. Melander, R. W. Doane, and C. T. Brues, receiving his B.S. from Stanford in 1911. Later he studied ants with William M. Wheeler and zoology with Thomas Barbour at Harvard University, where, in 1915, he received an Sc.D. for his work on ants. Wilson (1959) writes that most of Mann's entomological field work was conducted while he was a Sheldon Travel-

Figure 122. W. M. Mann. Courtesy of the Smithsonian Institution

ing Fellow of Harvard University during 1911–1917. Mann collected ants and animals of all kinds in Cuba, Haiti, Mexico, Turkey, Asia, the South Sea Islands, British Guiana, Argentina, and along the Amazon River in South America.

From 1917 to 1925 he served as a specialist on ants in the U.S. Bureau of Entomology. In 1925 he became director of the National Zoological Park in Washington, D.C. According to Wilson, "A common story has it that Mann's faculty sponsor at Harvard, William Morton Wheeler, was at first keenly disappointed when he abandoned a full-time career as entomologist for zoo-keeping but soon became completely reconciled by his former student's obvious genius in the latter role."

Snyder *et al.* (1961) provide the following description: " 'Doc' Mann was a colorful and distinguished personality, small in stature, wiry, careless of dress, with a puckish expression, and a keen sense of humor. Kindly and generous, he liked people and had friends in many walks of life. As director of the National Zoo he met presidents, congressmen, reporters, circus clowns, executives, 'show people,' and children, and created great interest in all. He encouraged interest in animals, especially by children whom he loved, and who came constantly to see him at the zoo or at home. . . . Bill wanted people around him; he liked to give food and drink. He was sensitive and emotional. When any of the baby animals neglected by the mother were taken home to be cared for by Lucile [his wife] and died, Bill took it very hard."

Besides many scientific papers, Mann wrote two popular works that are highly regarded: *Ant Hill Odyssey* (1948), which is biographical and describes his many explorations; and "Stalking Ants, Savage and Civilized," in the August 1934 issue of *The National Geographic.*

Mann died on October 10, 1960, after a long illness. His collection of ants and ant guests are in the Museum of Comparative Zoology at Harvard and in the U.S. National Museum.

GRACE ADELBERT SANDHOUSE (1896–1940)

Miss Sandhouse was born on June 1, 1896, at Monticello, Iowa. She graduated from the University of Colorado in 1920, where she studied entomology and zoology with Professor T. D. A. Cockerell, who led her to the study of bees. She received her Master's degree from that University in 1923, and her Ph.D. from Cornell in 1925, where she held a teaching fellowship for 1924–1925. She joined the Division of Insect Identification of the U.S. Bureau of Entomology in 1926. There she rapidly became an authority on the Hymenoptera, especially on the bees of the genus *Halictus* and *Osmia,* and published important monographs

Figure 123. G. A. Sandhouse

on them. She died in a hospital in Denver, Colorado, on November 9, 1940, after a long illness.

Cushman and Russell (1940) wrote her obituary.

VERNON SENNOCK LYONESSE-LIANCOUR PATE (1903–1958)

Vernon Pate was born in Philadelphia on August 31, 1903. He received his A.B. from Cornell in 1928 and his Ph.D. in 1946. During the summers between 1927 and 1933 he worked on stream surveys of the aquatic insects of New York. He was an instructor in taxonomy at Cornell from 1932 through 1947 and assistant professor from 1948 to 1952, when he resigned.

From 1929 to 1948 Pate published ninety-two papers on sphecoid wasps. While at Cornell he studied the Hymenoptera with J. C. Bradley, an authority on the order, and received his Ph.D. on the reclassification of the crabonine wasps. He had a thorough knowledge of the zoogeography, paleontology, and evolution of wasps. Although he was greatly interested in wasp biology and behavior, he never published anything on the subject.

Krombein (1961) gives a brief glimpse of him at Cornell: "A generation of students will remember Pate's combined office and laboratory. One had to thread his way to the inner sanctum through a maze of insect storage cabinets and bookcases arranged in baffles. The air was blue

Figure 124. V. S. L.-L. Pate

from his chain-smoking, and cigarette ashes were dribbled liberally over the floor, tables and shirtfront of the occupant. These students also remember the stimulating classroom and laboratory lectures that were made so vivid and meaningful by Pate's broad knowledge of biological fundamentals and his ability to relate these to the particular subject of discussion."

Pate died in Philadelphia on October 30, 1958.

12 Notable Dipterists

> How fares it sadly with the Man
> Whose soul doth patience lack
> When he to smite fugacious flies
> Himself doth fiercely whack.
> *Entomological News,* 2 (7): 140 (1891)

Flies have plagued man from the very beginning of their long acquaintance. A few, and fortunately, relatively few, are important carriers of disease or are plant pests. Some entomologists study flies as disease vectors or as crop-feeders and others just because they are there. In this chapter I tell the story of entomologists who have made important contributions to our knowledge of flies.

CHARLES ROBERT OSTEN SACKEN (1828–1906)

C. R. Osten Sacken, also known as Karl Robert Romanovich baron von Osten Sacken, once wrote an American dipterist, "As the Grandfather of American Dipterology, I am very much interested in the progress of my descendants." His work on North American flies will always be linked with the efforts of the famous German dipterist, Hermann Loew, a link which is recorded in Osten Sacken's book, *Record of My Life Work in Entomology,* published in 1903.

Osten Sacken was born in St. Petersburg (Leningrad) in August 1828. He began the study of insects at the age of eleven after a young Russian nobleman introduced him to the pleasures of collecting. He received his education in St. Petersburg. He collected in all orders except the Lepidoptera and while in Russia published on the insects of St. Petersburg.

In 1856, according to his own account, he

> was appointed Secretary of the Legation in Washington, and started for my destination in the first days of April of that year. During my journey, which lasted two months, I made the acquaintance of the principal entomologists and

Figure 125. C. R. Osten Sacken

zoologists in the cities I visited. At Konigsberg, Prussia, I met Dr. H. A. Hagen, and formed an acquaintance which ripened into a lifelong friendship, and became of great importance for American entomology. . . .

The *second* period of my entomological career embraces the *twenty-one* years of my residence in the United States (1856–1877), during which, until 1862, I was secretary of the Russian Legation in Washington; in that year I was appointed Consul General of Russia in New York, which thus became my residence between 1862 and 1871. I resigned my post in 1871 and made several journeys to Europe and back, until in the autumn of 1873 (this time as a private citizen), I again settled in the United States, where I remained till 1877. These twenty-one years were, as regards entomology, principally devoted, in collaboration with Dr. *H. Loew,* to the task of working up the *Diptera of North America* north of the isthmus of Panama. A great deal of my time, as will be seen, was spent in acting as a purveyor of material for Loew to work upon, and as a translator and editor of his manuscripts. . . .

I promised to send Loew as much material as I could, on the condition, however, that he should consider the collection thus gradually accumulating in his hands not as his property, but *as a trust.* My purpose was, by this means, to form a collection of North American Diptera containing the type specimens described by Loew, as well as specimens determined by him from earlier authors, and as the case might be, an abundant supply of as yet undescribed and determined specimens. Such a collection I expected, sooner or later, to be brought back to the United States, in order to form a solid foundation for the further study of the American fauna. This scheme enabled me to receive without stint the numerous contributions in collections and specimens, which were, *most generously,* put at my disposal by different collectors during my long residence in the United States. As will be shown in the sequel, this scheme came to a successful conclusion, principally in consequence of the generous intervention of Prof. Louis Agassiz. In 1877 this collection, containing (according to Loew's estimation) about 1350 species described by himself, 330 species described by earlier authors, and a large number of undescribed species, forming a total of about 3000 species, came back to the United States and was safely housed in the *Museum of Comparative Zoology* in Cambridge, Mass. At the same time Loew received from the Museum a liberal remuneration for his work on the collection.

Loew published a number of papers in German periodicals on the Diptera he received from Osten Sacken. The first of three volumes by the Smithsonian Institution entitled *Monographs on North American Diptera,* which Osten Sacken edited, appeared in 1862. Osten Sacken himself worked on the Tipulidae, Tabanidae, and Cecidomyiidae, and on the Cynipidae in the Hymenoptera. In 1858 the Smithsonian Institution published his *Catalogue of the Described Diptera* and in 1878 his *Catalogue of North American and Western Diptera.*

In 1873, when Osten Sacken returned from a stay in Europe, he settled in Cambridge to pursue his studies of the North American Diptera at the Museum of Comparative Zoology as one of the celebrated group of entomologists associated with Dr. Hagen.

Between December 1875 and September 1876 Osten Sacken travelled and collected throughout California, the Sierra Nevadas, and the Rocky Mountains; and some of his Diptera were described in a paper published in 1877, after which he wrote "I bade farewell (and this time for good) to the United States and sailed in June for Europe."

Once in Europe again, Osten Sacken went to Guben, Germany, where Loew, then in ill health, was living and undertook the "delicate negotiations, by which Loew was reimbursed for his labors on the North American material and surrendered it all to the Museum of Comparative Zoology . . ." (Aldrich, 1906) Osten Sacken supervised the packing and shipping of the collection, which fortunately arrived safely in Cambridge.

Within a few months Osten Sacken settled in Heidelberg. Some time after 1877 he acquired Philipp C. Zeller's collection of Diptera and began to work on a catalogue of the Diptera of the world, excluding Europe, studying the entomological collections of museums throughout the Continent.

Then in 1894 he returned to Russia to serve in the Imperial Foreign Office.

There can be no doubt about Osten Sacken's help to the newfledged dipterists in the United States by corresponding with and naming flies for them, for he said that between 1856 and 1872 he received 617 letters from ninety-nine American correspondents. Out of gratitude to Professor Agassiz for purchasing the Loew collection, Osten Sacken presented his own collection of Diptera to the Museum of Comparative Zoology.

Aldrich (1906) says that Osten Sacken "wrote in Russian, German, French, Italian, English, and on occasion in Latin; he preferred English, in which he had a literary style distinguished for clearness, force, and accuracy. The striking qualities of his character were energy, farsightedness, persistence, keen discrimination and conscientiousness. No pecuniary consideration ever lessened the completeness of his devotion to the Diptera." However, he had the unfortunate habit of thinking that those who disagreed with him, including Loew, did so, not because they had their own opinions, but because they were mildly insane or had a personal animus against him.

Osten Sacken died in Heidelberg on May 20, 1906.

SAMUEL WENDELL WILLISTON (1852–1918)

Although entomologists honor Williston for his work in dipterology, he was even better known as a paleontologist. Among those who have written about him are Osborn (1919), McClung (1919), Aldrich (1919), and Lull (1924).

Williston was born in Roxbury, now a part of Boston, on July 10, 1852. His father was an intelligent but illiterate blacksmith, his mother a woman of sound education. In 1857 the family emigrated to Kansas and settled near the town of Manhattan in a small log cabin "about 15 feet square, containing a single room below and a loft above where the four boys slept. Indians were numerous and troublesome." (Osborn)

His mother was determined that her boys should not be handicapped by the illiteracy that hobbled her husband. Since Samuel was less robust than his three older brothers, he made up his mind to excel them in learning. He became a voracious reader and by the age of seven had consumed such books as Stevens' *Antiquities of Central America* and William H. Prescott's *Conquest of Mexico.*

Samuel first collected fossils from a bluff nearby, which, he was told, were left by the "great deluge." In 1866 at the age of fourteen he was admitted to the Agricultural College of the Kansas State University in Manhattan. He paid for his education by means of hard manual labor and also learned the printer's trade. He took every course given by Professor Benjamin F. Mudge, including "Natural Philosophy, Chemistry, Botany, Geology, Zoology, Veterinary Science, Mineralogy, Surveying, Spherical Geometry, Conic Sections, Calculus, etc." (Osborn)

At the age of fifteen, after reading Charles Lyell's *Antiquity of Man,*

Figure 126(A).

Figure 126(B). (A) S. W. Williston. (B) S. W. Williston and J. M. Aldrich; courtesy of J. J. Davis

Williston became a disciple of Darwin. Some seven years later, says Osborn, he "delivered in the local Congregational Church what he believed was the first public lecture given west of the Mississippi River in favor of evolution." For this he was severely criticized by members of the church.

His college education was interrupted when at eighteen he ran away from home to work on a railroad. Before he was twenty he became a transitman at a fine salary, but after a bout with malaria he returned to college, from which he graduated in 1872. In the panic of 1873, railroad engineering was a deflated profession, so he began to "read" medicine in the office of a local physician. At this time another medical student got him interested in beetles, and he became an ardent collector. During the summers of 1874 and 1875 he assisted Professor Mudge on fossil-collecting expeditions in western Kansas for Professor O. C. Marsh of Yale University. During the winter of 1875–1876 he studied at the medical school of the University of Iowa and in the spring of that year was invited to become Marsh's assistant at a salary of $40 a month.

One of his first jobs with Marsh was to study bird skeletons using Richard Owen's *Comparative Anatomy* as a guide. Later he was sent by Marsh to collect the bones of giant sauropods (herbivorous dinosaurs) in Colorado and Wyoming. He published a brief account of his observations in 1878, but Marsh would not permit him to publish anything further on paleontology during the nine years he worked for him. This led Williston to study the Diptera, which Marsh did not consider part of his province. Williston received his M.D. from Yale in 1880 and then pursued a Ph.D. under Marsh, which was awarded him in 1885. The next year he served as a demonstrator of anatomy at Yale and an assistant editor of *Science* under Professor S. H. Scudder. This meant editing in Boston during the day and teaching anatomy in New Haven at night. The following year Yale University recognized his ability and appointed him a full professor of human anatomy. C. V. Riley offered him the first assistantship in the Division of Entomology, but Williston declined despite the higher salary, because he was aware of the difficulty of working under Riley.

In 1890 after three years as a professor at Yale, he was appointed Professor of Historical Geology and Palaeontology at the University of Kansas. "Twelve years of his prime were spent in this institution," writes Aldrich, "years crowded with productive labor. He helped organize the medical department of the University and took on the deanship of it along with his other work; this almost broke his vigorous health, and he had to slacken his pace—perhaps never again quite regained it [his health]."

Aldrich, who had studied under Williston and was himself an eminent dipterist, sums up Williston's contributions to entomology as follows: "Williston never held an official entomological position. But he found time to do much valuable work as a pioneer in dipterology. . . . At this time Osten Sacken had returned to Europe, and there was not a single American student of the order but Edward Burgess, the Boston yacht designer, who published only one small paper. In the absence of guidance, he plowed his way by main strength (as he often narrated to the writer) through descriptions of species until here and there he made an identification, which served as an anchor point for a new offensive." His first paper on flies dealt with the wing venation of the Bombyliidae and appeared in the *Canadian Entomologist* for November, 1879. He was greatly influenced by Schiner's *Fauna Austriaca,* with its keys to genus and families, and he decided to emulate it with a similar work on American flies.

During the following decade he published what Aldrich called "tentative papers analyzing the American families and genera of the flies. These he extended and enlarged in a pamphlet in 1888, and again in a bound volume in 1896; and in 1908 published a third edition still more complete with 1,000 figures, his well-known Manual of Diptera. This third edition is his main contribution to entomology. It is a handbook unapproached by anything else dealing with a large order of insects. From necessity he published it at his own expense; it was eight years before the receipts from sales covered the cost of printing, but happily he lived to see this consummation."

He also worked on the Asilidae, Conopidae, Tabanidae, and tropical Diptera. "As his official duties grew more exacting," Aldrich continues, "he gradually abandoned entomology, but he had as many farewell appearances as an opera singer, for he could not resist the temptation to come back again and again." After 1896 his other duties prevented him from spending much time on flies, except for the two years he spent preparing the third edition of the *Manual of Diptera,* which he illustrated with some eight hundred of his own figures.

Williston's types are scattered about in a number of museums, the National Museum, the Museum of the University of Kansas, the British Museum, and the American Museum of Natural History. "Williston did not believe in designating a single type specimen, hence in some cases his types of the same species are in two museums. He had no collections of Diptera in his last years, although he still retained his fine library in the order." (Aldrich)

Although Williston never gave formal entomology courses he assisted and helped train such dipterists as W. A. Snow, Hugo Kahl, C. F. Adams, A. L. Melander, and J. M. Aldrich, among others. Aldrich says that it was Williston's presence at the University of Kansas that drew him there in January, 1893. "He received me with open arms, and helped me in every way possible until I left in July to take up my work in Idaho. Then I saw him only a time or two in twenty years, and had few and short letters from him, for he was a notably poor correspondent. After coming to Indiana in 1913 I was so near that we were frequently together. My sketch would be entirely inadequate without some acknowledgement of my personal obligation. In Kansas he lent me money; he wanted me to live in his house; he could not do enough to further my scientific aspirations." Williston told Aldrich that some of the happiest hours of his life were spent working on the Diptera.

When Williston went to the University of Kansas in 1890 as a professor of geology he also taught classes in anatomy, physiology, histology, embryology, evolution, and meteorology. According to McClung, who studied under him, Williston was a "born" teacher but did not care for large classes. McClung cites one of his students as writing: "I believe that the 'Mark Hopkins and a log' idea of a university was never more nearly realized than in Dr. Williston. His knowledge of men and things were so wide and his acquaintance with many branches of science so intimate that in the heart of a barren fossil field, or under the stars at night by the side of a camp fire, some bird, or flower, or fossil, some insect—'one of mine, I named it in 1870-odd'—would start a talk that held his little band of student assistants enthralled until hunger, thirst, or sleep were forgotten."

While at the University of Kansas he made many forays into fossil fields and published studies of Cretaceous reptiles, the plesiosaurs, mosasaurs, sea turtles, and pterodactyls, as well as in anatomy and on the Diptera. McClung adds that he, "more than any other man in the institution, understood and lived the life of a real investigator, and the students who went out from his laboratory carried the inspiration and methods of his life with them." It is not at all unusual that a man of his abilities should have aroused the jealousy of lesser men on the University staff. Conditions finally became so unbearable that with great reluctance he accepted a position at the University of Chicago; the post was one of the most distinguished professorships in vertebrate paleontology in the United States.

At the University of Chicago he lectured and continued his investigations of Cretaceous sea reptiles, besides completing the third edition of his *Manual of North American Diptera*. In 1907 he began to publish studies of the reptiles of the Permian period, based partly on the excellent specimens he collected in Texas. These studies were summarized in 1914 in his *Water Reptiles of the Past and Present*. According to Osborn his work "marks the transition period between the work of the founders of American paleontology—Joseph Leidy, E. D. Cope and O. C. Marsh—and that of the large and increasing younger school of men who are taking up this wonderful subject and who may well follow his high example of unswerving integrity as an observer and broad philosophy as a generalizer." Osborn speaks of Williston's fine relationships with colleagues, assistants, and students, which were not marred by the jealousies prevalent among some of his famous predecessors.

Aldrich remembers Williston as a large and vigorous man of great mentality, who had positive opinions but who did not indulge in controversy or personalities. "My last mental picture of the man represents him on a day last winter, sitting at a table before a window in his study at home, in one hand a long-snouted reptilian skull, in the other a drawing pen with which he was rapidly making a sketch of it."

A man of Williston's capabilities was naturally the recipient of many honors. Yale gave him an honorary D.Sc. and he was a member of the National Academy of Science. He was a very active and prominent member of Sigma Xi.

He died in Chicago on August 30, 1918, leaving a wife, three daughters, and a son.

DANIEL WILLIAM COQUILLET (1856–1911)

Coquillet was not only a dipterest but also an outstanding economic entomologist and taxonomist, and a pioneer in biological control of harmful insects. Banks (1911), Cresson (1911), Walton (1914), Essig (1931), Doutt (1958) supply the information for this sketch, though others have written about him.

Coquillet was born on a farm in Pleasant Valley near Woodstock, Illinois, on January 23, 1856. As is indicated by his name, he was of French origin. Coquillet helped on the farm, and received enough education locally to teach for a term or two. He also collected insects, especially moths and butterflies, which were sent to A. R. Grote of Buffalo for identification. Coquillet and his brother were also interested in birds, and together on a small hand-press they printed a booklet on *The*

Figure 127. D. W. Coquillet

Oölogy of Illinois, which described the eggs and habits of the birds in the state.

His first paper on entomology, "On the Early Stages of Some Moths," published in the *Canadian Entomologist* for 1880, Vol. 12, No. 3, caught the eye of the entomologist Cyrus Thomas. Later he prepared more extensive articles on the larvae of the Lepidoptera for Thomas' *Tenth Annual Report* (1881) and contributed other articles to the succeeding reports, including one on the army worm.

When in 1882, Coquillet developed incipient tuberculosis, his parents moved to Anaheim in Southern California. There he continued to collect insects, specializing in flies, especially bee flies. He also studied the beetles of California. C. V. Riley became aware of Coquillet's entomological publications and in 1885 appointed him a field agent of the U.S.D.A. Division of Entomology. In the same year Coquillet learned of the success some farmers were having with a poison mash for locusts. This mash consisted of, by weight, one part arsenic (Paris green or white arsenic), one part sugar, and six parts bran, with enough water to make a wet mash. Coquillet made this formula known to C. V. Riley, and until World War II this bran mash was one of the most successful methods for controlling grasshoppers, as well as crickets, earwigs, cutworms, snails, and other pests.

Some of Coquillet's studies on the poison bran mash were made on the Buhach Farm in Merced County, California, where he was studying the cultivation of pyrethrum, the insecticidal agent so commonly used in household sprays and aerosols. These studies appeared in 1886 as U.S.D.A. Bulletin 12, *Report on the Production and Manufacture of Buhach.*

Poison bran mash and Buhach were incidental to Coquillet's main job in Southern California, which was to find ways to cope with the destructive cottony cushion scale, *Icerya purchasi* Maskell, then devastating the citrus groves of the state. Albert Koebele, another U.S.D.A. entomologist situated in Northern California, was sent South to work with Coquillet on this pest. As we have seen in the sketch of Koebele, where a fuller account of the Vedalia affair is given, the pair did not get along, and when Coquillet was laid off for about a year, supposedly because of insufficient funds, he pursued his studies on the control of citrus pests and, working with J. W. Wolfskill and Alexander Craw on the Wolfskill ranch near Los Angeles, discovered a method of controlling scale insects with hydrocyanic acid gas (HCN). Coquillet and his associates attempted to obtain a patent on the HCN fumigation of citrus trees, but when rumors of their success came to the ears of other growers, the University of California had their chemist F. W. Morse study the problem; in six months' time he made an independent discovery of the efficacy of HCN fumigation.

Riley rehired Coquillet in 1887 even though chagrined that Coquillet had made his HCN discoveries on his own time rather than as a U.S.D.A. employee. Coquillet then continued his HCN studies as a member of the Division of Entomology. In the same year, Koebele was sent to Australia as a quasi member of a commission to the International Exposition to Melbourne with the actual duties of studying the parasites of the cottony cushion scale. As noted earlier this was the occasion of his introduction in this country of dipterous parasites of this scale and of larvae of the Vedalia beetle, which likewise preyed on the scale. Despite Coquillet's success on the Wolfskill Ranch with the Vedalia sent him by Koebele, he mostly received complaints and criticism from both Koebele and Riley.

Fortunately, this was not the end of the trail for Coquillet. He continued his economic work in California and also became an authority on bee and robber flies. In 1893, Riley's relations with the California growers had become more than strained, and he recalled Coquillet to Washington. For a time, Coquillet worked on the San Jose scale in the East, and in 1896 he became an honorary custodian of the Diptera in the U.S. National Museum. He published extensive works on the taxonomy of the Tachinidae and shorter works on buffalo gnats, mosquitoes, and other Diptera. His *Type Species of North American Genera of Diptera* (1910) has been very helpful to American dipterists.

Coquillet was a "lumper," not a "splitter," in taxonomy and he apparently paid little attention to slight differences between some species. All in all, he described about a thousand species of Diptera. According to Walton, "The work of determining the great mass of material received by the Museum, gave him little time to devote to the descriptions of new forms, so that we are now complaining of his short diagnoses."

Essig describes him as "a tall, quiet, but energetic man who was well liked and who did a great deal to assist the younger entomologists of his time," but he was also shy and ascetic.

When he died in Atlantic City, New Jersey, on July 7, 1911, his excellent collection of Diptera became part of the U.S. National Museum.

CHARLES HENRY TYLER TOWNSEND (1863–1944)

According to James (1945), Townsend was born in Oberlin, Ohio, on December 5, 1863, and attended high school in Michigan. From 1887–

Figure 128. C. H. T. Townsend

1891 he took a medical course at Columbia College in South Carolina while working as an assistant to C. V. Riley. He was one of the first to study the habits of the cotton boll weevil in Texas, to appraise it as a potentially injurious pest, and to recommend methods for its control.

From 1891 to 1893 he taught at the New Mexico College of Agriculture and Mechanical Arts in Las Cruces. In 1893, in an exchange of positions with T. D. A. Cockerell, he became Curator of the Public Museum in Kingston, Jamaica. He worked for the New Mexico Agricultural Experiment Station in Las Cruces from 1898 to 1899, and taught in the Philippines from 1904 to 1906. He received a B.S. and a Ph.D. from George Washington University in 1908 and 1914, respectively.

In 1909 Townsend was appointed Entomologist and Director of the Experiment Station of Peru. After four years in that position, he went to work in the U.S. Gypsy Moth Parasite Laboratory in Melrose Highlands, Massachusetts; and then, after D. W. Coquillet's death, succeeded him as a systematic entomologist in the Bureau of Entomology.

Beginning in 1919 he worked for ten years in Peru and Brazil, making important contributions in the fields of agricultural and medical entomology. Wilson (1942) describes how Townsend discovered the insect vector of verruga: "There was rather impressive evidence that the disease was being carried by insects—perhaps nocturnal insects, since Andean folklore has it that the affliction is acquired at night. The fact that the disease is segregated in particular highland valleys was another significant hint. Charles H. T. Townshend [*sic*], an American entomologist in Peru, began work on the insect theory, suggesting first the tick as carrier, then isolating approximately fifty blood-sucking insects common to the region where the disease is most frequent (9 to 15 degrees south latitude). One by one, Townshend eliminated the suspects until finally he had reduced the eligibles to the horse fly, buffalo gnat, and phelebotomus—the 'vein opener.' The first two he eliminated because they bite only during the daytime. The third he chose because it is nocturnal."

According to James (1945), Townsend was a controversial writer. Perhaps this was why he founded his own publishing company in São Paulo. However that may be, his subjects included, besides insects, biological control, economic biology, taxonomy, the theory of gravity, and the theory of the moon's origin. He worked on the Diptera for more than sixty years, naming over a thousand genera and numerous species of flies. His *Manual of Myiology* was 3,760 pages long and had ninety-four original plates. Much of it is devoted to the muscoid and oestroid genera of the world.

Townsend died quietly in his sleep on March 17, 1944, in a suburb of São Paulo.

FREDERICK KNAB (1865–1918)

Knab was born in Wurzberg, Bavaria, on September 22, 1865. When he was eight years old, he moved with his parents, Oscar and Josephine

Figure 129. Frederick Knab

Knab, to the United States. They settled in Chicopee, Massachusetts, in 1873. Oscar Knab was an engraver and a painter, and one of his brothers served as court artist to the King of Bavaria. Frederick inherited the family's artistic bent and was sent to Munich for two years to study art. When he returned to Chicopee he took up landscape painting.

In 1885–1886 he had an opportunity to pursue his boyhood interest in natural history and insects along the Amazon River. In 1903 he studied the biology of mosquitoes of New England for L. O. Howard. From 1903 to 1904 he worked as an entomological artist for S. A. Forbes of Illinois. He then began to collaborate with Howard and Dyar on mosquitoes. In 1906 he became an assistant in the U.S. Bureau of entomology, where he continued his study of mosquitoes and other Diptera as vectors of disease. When D. W. Coquillet died in 1911, Knab succeeded him as Custodian of Diptera in the U.S. National Museum.

Knab is best known for his collaboration with Howard and Dyar on the prestigious four-volume work, *The Mosquitoes of North and Central America and the West Indies,* which was published by the Carnegie Institute from 1912 to 1917. Knab and Harrison Gray Dyar, at their own expense, had conducted biological studies and collecting trips in subtropical and tropical countries, and the former prepared many excellent illustrations for this work.

According to Caudell *et al.* (1919), Knab died in Washington, D.C., on November 2, 1918, as a result of "an insidious, lingering disease, probably insect-borne, which he contracted during his expeditions to Brazil. Its nature baffled the medical specialist until Mr. Knab himself correctly diagnosed it through the diligent study of the South American medical literature." We do not know what the disease was.

JOHN MERTON ALDRICH (1866–1934)

Aldrich's life is a recapitulation of many others in this book, of the farm boy interested in natural history and insects who, through intelligence and perseverance, makes the most of his opportunities to rise to the top of his profession. His story emerges from his own reminiscences (1930) and from articles by Wade (1930), Walton (1934, 1935), and Melander (1934).

Aldrich was born January 28, 1866, in Rochester, Minnesota, then an obscure village untouched by the fame of the Mayo brothers. When he was fifteen, his family moved to a farm in South Dakota and John attended the high school nearby for two years. It was probably in 1888 that he attended the State Agricultural College at South Dakota State University in Brookings, working his own way through. He received a degree

Figure 130. J. M. Aldrich

in three years because the president of the University was eager to have a graduating class. In his last term of school he studied entomology in a course on zoology presented by I. H. Orcutt, M.D.

After graduating he returned to his father's farm, then decided to apply for the job of assistant to Dr. Orcutt, who, in the interim had become Entomologist to the South Dakota Agricultural Experiment Station in Brookings. Dr. Orcutt wrote that if he would take a course in entomology during the winter, he would make an effort to employ him in some entomological capacity the following year.

Aldrich made a little money teaching one term in a local school and then applied to the University of Minnesota for instruction in entomology. The University gave no formal course but Otto Lugger, a man well qualified to accept a special student, was then Station Entomologist at St. Anthony's Park. An enthusiastic entomologist, he had worked for C. V. Riley in Missouri and, besides being a fine economic entomologist, was well acquainted with the Coleoptera. Aldrich studied J. L. LeConte's and G. H. Horn's *Classification of the Coleoptera of North America,* and through his brief encounter with Lugger's teachings absorbed some of his great enthusiasm for insects.

In the spring he was given a three-months' job at $40 a month in the South Dakota Experiment Station, and in the fall he was placed on the staff at $500 a year with the understanding he would devote his winters to study.

In the autumn of 1889 Aldrich went to Michigan State University at East Lansing to study entomology under Professor A. J. Cook. Aldrich

(1930), curiously speaking of himself in third person, writes: "Professor Cook was described by the speaker [Aldrich] as an excellent teacher and a keen and practical man of affairs, with tremendous energy. He advised young Aldrich to select a single order as a speciality, and to proceed at once to get together a library and collection; he also suggested the Diptera as a large order in which there were but two workers ([S. W.] Williston and [D. W.] Coquillet) at the time in the country. The advice was accepted, and the library and collection duly begun in the spring of 1890. Aldrich became a subscriber of *Entomological News* before the second number of Volume 1 was issued. Williston sent him separates, as did Coquillet and Osten Sacken, and he began buying at an alarming rate out of his small salary."

The following November Aldrich started on a journey to seek a job under Dr. H. A. Hagen and see the insect collection at the Museum of Comparative Zoology at Harvard. Along the way he stopped at Ames, Iowa, to meet Professors Herbert Osborn and C. P. Gillette, and they invited him to accompany them to the meeting of the Association of Economic Entomologists at Champaign, Illinois. "Professor Osborn shared his sleeping-car berth en route with the impecunious student, a characteristic act of kindness never forgotten," Aldrich writes. When he arrived at Champaign, he found that through Professor Cook's efforts he had been elected to the Association. Here, also, he had the opportunity to meet C. V. Riley, L. O. Howard, J. B. Smith, and other well-known entomologists.

When Aldrich arrived at Harvard, his hopes of working under Dr. Hagen were completely shattered, for Hagen had just suffered a stroke and the department and collections were closed. Despite the intervention of a sympathetic secretary, Alexander Agassiz, Louis Agassiz' son, had no interest in the hard-pressed and disappointed youth and made no attempt whatsoever to help him. Then Aldrich set out for Washington, but stopped off at Brown University in Providence, to see Dr. A. S. Packard, who welcomed him warmly and invited him to his home for dinner, where he "spent the evening showing the young entomologist his library and unpublished manuscripts thereby making a most favorable and lasting impression on him." (Walton, 1934)

Aldrich reached Washington in late November. He pawned his watch for five dollars in order to pay for a breakfast, then he went to see L. O. Howard. His faith in Howard was not misplaced: he reports that the latter "took him in charge, suggested a place to stay and arranged about his work in the National Museum where the collection of insects was under the direction of Martin Linell." Aldrich spent about three weeks studying the Diptera, especially Williston's collection of Syrphidae, and attended a meeting of the Washington Entomological Society at Howard's invitation, which, Aldrich says, "was in a private house, and the small room was soon so full of tobacco smoke that at the conclusion the visitor [Aldrich] was obliged to seek fresh air without sharing in the social hour which was then an outstanding feature."

In Washington he became acquainted with Dr. C. H. T. Townsend, a specialist in the Tachinidae, and, as a result, Aldrich became interested in this family and thereafter sent Townsend many of his Diptera for identification.

For want of money, Aldrich was soon forced to return home, where he spent the winter classifying his Diptera by means of his newly assembled library. He returned to Brookings in 1892 to work in the South Dakota Experimental Station and in the summer made his first real collecting trip. He joined Lawrence Bruner, Professor of Entomology at the University of Nebraska and a party, including Bruner's family, on a trip through the Black Hills.

Faculty infighting at Brookings ousted both Dr. Orcutt and Aldrich from the Experiment Station in 1892. Aldrich then headed for the University of Kansas, where, according to Walton (1935), Dr. S. W. Williston lent him money and pressed him to come and live in his home. From that time on Aldrich held Williston in veneration and even tried to model his life and work after him. The University gave Aldrich an M.S., his second, for he had already earned one from South Dakota State College in 1891. In the summer of 1893, he left Kansas to become the first Professor of Zoology at the University of Idaho.

At the age of twenty-seven, Aldrich, as Professor of Zoology and Entomologist in the Experiment Station of the University of Idaho, was ready to plunge into his life's work. He investigated apple insects, grasshoppers, and other phases of economic entomology. He studied the Diptera, especially the Dolichopodidae, and began his monumental *Catalogue of the North American Diptera,* which was published by the Smithsonian Institution in 1905.

In regard to the *Catalogue,* Walton (1905) says: "Too high a figure can hardly be set in estimating the value to students of American Diptera of this volume of nearly 700 pages. This catalogue not only lists 8,300 nominal species, with at least twice that number of references to their literature, but contains innumerable data on distribution and the host relations of the included species. It is safe to say that no other single publication, save perhaps Williston's *Manual of the North American Diptera,* has done so much to stimulate and facilitate the study of the order in America."

It was in 1893 that Aldrich married Miss Ellen J. Roe of Brookings,

South Dakota, a marriage that ended tragically in 1897 with her death and that of their infant son. Eight years later he married Miss Dell Smith of Moscow, Idaho.

In 1906, he took a sabbatical leave to attend Stanford University, where he received a Ph.D. His *Catalogue* was accepted as his thesis. During his absence, the University of Idaho burned to the ground. Fortunately he had moved his library and collection of Diptera to his father's house.

In 1913 Aldrich's association with the University of Idaho was terminated in a fashion that shocked the academic world. Of the circumstances we know only what Melander (1934) reports: "It is unnecessary now to reopen the sorry case and discuss the vagaries of an incompetent administration other than to recall that those of us who knew the situation well regarded the dismissal as an outrageous and unwarranted interference."

Howard immediately appointed him Entomological Assistant in the U.S. Bureau of Entomology under F. M. Webster, who headed the Cereal and Forage Crop Insects section, and he was stationed in West Lafayette, Indiana, for the next five years, investigating the life histories of the stem flies whose maggots bore into the stems or mine the leaves of grasses and cultivated grains, and of other flies as well. Webster suggested that he work on a revision of the North American sarcophagid flies, and in 1916 the Thomas Say Foundation published his monograph *Sarcophaga and Allies in North America.*

J. J. Davis (1958) says a neighbor, who saw Aldrich go out day after day with his net, once remarked to his wife "that he sure would like to have a job where he could go fishing every day."

Shortly after Frederick Knab died in 1918, Howard transferred Aldrich to the National Museum, where in 1919 he was appointed Custodian of the Diptera and Associate Curator, a position he held until his death. He became President of the Entomological Society of America in 1921 and received various other honors in his lifetime.

Walton (1935) describes Aldrich as a genial, kindly man, short in stature, and in his later years inclined to rotundity. He spoke rapidly, in a rather high-pitched voice, and had a fund of dry humor. "He was especially fond of children, and nothing pleased him more than to fill his automobile with underprivileged youngsters just before Christmas and, after having provided each with a bit of money, to haul them about Washington on a shopping tour."

As Custodian of Diptera he completely reorganized the collection and kept up a large correspondence on flies with people all over the world. According to Walton (1934), "His descriptive work possessed a quality that is exceptional and his command of good English was indeed remarkable. All who knew him well will remember with wonder his amazing ability to converse on his beloved order almost 'ad infinitum' and at great speed, without faltering an instant for a shade of expression or a technical name. Dr. Aldrich to the end, maintained his boyish enthusiasm . . ."

Aldrich collected a great deal, especially in the Western states. He also made trips to Canada, Alaska, Guatemala, and Europe. In 1928 he presented his collection of Diptera, consisting of 45,000 specimens and 4,000 named species, to the National Museum. Along with this went his extensive card catalogue and his fine library on the Diptera.

Aldrich died on May 27, 1934.

EPHRAIM PORTER FELT (1868–1943)

For much of the first half of this century, Dr. Felt was one of the most active and respected entomologists in the United States. As State Entomologist of New York, as Editor of the *Journal of Economic Entomology,* as an authority on shade-tree insects and gall insects, and as an author and lecturer he had a profound influence on his colleagues and the lay public. Bromley (1944), Burgess (1943), and Muesebeck and Collins (1944) wrote obituaries of him.

Felt was born in Salem, Massachusetts, on January 7, 1868, the son of Charles W. and Martha Seeth Felt. He attended local schools and worked when he could on nearby farms. He began his higher educa-

Figure 131. E. P. Felt

tion at the Massachusetts State Agricultural College (now the University of Massachusetts) though he had the ministry in mind. However his interest in insects and the influence of Professor C. H. Fernald led him to take up entomology. He received a B.Sc. from Massachusetts State College and also from Boston University, both in 1891. Upon graduation, he became an assistant to Fernald and began to study the gypsy moth, which was accidentally imported into Massachusetts and was causing great injury. Awarded a fellowship to Cornell University, Felt studied under Professor J. H. Comstock during the summers of 1892 and 1893 and received a D.Sc. from that university in 1894. He taught natural science at the Clinton Liberal Institute at Fort Plain in New York from 1893 to 1895. In 1895 he was appointed assistant to the State Entomologist of New York, Dr. J. A. Lintner, succeeding Lintner when the latter died in 1898.

Felt married Helen Maria Otterson in 1896 and made his home at Nassau, near Albany, where he became very active in school, church, and civic affairs. They had three daughters and a son.

His work, *The Mosquitoes or Culicidae of New York State* was published in 1904; his two-volume *Insects Affecting Park and Woodland Trees* appeared in 1906. The latter work contained hundreds of illustrations, including many fine colored plates. Felt also worked on dipterous gall insects, the midges of the family Cecidomyidae; in his lifetime he described 1,060 species in this family. His *Key to the Insect Galls* appeared in 1918, a revised edition in 1925. His *Plant Galls and Gall Makers* was published in 1940.

According to Bromley (1944), "he frequently presented papers at the New York Entomological Society meetings, which included subjects of a varied nature from gall insects to poetry concerning insects, as well as numerous valuable contributions on shade tree insects." He was also well known for his studies on wind-borne insects, and often collected them on tall buildings, including the Empire State. Needless to say, this created a great deal of interest in newspapers and other news media. By these studies he was able to ascertain how certain insect pests spread, for example, the European elm bark beetles, carriers of Dutch elm disease. He was also greatly concerned with the part played by flies, mosquitoes, and other insects as vectors of disease organisms.

Burgess (1943) explains how the *Journal of Economic Entomology* came into being under the editorship of Felt:

> In 1907 the Association [American Association of Economic Entomologists] faced one of the most severe crises in its history. The United States Department of Agriculture had, up to that time, published the annual reports of the Associa-

tion as bulletins of the Bureau of Entomology. Conditions arose so that this arrangement was discontinued and, as the Association at that time had a balance of only $50.12 in the treasury, the outlook for publishing the annual report was most discouraging. However, at that meeting a publishing company was organized among the members to which non-interest bearing stock was sold at $10.00 a share to finance a publication to take care of the annual reports of the meeting and to publish a limited number of important papers which might be presented during the year. The company agreed to publish six numbers annually at a subscription price of $1.00 per annum to each member and $2.00 to non-members, the publication to be the official organ of the Association. This proposal was made by Professor E. D. Sanderson and arrangements were perfected so that on repayment to the stockholders all property belonging to the company, including back numbers, would revert to the Association.

Dr. Felt was elected Editor, the writer, Associate Editor, and through the energy of Professor [E. D.] Sanderson, who was made business manager and Wilmon Newell, who consented to become advertising manager, together with L. O. Howard, James Fletcher, Henry T. Fernald, S. A. Forbes, H. A. Morgan and Herbert Osborn, who were elected as the advisory board, the publication was launched as the Journal of Economic Entomology and the first number published in February, 1908. In 1911 the indebtedness incurred had been paid from Journal receipts and the publication and its assets turned over to the Association. Dr. Felt continued as Editor until 1935 when he retired at his own request and was made honorary editor which position he held until the date of his death.

From 1898 to 1911 he was the entomological editor of the *Country Gentleman* and frequently contributed to the *New York Times*. Five volumes of the *Index to the Literature of American Economic Entomology*, compiled by Miss Mabel Colcord, appeared under his editorship from 1915 through 1942. He often wrote for other nonscientific as well as scientific periodicals; he lectured to groups of all kinds and spoke on the radio. During his thirty years as Entomologist of New York State he issued twenty-five important official reports. In addition to being connected with the New York State Museum as State Entomologist he co-operated with the State Department of Agriculture and worked closely with Cornell and Syracuse Universities as a consultant.

In 1923 and 1924 he continued his studies of the gypsy moth for the State Conservation Commission. He conceived the idea of creating a treated barrier zone to prevent the moth from spreading westward to uninfested parts of the country. In 1928 he retired from his post as Entomologist for New York State to become Director of the Bartlett Tree Research Laboratories; he was recognized as a world authority on the care of shade trees. Three hundred newspapers subscribed to his weekly syndicated newspaper article "Talks on Trees." His *Manual of Tree and Shrub Insects* was published in 1923, and in 1930 he was coauthor with W. H. Rankin of *Insects and Diseases of Ornamental Trees and Shrubs.* He published *Our Shade Trees* in 1938, *Pruning Trees and Shrubs* in 1941, and *Shelter Trees in War and Peace* in 1943.

Bromley (1944), who knew him well and was quite fond of him, writes that he was honest, sincere, upright, and a great scientist. He did not permit trivialities to upset his tranquil nature.

> Since 1929, it was my privilege and honor to work with him side by side on important research and I have never ceased to marvel at his deft ability to size up and conquer a problem and at his unruffled composure in the face of trying obstacles.
>
> He was an avid reader, and enjoyed the radio from the political speeches of the great to the humor of Charlie McCarthy. He had a deep and abiding sense of humor and was quick and accurate at repartee. A remark was once made on his winged collar, so characteristic of his dress. Dr. Felt immediately replied "What could be more fitting to my profession?" pointing to the two-winged insects he was studying.
>
> He was faithful to the minute to his commitments. Many were the occasions when he accepted an invitation to talk at some humble, out-of-the-way garden club or other meeting. . . .
>
> His appearance as well as his character was left unchanged by the surging tide of years and he was until the last the same steadfast, striking figure. His carefully trimmed white hair and Van Dyke beard, his calm, upright bearing, his impeccable dress, his gentle but firm voice, all reflected the imprint of his great character.

Professor S. W. Frost, who was well acquainted with Felt and wrote me his recollections of him, which in general confirm Bromley's, adds that, though "opinionated, his judgement was sound," that he was full of enthusiasm, and that he was much in evidence at "all entomological meetings" because of his frequent attempts to straighten out minor disagreements. He served as secretary and in 1902 as president of the American Association of Economic Entomologists and took part on many important committees.

"He was a friendly person," said Frost, "always willing to assist young entomologists. I was certainly in that category when I knew him best."

On December 14, 1943, Dr. Felt died of a heart attack in his office, shortly after telling his secretary that he felt as though he was coming down with the flu. "With a twinkle in his eye, he said that he had such symptoms before and that nothing came of them. 'You see,' he said, 'I am still pretty tough.' " (Bromley) These were the last words his secretary heard him speak.

JOHN RUSSELL MALLOCH (1875–1963)

Born in Scotland on November 16, 1875, Malloch received a B.S. from the University of Glasgow in 1897, and emigrated to the United States in 1909. He served as Scientific Assistant in the Bureau of Entomology in Washington, D.C., in 1912 and 1913 and as Entomologist in the Illinois Natural History Survey from 1913 through 1921. In 1917 his monograph, *A Preliminary Classification of Diptera, Exclusive of Pupipara, Based on Larval and Pupal Characters,* was published by the Survey. Then from 1921 to 1934 and 1936 to 1938 he worked as a Biologist with U.S. Biological Survey in the City of Washington.

Although interested in the Lepidoptera, Hymenoptera, and Hemiptera, he was primarily a dipterist, and an authority on the acalyptrate

Figure 132. J. R. Malloch

Diptera and Muscidae of the world. He published numerous papers on exotic Diptera and described 114 genera and 851 species of Diptera north of Mexico.

Sabrosky (1963), whose one-page obituary is our only source of information about Malloch, writes that he had a phenomenal memory and carried a "card catalog" in his head, which resulted in some errors. "He must unquestionably be judged one of the keenest, most perceptive, and most prolific of dipterists . . ."

Malloch died in a nursing home in Tampa, Florida, on February 18, 1963.

Figure 133. C. L. Fluke

CHARLES LEWIS FLUKE (1891–1959)

Fluke was born on a farm in Grand Junction, Colorado, on August 7, 1891. He majored in entomology at Colorado Agricultural College, receiving his B.S. in 1916. He did his graduate work at the University of Wisconsin and obtained his M.S. in 1918 and his Ph.D. in 1928. As a graduate student he completed a thesis on the biology and systematics of the Syrphidae; in time he became a world authority on the North and South American members of this family. First an assistant, then a professor, he was chairman of the Department of Entomology at the University from 1942 through 1946. He excelled as an instructor of both undergraduates and graduates and conveyed his enthusiasm to his students. During the summer months he devoted himself to research on insect pests of fruits and vegetables.

Fluke reached emeritus rank at the University of Wisconsin in 1958 and died after a long illness on February 11, 1959.

Allen (1959) wrote his obituary.

13 Notable Arachnologists

twas an elderly mother spider
grown gaunt and fierce and gray
with her little ones crowded beside her
who wept as she sang this lay
curses on these here swatters
what kills off all the flies
for me and my little daughters
unless we eats we dies

don marquis

To the housewife few creatures are more odious than spiders, scorpions, and centipedes, while to the scientific observer few are more interesting. Starting with Nicholas Marcellus Hentz, an early specialist in spiders, our arachnologists are as lively a group as any in this book. John Henry Comstock, an arachnologist of distinction, was considered in an earlier chapter.

NICHOLAS MARCELLUS HENTZ (1797–1856)

One of the treasures of the library of the Boston Museum of Science are several volumes containing original drawings and paintings of spiders by Nicholas Marcellus Hentz, the first authority on spiders in the United States. When Edward Burgess (1875) edited Hentz's *The Spiders of the United States,* he also prepared some notes on its author, from which the following quotations are taken.

Nicholas Marcellus Hentz was born in Versailles, July 25, 1797. His father, an advocate by profession, was actively engaged as a politician at the time of Hentz's birth, and had been, shortly before this event, obliged to flee from his home in Paris, and to conceal himself in Versailles under the assumed name of Arnould. To the agonizing fears and alarms which his mother was obliged to undergo during this period, Hentz was wont to attribute the peculiarities of his nervous system, which were, as will be seen very remarkable.

Figure 134. N. M. Hentz. From H. C. McCook, *American Spiders and Their Spinning Work,* Vol. 3

At the early age of between twelve and fourteen years he began the study of miniature painting, for which he showed great talent and [in which he] became highly proficient. He soon, however, became interested in medicine and entered the Hospital Val-de-Grâce as a student. His son still possesses, in an old parchment-covered memorandum book, the following record in Hentz's then boyish hand-writing, "le vendredi 22 octobre 1813, j'ai été au Val-de-Grâce, M. Hentz." There he remained, busied with his studies and duties as hospital assistant, until the fall of Napoleon, when his father was proscribed and obliged to flee to America, wither Nicholas and one of his brothers accompanied their parents.

The party sailed from Havre-de-Grace, in the bark "Eugene," January 22, 1816, and arrived in New York City on March 19. Here and in Elizabeth Town they spent a few weeks in collecting their personal effects and making arrangements to move into the interior, an undertaking which was then quite formidable. They arrived in Wilkesburg, Pennsylvania, in the latter part of April, where it is probable that Hentz's parents finally settled.

Hentz himself for several succeeding years lived in Boston and Philadelphia, where he taught French and miniature painting. He also passed a short time on Sullivan's Island, near Charlestown, S.C., as a tutor in the family of a wealthy planter, a Mr. Marshall. All this time, whenever leisure hours allowed it, he was engaged in entomological studies, directing his special attention, as has already been said, to the spiders. While in Philadelphia he became intimate with the naturalist [Charles Alexander] Le Sueur. Le Sueur was accustomed to etch his own drawings, and having the use of his press, etc., Hentz made etchings of some of his spiders, as well as an alligator, which he had dissected to study the nature of its circulation.

In the winter of 1820–21, he attended a course of medical lectures in Harvard University, but finally abandoned the study of medicine, and engaged himself as a teacher in a school for boys at Round Hill, Northhampton, Mass., where Bancroft, the historian, was also employed. Here he was married to Miss Caroline Lee Whiting, the daughter of Gen. John Whiting, of Lancaster, on Sept. 30, 1824, and Mrs. Caroline Lee Hentz became afterwards well known as a poet and novelist.

Shortly after their marriage the couple moved to Chapel Hill, North Carolina, Hentz to teach modern languages at the University of North Carolina. Cobb (1932) says they enjoyed the "kindness, warm feeling, hospitality, and union of Chapel Hill." While Hentz collected spiders as an avocation, Mrs. Hentz wrote verses and a play, and both painted in water colors and drew.

Every few years they moved from locality to locality to establish female seminaries and other schools. In 1830 they went from Chapel Hill to Covington, Kentucky, and in 1834 to Florence, Alabama, in 1842 to Tuscaloosa, and 1846 to Tuskegee, Alabama, and in 1847 to Columbus, Georgia.

In Columbus, Hentz became ill, suffering a nervous disorder which became progressively worse. He then became addicted to morphine.

Hentz was "a small, spare man, about five feet and a half in height, and weighing only one hundred and ten or one hundred and fifteen pounds. Although of a genial, affectionate, and generous nature, his peculiarly nervous organization made him often morbidly sensitive and suspicious, and a prey to groundless fears, which not a little marred his enjoyment of life." (Cobb) He had the remarkable habit of suddenly dropping to his knees or taking off his hat and ejaculating his prayers, regardless of time or place. He enjoyed long walks with his sons in the woods, collecting and observing insects, and delighted in fishing and hunting.

He was a good friend of Dr. T. W. Harris', and their correspondence, at great length from 1824 through 1839, is preserved in S. H. Scudder's *Entomological Correspondence of Thaddeus William Harris, M.D.* Hentz, who was interested in both beetles and spiders, received help in classifying beetles from both Harris and Thomas Say, and in turn provided both of them with many specimens and much information.

In a letter dated January 1, 1826, from Northampton, Massachusetts, Hentz ends one of his letters to Harris with regrets that they lived so far from one another. "What pleasures we might enjoy together. I feel the want of books still more than you do. You have access to libraries, and can consult Olivier's valuable work." It is apparent from Harris' correspondence to Hentz that in Hentz he found a kindred spirit.

In some of his earlier papers published in the *Journal of the Academy of Science* in Philadelphia Hentz described some newly discovered beetles from Massachusetts and Pennsylvania. He published several other papers on beetles, but almost all his later works were devoted to spiders. These papers on spiders appeared in the *American Journal of Science* and the *Journal of the Boston Society of Natural History*. By 1841 he had described 141 species of spiders, most of which he also illustrated.

The Boston Society of Natural History undertook the publication of Hentz's book on spiders but was delayed for years in this project because of the expense of the illustrations. It appeared, with twenty-one plates, in 1875, long after Hentz had died. It was the first noteworthy work on American spiders.

According to Woodson (1950), although the book is scientific in its attitude, there is much in it written in a popular style that should appeal to the lay reader. He describes a certain cocoon as "large as a small plum, like a pear hanging down," the abdomen of a certain spider as similar to a "bishop's mitre," a male copulating with a female spider "like a pygmy upon a mountain, or rather under a mountain." And he notes that despite the unsavory reputation of some female spiders for digesting their true loves on their honeymoon, some of the females "are so gentle that I have seen several allow the males to dwell in the same tent with them, the pair living decently together as husband and wife should among Christian people." Although their pictures survive, Hentz's specimens of spiders have long since disappeared.

He died in the home of his son Charles in Mariana, Florida, on November 4, 1856.

GEORGE MARX (1838–1895)

Dr. George Marx and James H. Emerton (the latter did notes and descriptions for N. M. Hentz's *The Spiders of the United States*), both early and outstanding arachnologists, were also superb scientific illustrators. Much of the following material is derived from an obituary on Marx by Riley *et al.* (1896).

George Marx, son of a court chaplain at Laubach of the same name, was born in the Grand Duchy of Hesse on June 22, 1838, and he spent his boyhood there and in Leeheim. At the age of fourteen he entered the gymnasium at Darmstadt to prepare for the ministry, and became so proficient in botany and showed such skill as an artist that he was soon engaged to illustrate a book on the local flora. Shortly thereafter, against

Figure 135. George Marx. Courtesy of the Academy of Natural Sciences

his father's advice, he decided to change his career from the ministry to pharmacy, because this would permit him to delve more deeply in botanical studies. After completing his pharmaceutical studies in Giessen, in 1860 he moved to the United States.

Shortly after arriving in the States, the new immigrant enlisted in the Union Army as a private in Company K, 8th New York Volunteers. He remained with this company until after the battle of Bull Run on July 21, 1861, when because of his pharmaceutical and medical knowledge he was transferred to the medical corps as an assistant surgeon. His letters to his parents describing army movements and army life were printed in German newspapers and created a sensation. Because of illness and a severe wound he was honorably discharged from the army in July 1862 and returned to New York to take up pharmacy.

Marx moved to Philadelphia in 1865, began a business, and married Miss Minnie Maurer. He also began to collect and study arachnids. In 1878 he accepted a position in Washington in the Department of Agriculture as a draftsman and was attached to the section on Entomology, which became the Division of Entomology in 1881. Many of the plates and figures published by the Division betwen 1878 and 1883 were from his talented hand. C. V. Riley used many of them in his works. In later years Marx devoted more of his time to requests from other divisions, and in 1889 he was named chief of the newly established Division of Illustrations.

In Washington he became an authority on spiders and other arachnids.

American arachnologists such as H. C. McCook, James Emerton, G. W. and E. G. Peckham, as well as European arachnologists were indebted to Marx for much of their information about these insects. He wrote over thirty papers on the arachnids, and, of course, illustrated many of them with his excellent drawings. The titles of a few of his publications follow: "On the Morphology of Scorpionidae" (1888), "Catalogue of the Described Araneae of Temperate North America" (1889–1890), and "On the Effect of the Poison of *Latrodectus mactans* Walck. upon Warm-Blooded Animals" (1891). During the last few years of his life he began to study ticks, and in 1892 published his "On the Morphology of Ticks." Marx completed and edited E. Keyserling's *Die Spinen Amerikas* (1891) after the latter's death. Despite all of these interests, Marx kept up the medical skills he acquired on the battlefield and received an M.D. from Columbian University in Philadelphia in 1885.

Mrs. Comstock in *The Comstocks of Cornell* (1879) writes of him: "Dr. George Marx, artist for the scientific divisions of the Department of Agriculture, was a striking character. He . . . was German, and was tall, blonde, handsome, and dignified. He had studied medicine but liked scientific work better. He was an excellent, painstaking artist. Not only were his pictures accurate, but with the feeling of an artist he made them beautiful whenever possible. His wife was a typical German *frau* of the upper class, devoted to her home, to her husband, and to music. Dr. Marx was an authority on spiders. He was witty and often entertained us when there was a lull in the work. The systematists had wrought chaos in the genera of spiders; there was a tangle of synonyms hard to unravel. In recounting his perplexities, Dr. Marx said: 'I shall tell you about it. The man who studies spiders stays out with his friends an evening, drinking much beer, comes home late, wakes up next morning with katzenjammer; his breakfast is late, his coffee muddy, his eggs bad, his wife cross, and he says "God damn" and goes upstairs and erects a new genus.' When one of us remarked, 'The spirit is willing but the flesh is weak,' he repeated it in German and translated it into English as 'The ghost is willing but the meat is weak'—a saying we long remembered and which relieved many a trying situation."

Marx was a founder and a very active member of the Entomological Society of Washington, where he was admired for his genial nature, natural wit, broad learning, and artistic skill.

When Marx died in Washington, on January 3, 1895, he left his collection of more than a thousand species of native and exotic spiders to the U.S. National Museum. His other arachnids and his library were offered for sale for $1,500.

Anon. (1895) has further biographical notes on Marx.

Figure 136. Rev. H. C. McCook

HENRY CHRISTOPHER McCOOK (1837–1911)

Anyone who browses often in used bookstores is likely to come across the name of the Reverend Henry C. McCook, D.D., as an author of scientific and popular works on ants, spiders, and related animals. A first-rate naturalist, McCook has contributed a great deal of fascinating information on these subjects. The following sketch is based on obituaries by Calvert (1911) and Skinner (1911).

Henry was born on July 3, 1837, in New Lisbon, Ohio, the son of John and Julia Sheldon McCook. John McCook, a physician, was of Scotch-Irish descent, his wife was a New Englander. Henry was educated in public schools in Ohio and attended Jefferson College in Canonsburg, Pennsylvania, from which he received his A.B. in 1859. (This college is now part of Washington and Jefferson College at Washington, Pennsylvania.) He married Emma C. Herter, from New Lisbon, Ohio, in 1860. After a brief period as a printer's apprentice, then as a law student, he enrolled at Western Theological Seminary, Allegheny, Pennsylvania. When the Civil War broke out in 1861, he assisted in raising the Forty-first Regiment of Illinois Volunteers, in which he did service as a lieutenant and chaplain for two years.

According to Wadsworth (1911), McCook was ordained as minister by the Presbytery of Steubenville, Ohio, in 1861 and served in Clinton, Illinois, in 1862–1864 and St. Louis, Missouri, in 1864–1869. In November 1869 he became pastor of the Seventh Presbyterian Church of Philadelphia, continuing in this capacity until ill health forced him to resign in 1902.

The dedication in McCook's *The Natural History of the Agricultural Ant of Texas* reveals that his parishioners' "kindness to their pastor" allowed him "the annual summer vacations which have made possible the prosecution and publication of the following and other nature-studies."

In August 1876 McCook camped on Brush Mountain, near Hollidaysburg, Pennsylvania, to study the mound-building ants, *Formica exsectoides* Forel, and in 1877 he published *Mound-making Ants of the Alleghenies, Their Architecture and Habits.*

During three weeks of the summer of 1877 he investigated the agricultural ant, *Pogonomyrmex barbatus* (F. Smith), in Austin, Texas, studied their habits and also checked the accuracy of observations previously made by S. B. Buckley and Dr. Gideon Lincecum. In addition, he observed the leaf-cutting ant, *Atta texana* (Buckley). He pitched tent in a young live-oak grove and remained for several weeks minutely watching these interesting insects where, says Skinner, the "intense heat and sandy wastes made a trying combination." He was stooped over "on hands or knees or prone upon the face crawling slowly along, with eye fixed on the eager insect," sometimes being led a tiresome chase for many feet by a hurrying emmet.

The result of these careful observations was his *The Natural History of the Agricultural Ant of Texas, A Monograph of the Habits, Architecture and Structure of Pogonomyrmex barbatus* (1879). Unlike Lincecum, he concluded that the ants do *not* deliberately sow a crop of seeds around the periphery of the mound of the nest. The planting is the result of seeds cast out or accidentally dropped by the ants.

McCook set out in July 1879 to inquire into the behavior of the honey ants of the genus *Myrmecocystus,* and in the Garden of the Gods in Colorado he uncovered many of the mysteries concerning these remarkable ants with their strange "honey casks." In 1882 he published his observations on the honey ants and Western agricultural ants under the title *The Honey Ants of the Garden of the Gods and the Occident Ants of the American Plains.* By following the movements of honey ants at night with a lantern he had discovered that the sources of the honey they collected were the oak galls of cynipid wasps.

If Dr. McCook is well-known for his studies on ants, he is equally famous for his investigations of the behavior of spiders. These again were done during his summer vacations and other spare time. Toward the back of his book on the honey ants and the occidental ants, he placed a prospectus of his work on American spiders. He devoted ten years to its preparation and the work was to contain the best paper,

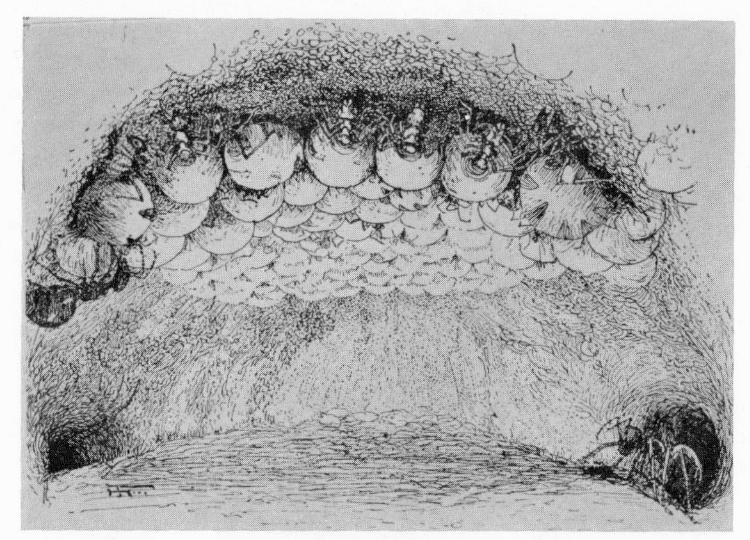

Figure 137. A drawing of honey ants from William Morton Wheeler, *Ants*

typography, binding, cuts, and colored lithographic plates. "It will be issued in three parts, each part *complete in itself*." The "Authors Edition" numbered one hundred and fifty copies and sold for $15 for the three parts.

This excellently illustrated work, of which a total of only 250 sets were printed by the author, is now rare and worth hundreds of dollars a set. The three parts of McCook's *opus magnum, American Spiders and Their Spinning Work—A Natural History of the Orbweaving Spiders of the United States with Special Regard to Their Industry and Habits* appeared seriatim in 1889, 1890, and 1893.

In the preface of one of his books, which I have been unable to locate, Dr. McCook wrote that "The duties of my calling in a large city have held me rigorously away from the open country except during two months of the year. Summer vacations, and leisure hours as a most busy life would allow, have been given to the pleasant task of following my little friends in the aranead world into their retreats, and watching at the doors of their fragile domiciles for such secrets of their career as they might happen to uncover. Occasional excursions at other times were unavoidably brief, and often broken off at the point of promised discoveries. I have, in part, indeed, overcome this obstacle by transporting and colonizing specimens, and by directing the observations of others. But, at the best, artificial conditions fall short of Nature's fulness, and no faithfulness of assistants can quite equal personal investigations."

Besides his scientific publications, McCook wrote a number of popular works on insects and spiders such as *Tenants of an Old Farm; Leaves from the Note Book of a Naturalist* in 1885, *Nature's Craftsmen, Popular Studies of Ants and Other Insects* in 1907, *Ant Communities and How They are Governed, A Study in Natural Civics* in 1907. He was a humorous and eloquent lecturer.

Colonel O. C. Bosbyshell (1911) portrays Dr. McCook as chaplain of the Second Regiment in the National Guard of Pennsylvania as follows:

> I recall the encampment of the entire National Guard of the State in Fairmount Park, Philadelphia, in the summer of 1879. The Second Regiment occupied Chamounix Heights and the ridge poles of the tents at nights became the resting places for hundreds of spiders—a most fruitful place to please the Chaplain. He approached a coterie of officers with his hands closed, exclaiming, "Who wants it?" Knowing the doctor's predilection for uncanny insects, no one was anxious. He opened his hands, revealing an ugly, vicious-looking spider, which he assured us was harmless, and then followed a most interesting talk on the habits of the species, as well as other kinds in the camp.
>
> His dignity of manner forebade anything like license or liberty of talk in his presence, and his readiness to lead to topics interesting and instructive made the leisure hours of camp life most enjoyable.
>
> The "boys" as the men were familiarly called by him, never forgot to do him honor and respect. Unlike some chaplains of the war and of the guard, he never once was referred to as their "Holy Joe." He was a welcome officer in camp to those of all ranks. Ever alert to improve conditions, he sought all means to accomplish such an end. He was a rigid disciplinarian, ever upholding those in authority.

When the Second Regiment was sent to Cuba during the war with Spain in 1898 he went with it. He also founded and served on the National Relief Commission in the same war, reorganized a hospital in Havana, rearranged and marked the graves of American soldiers in Cuba, and served as chaplain in other semimilitary organizations. He took an active part in the establishment of the Philadelphia City flag, participated in political reform and philanthropic movements, and designed the seal of the Presbyterian Church in America.

McCook was a very active member of the Academy of Natural Sciences in Philadelphia, where he and Edward D. Cope, a well-known paleontologist who supported Darwin's evolutionary theory, staged some interesting arguments. He joined the American Entomological Society in 1877 and was its president from 1898 to 1900. He received honorary degrees from Lafayette College and Washington and Jefferson College.

He died in Devon, Pennsylvania, on October 31, 1911 and was survived by a son and daughter by his first wife, who died in 1897.

JAMES HENRY EMERTON (1847–1931)

Emerton was a skillful illustrator of insects, spiders, plants, and other living things, a sculptor of octopuses and anatomical models, a naturalist, and a notable authority on spiders. Nathan Banks thought highly of him and of his contributions to natural history, and the following is chiefly drawn from two articles by him (1931, 1932).

Emerton was born in Salem, Massachusetts, on March 31, 1847. When a schoolboy, he was introduced to the study of outdoor life by George F. Markoe, a young helper in the drugstore owned by James Henry's father. "They collected plants, insects and shore invertebrates and at the age of fifteen [Emerton] was frequently visiting the Essex Institute, where he became acquainted with A. S. Packard, F. W. Putnam, John Robinson, Caleb Cooke, and others who later became more or less prominent students of Natural History." (Banks, 1932)

James had an innate ability to sketch and model natural history objects; he took no lessons. By the year 1868 he felt professional enough to advertise his work in *The American Naturalist*:—"James H. Emerton, Zoological and Botanical Draughtsman, Salem, Mass., is prepared to execute drawings on paper or wood for Zoological Subjects. Especial attention given to the delineation of Insects. References: Editors of American Naturalist."

As a natural history artist he worked with various authorities in museums and educational institutions. He made many of the drawings

Figure 138. J. H. Emerton

for A. S. Packard's *Guide to the Study of Insects* and for the *Botany of the Fortieth Parallel* by Watson and Eaton. He was elected to the Boston Society of Natural History in 1870 and was an assistant in the museum of the Society from 1873 to 1874; during this time he also contributed notes to N. M. Hentz's work, *The Spiders of the United States.*

In the meantime he was collecting and studying spiders in New England, and in 1875 he went to Europe with his collection, where he spent a year comparing his specimens with those in collections in London, Paris, Berlin, Jena, Leipzig, and Stockholm and met such European arachnologists as O. P. Cambridge, E. Simon, L. Koch, T. Thorell, and E. Ohlert. Later he prepared a paper on his European studies.

During the summers of 1877 and 1878, Emerton presented a series of lectures on zoology and spiders at the Summer School of Biology in Salem. He then became curator in the Museum of the Peabody Academy of Science, where he prepared a guide for visitors and spent one hour each day conducting them through the museum. In 1880 he began to illustrate for Professor A. E. Verrill of Yale and prepared models of the great squid and octopus in the Museum of Comparative Zoology at Harvard and the National Museum in Washington.

In 1884 Emerton married Mary A. Hills. They made their home in Boston.

He made anatomical models for medical colleges and scientific drawings for a great variety of disciplines. Some of the well-known books illustrated by him were C. S. Minot's *Textbook of Embryology,* A. E. Verrill's *Marine Invertebrates,* S. H. Scudder's *Butterflies of North America,* G. W. and E. G. Peckham's *Wasps—Social and Solitary,* D. G. Eaton's *Ferns of North America* and A. S. Packard's *Monograph of the Geometridae.*

Emerton collected spiders throughout much of the United States and Canada. He went with A. P. Morse to the South, with Alexander Agassiz to the West Indies, and to the Sierras with G. W. Peckham, and even collected around Hudson Bay.

He was the author of many scientific papers and monographs on spiders, being principally interested in the taxonomy and distribution of the spiders of New England and Canada. Some of his more important works were *The Structure and Habits of Spiders* (1878), *Life on the Seashores, or Animals of our Coast and Bays* (1880), "New England Spiders of the Family Theridiidae" (1882), and *Common Spiders of the United States* (1902). All in all, he described about three hundred and fifty species of spiders, and most of his types are in the Museum of Comparative Zoology.

He was very active in a number of natural history societies in Boston and was Secretary and a founder of the Federation of New England Natural History Societies.

According to Nathan Banks, Emerton was a founder of the Cambridge Entomological Club, for many years an officer, and one of the most regular attendants, often speaking and exhibiting specimens. At times he lectured publicly, and he collected spiders until a few months before his death on December 5, 1931. He willed his library and collection to Banks.

"Aside from being a naturalist he was an artist for the sake of art. He painted hundreds of water colors, often depicting the sea, the shore, or ships. For several seasons this was done at Ipswich, and in later years he went regularly in July to Gloucester for painting. He frequently exhibited before art societies . . ." (Banks, 1932)

Banks (1931) says of him: "To one who knew him for many years, his independent and adventurous spirit, his directness and simplicity, his kindly interest in the scientific work of others, and his continuous endeavors to attract the young to the study of Natural History overtop much of his scientific achievement."

ALEXANDER PETRUNKEVITCH (1875–1964)

Petrunkevitch was born in December 1875 at Pliski, Russia, of parents of noble birth, his father being the equivalent of a baronet and, according to Hutchinson (1945), "a prominent liberal statesman and

Figure 139. Alexander Petrunkevitch

political writer, [who] gave his son that love of freedom that was to bring him, with so many of his compatriots, to the New World."

Petrunkevitch was greatly interested in natural history and collected and studied beetles from his youth. His early academic training was at the University of Moscow, where he was simultaneously attracted to zoology and literature, publishing in both fields. He translated Byron's *Manfred* into Russian and Russian works and poems by Pushkin into English.

His liberal convictions and enthusiasm for social change swept him into the political torrents of the times and he was forced to leave Russia to escape arrest. He made a fortunate choice, when he took refuge in Freiberg, Germany, for there he came under the influence of the great German zoologist August Weismann. One of Petrunkevitch's first papers was on the development of the heart of a chrysomelid beetle, which foreshadowed his interest in the circulatory system of arthropods, and his Ph.D. thesis in 1900 was on the cytology and embryonic development of the honey bee, in which he showed that the workers and queen bees come from fertilized eggs and the drones from unfertilized eggs. A by-product of this study was the development of a fixing fluid known as Petrunkevitch Fluid. In Freiberg, he also served as a *Privatdocent* and married Wanda Hartshorn, an American student of philosophy, who bore him two children.

Petrunkevitch moved to America with his wife in 1903, where one of his bride's relatives kindly shaved off his beard to please the local gentry, since Petrunkevitch had no prior experience in the tonsorial arts. His first job in this country was as a lecturer on cytology at Harvard. In 1906 he became acting professor of zoology at Indiana University and regularly commuted to Massachusetts in order to be home with his wife, who was ill. When she developed tuberculosis in 1910, he accepted an instructorship in zoology at Yale in order to be closer to her. He became an assistant professor there in 1911 and a full professor in 1917.

For thirty-three years Petrunkevitch taught both graduate and undergraduate courses in invertebrate zoology, as well as in parasitology, entomology, histology, biological microtechnique, photographic technique, and, naturally, courses on spiders and other arachnids, such as scorpions, pseudoscorpions, solpugids, harvestmen, and mites.

According to Woodruff (1945): "Large as has been Petrunkevitch's influence on the students in his formal courses, it was exceeded in the case of the goodly number who chose him as a director of their research toward the dissertation. Here they profited not only by his breadth of view, scientific ideals and enthusiasm, but also by his especial solicitude for their progress in zoological thought and technique. And his interest extended to all the graduate students of the Department as is attested by the informal weekly teas that he held for them in his study." For forty years tea, cookies, and ideas were served from four to six in his office; they were known as Pete's Tea, or, in his later years, as Pete's Tea-Emeritus.

His book, *Morphology of Invertebrate Types,* based on original research, served as a laboratory manual in some of his classes. However, all was not laboratory work with Petrunkevitch because he loved field work and collected throughout the United States, Panama, and Puerto Rico. On the way to the collecting ground, Petrunkevitch would talk, says Kinkead (1950) "with great warmth and animation about many things besides spiders. A few of his numerous interests are history, folklore, petrography, meteorology, and dendrology."

Petrunkevitch began to publish on spiders when he moved to the United States, and his first papers on them were concerned with the optics of vision. In all he wrote about thirty papers, chiefly on spiders, many of them profusely illustrated. A number of his publications were major works and monographs of great importance and interest to students of their subjects. Some of his papers were on the mating habits of spiders, spider silk, the behavior of tarantulas when under the attack of *Pepsis,* the tarantula hawk. Tarantulas were among his favorites; at one time he had about one hundred and eighty of them living in his laboratory.

Hutchinson (1945) lists some of his more important works as *A Synomic Index-Catalogue of Spiders of North, Central and South America,* published in 1911, *On Families of Spiders* in 1923, *Catalogue of American Spiders—Part I* in 1939, and *A Study of Amber Spiders* in 1942. He also made an important contribution to the classification of spiders in *An Inquiry into the Natural Classification of Spiders, Based on a Study of Their Internal Anatomy,* which appeared in 1933.

In the laboratory, Petrunkevitch was a man of many skills, an authority on histological technique, an expert photographer, and a master mechanic and carpenter whose inventions are widely used by biologists. In the classroom he was a stimulating teacher, with a reservoir of knowledge that amazed and delighted his charges. Being a very kindly man he "had difficulty censuring the most delinquent of students."

Petrunkevitch never lost his interest in Russian politics, and in 1918 he wrote a book entitled *The Russian Revolution.* Although he took no part in such politics he did aid émigrés from his war-torn country. He was described as a "philosophical socialist," who was opposed to fascism and communism.

According to Eugene Kinkead, Petrunkevitch "holds that a belief in the hereafter is unnecessary for an ethical code. Man, he thinks, should do what is right not for any future award but as an example to others."

Nor did he believe that science could be used as a substitute for religious faith. In his "Profile" in *The New Yorker* of April 22 and 29, 1950, Kinkead writes: "In June, 1944, a few weeks before Petrunkevitch retired, his colleagues at Yale honored him at a special meeting of the Connecticut Academy of Arts and Sciences, of which he had been president for fourteen years. In order to throw Petrunkevitch off the track, the event was announced as a symposium on the gypsy moth, but it actually was a surprise party devoted to champagne drinking, speeches about Petrunkevitch, and the presentation to him of a commemorative volume of forty-four manuscripts on spiders, Petrunkevitch himself, and allied subjects, written by his associates, pupils, and other friends, who had been quietly engaged on the project for months." Kinkead gives the following picture of him at a somewhat later period in life: "He is a tall, broad-shouldered, gaunt, distinguished-looking man of seventy-four, slightly stooped from long years at the dissecting table and the microscope. A few stiff gray hairs rise from his head, and old-fashioned gold-rimmed spectacles are perched on his prominent nose. His name is pronounced 'Pay-trun-kay-vitch.' Although he officially retired in 1944, he keeps regular hours each weekday at Yale's vine-covered Osborn Zoological Laboratory. His office, which he has occupied for thirty-eight years, is Room 201, a large high-ceilinged, disorderly chamber crowded with books, desks, chemical oddments, laboratory glassware, stands, tables, microscopes, assorted aquaria and terraria, and spiders."

Petrunkevitch was an Honorary Curator of Arachnida at the American Museum of Natural History and received many honors in his lifetime.

He died in New Haven on March 9, 1964, at the age of eighty-eight, "a very part of the warf and woof" of Yale University and America's greatest arachnologist.

CYRUS RICHARD CROSBY (1879–1937)

Crosby was born at Penn Yan, New York, on January 9, 1879. He graduated from Cornell University in 1905 and became assistant entomologist at the University of Missouri, Columbia. In 1906 he joined the staff of Cornell University and occupied several positions in entomology, including that of extension entomologist, which he held until his death.

He was the author of a number of experiment-station circulars and bulletins, and collaborated with M. V. Slingerland on a *Manual of Fruit Insects,* published in 1914, and with M. D. Leonard on a *Manual of Vegetable-Garden Insects,* published in 1918.

In his spare time he worked on spiders, a subject he became interested in under the expert tutelage of Professor J. H. Comstock, and he

Figure 140. C. R. Crosby

collaborated with S. C. Bishop on a number of important taxonomic studies on spiders.

Johannsen (1937) writes: "Though physically rather deliberate he exhibited an astounding intellectual alertness, a tireless industry, and an unbounded enthusiasm which was an inspiration to the young men with whom he was associated. The catholicity of his tastes was shown by his interest in philosophy, religion, anthropology, geology and other subjects outside of the field of his profession."

Professor Crosby died suddenly on the evening of January 11, 1937, at Rochester, New York, leaving a wife and son.

HENRY ELLSWORTH EWING (1883–1951)

H. E. Ewing was for many years in charge of the arachnid and parasitic insect collections in the U.S. National Museum. He was also an authority on the classification of such insects as lice, fleas, the Protura, and Japygidae.

He was born in Arcola, Illinois, on February 11, 1883. He first studied medicine but later changed to entomology, obtaining an A.B. in 1906 and an M.A. in 1908, both from the University of Illinois. In 1910 and 1911 he attended Cornell and obtained his Ph.D. under J. H. Comstock, who undoubtedly stimulated his interest in spiders. For a time he worked in the Oregon Agricultural Experiment Station at Corvallis and later was a professor at the University of Iowa. In 1919 he joined the staff

Figure 141. H. E. Ewing

of the U.S. National Museum as a specialist on Arachnida. He remained on the staff until his death.

Ewing first began to publish in 1907, writing on oribatid mites, plant mites, and chiggers as well as other mites. He was greatly interested in ectoparasites, and in 1929 there appeared his well-known work, *A Manual of External Parasites.* Ewing was also the author of a very popular illustrated article on spiders, "Afield with the Spiders," which appeared in the *National Geographic* for August 1933. He became greatly interested in the parasites of reptiles and amphibians. For many years he maintained a colony of box turtles in his back yard and wrote on their biology.

Some later workers believe Dr. Ewing spread himself somewhat thin in trying to cope with the taxonomy of so many different orders. As he grew older he became somewhat withdrawn and unco-operative with other taxonomists. Nevertheless, he assembled large and important collections and wrote comprehensive works that were useful to later workers.

He died in Washington, D.C., on January 5, 1951. Baker and Gurney (1951) wrote his obituary.

14 Notable Entomologists of Divers Interests

There was a young farmer named Graham
Who, though bugs ate his crops, wouldn't spray them
He explained: "I've a droll
But effective control
I just catch all the females and spay 'em."

H. C. Crook

This chapter contains sketches of a distinguished group of entomologists, economic, medical, and industrial, as well as educators, administrators, and others with their roots deep in insect lore. They are considered in this chapter because of a variety of interests which do not readily fit them into some of the prior categories. Many other entomologists could be mentioned in this chapter but because of space considerations I have had to limit their number. I think most entomologists will agree that the workers discussed here have made many important contributions to entomology.

CHARLES NICHOLAS AINSLIE (1856–1939)

Ainslie was born October 13, 1856, near Amsterdam, New York. He graduated from Beloit College with an A.B. in 1877 and an A.M. in 1880. For twenty-two years he worked in a bank in Rochester, Minnesota, while collecting and studying insects in his spare time.

In 1906 he began to work for the U.S. Bureau of Entomology under F. M. Webster as an economic entomologist. During his twenty-four years of government service he dealt with a variety of agricultural problems in the Midwest and West. The insects he investigated included "greenbugs," the New Mexico range caterpillar, alfalfa weevil, Western grass-stem sawfly, pale Western cutworm, Hessian fly, chinch bug, grasshoppers, and many other cereal and forage pests. He was also greatly interested in the taxonomy of certain Hymenoptera of economic im-

Figure 142. C. N. Ainslie

portance, flowers, birds, and book collecting, and he was a talented musician.

Ainslie died in Sioux City, Iowa, on December 5, 1939. Walton and Caffrey (1940) and Packard and Wade (1940) wrote obituaries of him.

PERCY NICOL ANNAND (1898–1950)

Annand was born in Telluride, Colorado, on November 16, 1898. He earned his B.S. from Colorado State College in 1920, and his M.S. in 1922 and Ph.D. in 1928, both from Stanford University. After obtaining his M.S. he was made head of the Department of Biology at the Junior College of San Mateo, California.

Dr. Annand became associated with the U.S. Bureau of Entomology in 1929 in order to study the sugar beet leafhopper in the West. While so engaged he directed a staff of young entomologists and proved to be an excellent administrator and research director. He was then transferred to Washington, where in a few years he became a Division Leader in charge of cereal and forage crops. While in this position he directed a large-scale grasshopper control campaign.

Annand was appointed Assistant Chief of the Bureau in charge of research work in 1939; when the Chief, Lee A. Strong, died in 1941, Annand succeeded him. The war years weighed heavily on him. In 1943 he was elected president of the American Association of Economic Entomologists.

Figure 143. P. N. Annand

Death put an end to Annand's short but distinguished career on March 29, 1950. He was survived by his wife and two children. Cardon and White (1951) wrote his obituary.

ERNEST ADNA BACK (1880–1959)

Back was born on October 7, 1880, in Northampton, Massachusetts. He was orphaned as a small boy and was brought up by his father's sister. He graduated from Massachusetts Agricultural College (University of Massachusetts) in 1904, where he was guided by C. H. Fernald and H. T. Fernald, and received his Ph.D. in 1907.

Upon completion of his graduate work he joined the U.S. Bureau of Entomology, where, for a time, his job was to study the control of the citrus white fly in Florida, Cuba, and Mexico. From 1910 to 1912 he was entomologist and plant pathologist with the Virginia Agricultural Experiment Station in Blacksburg. In 1912 he rejoined the Bureau of Entomology and was involved in research on the control of fruit flies in Hawaii, Bermuda, and Spain.

In 1917 Back was made head of the new Division of Stored Product and Household Insect Investigations in the Bureau of Entomology and did fundamental research on clothes moths, carpet beetles, and other common household pests. He also contributed to the knowledge of the biology and control of numerous stored-product pests as well as to the knowledge of fumigants and fumigation techniques. From 1934 until his

Figure 144. E. A. Back

retirement in 1947 he served with the Division of Insects Affecting Man and Animals in the Bureau of Entomology and Plant Quarantine.

Back was a quiet, self-effacing man, well liked by his entomological colleagues.

He died at his home in Washington on May 21, 1959, leaving a wife and two sons. Reed *et al.* (1959) prepared his obituary.

ELMER DARWIN BALL (1870–1943)

E. D. Ball was born in Athens, Vermont, on September 21, 1870. He received a B.S. in 1895 and an M.S. in 1898 from Iowa State College. In 1907 he received a Ph.D. from Ohio State University, where he studied under Professor Herbert Osborn. They published many joint papers on leafhoppers and related Hemiptera. Ball taught and did research at a number of universities, including Iowa State, Colorado Agricultural College (Colorado State University), and Utah Agricultural College (Utah State University). From 1907 to 1916 he was Dean of the Utah Agricultural College and Director of the Experiment Station at Logan, Utah. During the years 1916 to 1918 he was State Entomologist of Wisconsin and organized the movement for the eradication of American foul brood. In 1920 and 1921, on leave from Iowa State College, he served as Assistant Secretary of Agriculture under Secretaries Edward T. Meredith and Henry C. Wallace; from 1921 to 1925 he was Director of Scientific Work in the U.S. Department of Agriculture.

Figure 145. E. D. Ball

Mrs. Blake (1951) describes Ball's entry on the Washington scene: "At about this time there came to the Department a director of scientific research in the form of an over-energetic man, E. D. Ball. He also was an entomologist, and possibly because of this, his interest in bringing the Bureau of Entomology up to his concept of full efficiency soon brought about many changes. It was the beginning of the end of the age of the individual and the real naturalist in the Bureau. Under Dr. Howard there had grown up and come to full maturity a group of scientists not to be seen there since. He had brought them together as a body of trained and enthusiastic men and then left them alone to work on their own projects in their own way. The result was outstanding fullness of production of high worth. What Ball and his successors did not realize is that you cannot inspire either an artist or a scientist by holding up before him schedules of efficiency. Real genius does not flourish under such conditions. Regimentation of any sort is a barren soil in which only red tape and officiousness grow."

While he was with the Department of Agriculture "one of his prominent contributions was a stubborn fight for a higher standard of training for scientific men. This included not only training but remuneration as well, and he is credited with organizing the Graduate School in the Department and serving as its director until 1925. He was also instrumental in the fight for establishing a Bureau of Home Economics and organized the Bureau of Agricultural Economics." (Osborn *et al.*, 1944.)

From 1925 to 1928 he was in charge of investigations on celery pests for the Florida State Plant Board, and from 1928 to 1931 he was Dean of the College of Agriculture and Director of the Agricultural Experiment Station at the University of Arizona in Tucson.

According to Vorhies (1944), "Dr. Ball was an enthusiastic teacher, helpful with counsel and financial assistance to worthy students. While he may have acquired some reputation for pugnacity in defense of his views (he was not one to compromise) he was at all times courteous and helpful to inquiring constituents, and a most loyal friend."

Vorhies claims it was Ball who "first pointed out the causal relationship of leafhopper infestation of sugar beets to curly top disease of that crop; and later repeated this achievement on tip burn of potatoes—the third disease proved to be insect borne. He put every possible hour into collecting and taxonomic study of Jassidae, Cercopidae, Fulgoridae, and especially Membracidae."

All in all he published about two hundred papers, both economic and taxonomic, and described some three hundred and eighty-five species of leafhoppers. His outstanding collection was acquired by the U.S. National Museum. In economic entomology he made notable contributions to the study of the codling moth, American foul brood, and the role of insects as transmitters of plant diseases.

During his life Ball received many honors; among them was his election as president of the American Association of Economic Entomologists in 1918.

He died in Pasadena, California, on October 5, 1943, after a long illness. Osborn *et al.* (1944) and DeLong (1944) as well as Vorhies wrote obituaries of him.

GEORGE WARE BARBER (1890–1948)

George Ware Barber was a devoted naturalist and a highly respected economic entomologist. He was born at Hyde Park, Massachusetts, on August 3, 1890. Outdoor life and natural history, especially insects and birds, interested him even as a boy. He graduated from Massachusetts Agricultural College with a B.S. in 1913, and received his M.S. and Sc.D. from Harvard in 1925 and 1927, respectively.

Barber entered the U.S. Bureau of Entomology in 1914 and worked on wireworms in Missouri, range caterpillars in New Mexico, and the Hessian fly in Kansas and Maryland. After serving as a Lieutenant of Cavalry in World War I, he returned to the Bureau.

In his new assignment he investigated the European corn borer in New England and Ohio, and the corn earworm in the East. While studying the corn earworm he discovered and developed the use of

Figure 146. G. W. Barber

oil-insecticide treatment of the corn silks to control this pest. This discovery was a boon to homeowners and farmers who tried to raise sweet corn without corn earworms or evidence of them. For this accomplishment he received official recognition by the U.S.D.A.

Barber retired from the U.S.D.A. in 1945 and became associated with Rutgers University, where he studied the behavior of the house fly and the effectiveness of such new insecticides as DDT. As Wade (1949) says, "His work at Rutgers University amply demonstrated his unusual qualities as teacher and as scientist. He had a profound influence on his associates and especially on the graduate students in entomology. His broad knowledge and experience, coupled with his success in aiding students and in winning their confidence, his jovial nature and insatiable capacity for work were a continuous source of wonder and admiration to his associates."

Barber was not only interested in entomology but also in birds and plants, and he was an accomplished musician.

He died in New York City on December 5, 1948, at the age of fifty-eight leaving a wife and son.

SHERMAN WEAVER BILSING (1885–1954)

For a large part of the first half of this century, Bilsing was one of the most prominent economic entomologists in the South. Martin *et al.* (1954) wrote his obituary, and the following is from this source.

Figure 147. S. W. Bilsing

Bilsing was born on December 8, 1885, on a farm near Crestline, Ohio. In 1912 he received a B.S. from Otterbein College and an A.B. from Ohio State University. He was awarded an M.S. in 1913 and a Ph.D. in 1924, both from the latter institution. Starting as an instructor of entomology, he joined Texas Agricultural and Mechanical College in 1913 and became head of his department in 1918. He served as head until 1947 and retired as Professor of Entomology in 1952. Highly respected as a teacher, he was also known for his generous loans to students to enable them to finish college. Bilsing was an authority on the life histories of many important insect pests in the South and was known for his studies on pecan insects, especially the pecan nut casebearer, on the cotton boll weevil, and the codling moth. His ability was recognized by his colleagues and he occupied numerous important positions in entomological and agricultural organizations.

Bilsing died on July 23, 1954; he was survived by a wife and son.

MAULSBY WILLETT BLACKMAN (1876–1943)

Blackman was born at Lawrence, Kansas, on March 26, 1876. He received his collegiate training at the University of Kansas, where he received a bachelor's degree in 1901 and a master's in 1902. He was an instructor of zoology and histology at the University from 1901 to 1904. He then studied at Harvard University on a fellowship in 1904 and 1905, when he obtained his doctorate. From 1905 to 1909 he was an instructor in histology at Western Kansas University and in 1909 became an

Figure 148. M. W. Blackman

assistant professor of zoology at Syracuse University and eventually Professor of Entomology there.

At the State College of Forestry of Syracuse University, he played a leading part in organizing a department of entomology with emphasis on forest insects. Craighead *et al.* (1944) write: "During the summers of 1925, 1926 and 1927 Dr. Blackman had appointments as field assistant in the Forest Insects Division in the Bureau of Entomology, and devoted his entire time to a detailed study of the Black Hills beetle and its aggressive outbreak in the Kaibab National Forest and surrounding regions." In 1929 he joined the Division of Forest Insect Investigations of the U.S. Bureau of Entomology as a specialist in the Scolytidae. "While with the Division of Forest Insects he acted, at times, as assistant chief, in which capacity the soundness of his judgements and recommendations reflected, no doubt, his broad biological background and remarkable grasp of the fundamentals of insect control." In 1937 he joined the Division of Insect Identification in the same Bureau. He was a highly regarded teacher and a discerning entomologist who often expressed himself with "disconcerting frankness." He died in Silver Springs, Maryland, on October 12, 1943.

ADAM GIEDE BÖVING (1869–1957)

Most entomologists find adult insects more attractive than the immature forms and consequently tend, to concentrate on the former.

Figure 149. A. D. Böving

Böving, to some extent, corrected this neglect of the larvae by devoting much of his life to their study.

He is said to have been born in "Saby," Denmark, on July 31, 1869, and he graduated from the University of Copenhagen in 1888. While attending college he supported himself by teaching courses in botany and zoology in the local schools. At the University Professor Frederik Meinert interested him in immature insects, and Böving received his Ph.D. on the chrysomelid genus, *Donacia,* the larvae of which live on aquatic plants below the water level. During his graduate studies he became connected with the Royal Zoological Museum where he was appointed Assistant Curator of Entomology after receiving his doctorate. He studied collections in other museums throughout Europe and became acquainted with entomologists there. In Denmark, he concentrated on the study of coleopterous larvae and the musculature of the male genitalia of dytiscids.

In 1913 Böving moved to the United States and was appointed an "Expert" in the Bureau of Entomology. He married for the second time in 1916, his first wife having died in 1911, and became a citizen of the United States in 1918. He worked for the U.S. Department of Agriculture until his retirement in 1939; then returned for another three years in 1942 to help out during World War II. During his years with the Federal Government he prepared many important papers, those on the larvae of the Japanese beetle, tobacco beetle, ladybird beetle, and the lead-cable borer being especially noteworthy. In 1930–1931 he and F. C. Craighead published an important monograph, *An Illustrated Synopsis of the Principal Larval Forms of the Order Coleoptera.* This was printed by *Entomologica Americana* in four parts by means of a personal subsidy from Böving.

Muesebeck *et al.* (1958) remark: "There can be no doubt of the stimulating effect that the Synopsis had on the study of beetle larvae in the United States as well as in other countries. For the first time a serious effort to arrange the larvae of this major order of insects in a natural or nearly natural system was successful."

Böving was a man of high ideals, free of professional jealousies, genial and courteous, of broad intellectual attainments, an energetic and careful worker, and naturally received many honors.

He died in Washington, D.C., on March 16, 1957, at the age of eighty-eight.

WILTON EVERETT BRITTON (1868–1939)

Britton was born in Marlboro, Massachusetts, on September 18, 1868. When he was a year old his family moved to a farm near Keene, New Hampshire. He received his B.S. from the University of New Hampshire in 1893 and then studied for a year at Cornell. In 1903 Yale University granted him a Ph.D. and in 1930 the University of New Hampshire conferred upon him an honorary Doctor of Science degree.

Britton began his career in 1894 as a horticulturist at the Connecticut Agricultural Experiment Station in New Haven. In 1898 he became in-

Figure 150. W. E. Britton

volved in nursery inspection. During the thirty-eight years he was State Entomologist and in charge of the Agricultural Experiment Station, many important studies of the insects of Connecticut appeared under his leadership, including monographs on the Orthoptera, Hymenoptera, Hemiptera, and Odonata. Britton was involved in economic studies of such insect pests as the San Jose scale, gypsy moth, Japanese beetle, oriental beetle, and the European corn borer.

He was active on the Eastern Plant Board and in the Crop Protection Institute and various other organizations, including the American Association of Economic Entomologists, of which he was president in 1909. He served as associate editor of the *Journal of Economic Entomology* from 1910 to 1929. He cooperated closely with the Federal Government on insect and quarantine problems. He had a beneficial influence not only on entomologists in his state, but throughout the country. Britton died on February 15, 1939, the year following his wife's death. Felt (1939) wrote his obituary.

CHARLES THOMAS BRUES, II (1879–1955)

Brues was born in Wheeling, West Virginia, on June 20, 1879. He attended high school in Chicago, where he and A. L. Melander (see pages 479–81) were introduced to the study of insects by H. E. Walter. Brues and Melander remained good friends for over sixty years, a friendship broken only by Brues' death.

Melander (1955) wrote: "We had prepared to enter Chicago University,

Figure 151. C. T. Brues

but on the way to matriculate we met Dr. W. M. Wheeler who advised us to go with him to the University of Texas where he had just accepted the headship of the zoology department. At Texas we wrote several joint papers, The Chemical Nature of Some Insect Secretions, and New Species of *Dolichopus* and *Hygroceleuthus,* as well as others with a single authorship. After a year we were given scholarships and later a class in entomology. On leaving Texas Brues went to Columbia University since his family had moved to New York, and I to Chicago University." Brues received an A.B. degree in 1901 and an M.S. in 1902, both from the University of Texas. After two years at Columbia University he returned to Texas as a special field agent in entomology for the U.S. Department of Agriculture. From 1905 to 1909 he served as Curator of Invertebrate Zoology at the Milwaukee Public Museum. In 1909 he became an instructor in Economic Entomology at Harvard under W. M. Wheeler; he was made Professor of Entomology in 1935. Brues was also an Associate Curator of Insects, at the Museum of Comparative Zoology there.

As a teacher, Brues was especially stimulating to his graduate students, in whom he had a friendly and sympathetic interest. His research was devoted to taxonomic and biological studies such as the phorid flies in the nests of ants, parasitic Hymenoptera, including fossil forms, the ecology of thermophilous insects, food and feeding habits of insects, medical entomology, tissue staining, and the like. Among Brues' publications are *A Key to the Families of North American Insects* (with A. L. Melander) in 1915, 2d ed. 1954; *Insects and Human Welfare* (1921, 2d ed. 1947); and *Insect Dietary* (1946).

Brues collected widely throughout the United States and in Jamaica, Peru, Ecuador, Cuba, Hudson Bay, the East Indies, and the Philippines. He was very active in the Cambridge Entomological Club and was the editor of *Psyche* from 1910 to 1947.

In 1946 he was made Emeritus Professor of Entomology at Harvard. He died in Crescent City, Florida, on July 22, 1955, leaving his wife, a son and a daughter, both of whom also became professors. Besides Melander (1955), Romer *et al.* (1955) wrote obituaries of him.

ALBERT FRANKLIN BURGESS (1873–1953)

Burgess was largely responsible for limiting the gypsy moth and the brown-tail moth to the New England and northeastern states for one hundred years. On an entirely different front, it was through the interest and direction of members like Burgess that the American Association of Economic Entomologists and the *Journal of Economic Entomology*

Figure 152. A. F. Burgess

became such practical and powerful organs for the applied entomologist.

Professor A. I. Bourne (1953), a friend of Burgess's, wrote his obituary, and the following is largely from his notes.

A. F. Burgess was born on October 2, 1873, in Rockland, Massachusetts, a member of an old New England family. He was educated in local preparatory schools until he entered Massachusetts Agricultural College at Amherst, from which he graduated in 1895, receiving an M.S. in 1897. Bourne notes that although Burgess had to spend much of his time working to pay necessary expenses at college he nevertheless managed to be an active member of the college band and of the musical club, to engage in several college sports, and to serve as president of his class.

Between 1899 and 1907 he served as an Assistant in Entomology at the University of Illinois, and as an inspector of nurseries and orchards of the Ohio Department of Agriculture. In 1907 he returned to Massachusetts under the U.S. Bureau of Entomology to take charge of the breeding experiments on the Gypsy Moth Project. "From that time to the date of his retirement in October 1943," writes Bourne, "Mr. Burgess' life work was devoted to the studies of the Gypsy Moth and particularly to the organization of the State and Federal programs for its control. . . .

"For many years he furnished the inspiration for and the driving force in most of the research conducted on all phases of the Gypsy Moth problem. He was the author of many scientific papers and stimulated the publication of many more by his subordinates. All of the publications for which he was directly or indirectly responsible were of the highest quality and could well be taken as models in accuracy of detail, thoroughness in coverage of subject matter and in excellence of presentation.

Together these papers on all phases of the Gypsy Moth and its control constitute what is probably the most comprehensive, concise, and complete coverage ever brought together on any single species of insect pest in the history of Economic Entomology."

Burgess believed in quarantines and used them as a tool or "supporting means" to keep pests from spreading from one area or from infesting a new area by being imported from abroad. "The formation and successful operation of the barrier zone in western New England and eastern New York were largely due to his initiative and ideas."

In 1916 he was placed in charge of the Federal efforts to control the gypsy and brown-tail moth. In 1928 he was transferred from the Bureau of Entomology to the Plant Quarantine and Control Administration with the grade of Principal Entomologist, retaining charge of the project of gypsy moth extermination.

Bourne says that "His work and his long association with entomological groups, together with the large number of men he trained in research and applied entomology, made him one of the most respected and widely known entomologists of the present day," and that "the high quality of his own work, his inspired leadership and his basic honesty and integrity" also won him international recognition in his field.

Burgess was secretary-treasurer, vice president, and in 1924 president of the American Association of Economic Entomologists and associate editor and business manager of the *Journal of Economic Entomology*. For many years he helped guide and shape the policies of both the Association and the *Journal*.

"Albert Burgess also had many very human traits," Bourne writes, "and peculiar qualities which made his friendship something to be enjoyed and always to be treasured. For example, he had a unique and refreshing sense of humor which he usually kept under strict control but which on appropriate occasions he could skillfully employ for his own enjoyment and the delight of his friends. One of his most conspicuous characteristics was his loyalty to his friends and his sympathetic understanding of their shortcomings."

Burgess died on February 23, 1953, after several years of illness; he was survived by his second wife and two sons from his first marriage.

HARRY E. BURKE (1878–1963)

Burke was a pioneer forest entomologist in the West and a specialist in the buprestids and other wood-boring insects. For over thirty years he was connected with the U.S. Bureau of Entomology and served the Government in a variety of capacities.

Figure 153. H. E. Burke

He was born on May 19, 1878, in Paradise Valley, Nevada. A raid by Paiute Indians persuaded his parents to move to California. In 1881 the family moved to the State of Washington, where Burke received most of his formal education. He obtained a B.S. in 1902 and an M.S. in 1908 from what is now Washington State University. In 1923 he earned a Ph.D. from Stanford University.

In 1902 he joined the Bureau of Entomology as one of A. D. Hopkins' two assistants in forest entomology. During his long career with the Federal Government he worked on such injurious insects as the tussock moth, mountain pine beetle, Western pine beetle, red turpentine beetle, Black Hills beetle, Pacific flatheaded borer, lead cable borer, and many others. For many years he was a specialist on shade-tree insects. He collaborated with Doane, Van Dyke, and Chamberlin on the well-known text *Forest Insects,* published in 1936. For a number of years he was concerned with insect problems in the national parks in the West.

Burke retired in 1934 at the age of fifty-seven to Los Gatos, California, and spent his remaining years advising on insect problems in the national parks and writing magazine articles on shade-tree insects. He died on March 26, 1963, at the age of eighty-four. Eaton and Keen (1964) wrote his obituary.

GUY CHESTER CRAMPTON (1881–1951)

According to Alexander (1952), Crampton was born in Mobile, Alabama, on September 21, 1881. He received an A.B. from Princeton in

Figure 154. G. C. Crampton

1904, an A.M. from Cornell in 1905 and from Harvard in 1920, and a Ph.D. from the University of Berlin in 1908, where he and James McDunnough, the Canadian lepidopterist, were classmates. He taught biology at Princeton University from 1908 to 1910 and at Clemson University in South Carolina from 1910 to 1911, when he joined the Massachusetts Agricultural College as Assistant Professor of Invertebrate Morphology. He became a full professor in 1915, a position he held until his retirement in 1947.

Crampton gained recognition for his studies on insect morphology and phylogeny, on which he wrote over one hundred papers. S. F. Bailey, a student of his from 1927 to 1929, has provided the following information:

He lived in a sort of an apartment in the attic of Fernald Hall where all entomology classes were held. I was never actually in his "quarters," not being one of the small group who used to philosophize with him into the wee small hours. He must have done a lot of his own cooking as the "egg on his vest" indicated. Frequently he would oversleep and come to class unshaven and obviously hurriedly dressed. He had a classy car and used to get a student to drive him on collecting trips in the summer. . . . Obviously he was a bachelor. I never saw him dressed in anything but a blue suit.

He was soft spoken and very patient in explaining the intricacies of the nerves and muscles of those pickled specimens. And, incidentally, many specimens he received from far away places *were* pickled in rum which you could smell intermingled with other hybrid alcoholic aromas in his office.

And that office! It was shelved high on all sides with books, pamphlets—uncatalogued—and the most fantastic array of glassware. The thousands of specimens of invertebrates were preserved in nearly every kind of jar, vial, bottle, reagent, container, square, round, oval, fluted, flared, squat, pyramidal, and of many colors. They all had a purpose and meaning—namely to illustrate comparative structures, their phylogeny and function—Lamarck not withstanding.

As you can gather the subject he taught (morphology) did not move *me* very much so perhaps I'm a poor choice to write this resume. At the moment I can recall only that he published relatively little and gave us very little in the way of mimeographs. Compared with Fernald and Alexander, he was not a top flight lecturer. His collection and anatomical studies appeared to be his whole life and he was a typical dedicated instructor of the old school—tolerant but never overly generous in handing out A's.

On October 31, 1951, four years after he retired, Crampton died from a heart attack in a hotel in Albany, New York.

ALFONSO DAMPF (1884–1948)

Dampf, one of Mexico's foremost scientists, was born in Esthonia on December 3, 1884. He studied at the University of Königsberg, received his Ph.D. in 1909, and worked in the Zoological Museum and Institute there from 1907 to 1912, and from 1920 to 1923; during the latter years he lectured on applied entomology. From 1913 to 1919 he was Government Entomologist in German East Africa, or Tanganyika Territory.

Dampf moved to Mexico in 1923, becoming a naturalized citizen in 1941. He served as Chief Entomologist in the Mexican Department of

Figure 155. Alfonso Dampf

Agriculture, Head of the Entomological Laboratory in the Mexican Public Health Service, and Professor of Entomology and Head of the Department of the National Agricultural College. He had a profound influence on Mexican entomology and did important work in all aspects of the subject, on the habits and distribution of migratory locusts, of black flies, fleas and other blood-sucking insects, and Lepidoptera. He sent Mexican insects to specialists throughout the world and was highly regarded by his colleagues, many of whom he met during his extensive travels throughout the world. C. P. Alexander (1948) writes: "It is difficult to conceive how one person could have accomplished the vast amount of work that Dampf has done. His collections of Mexican insects, generously distributed to many specialists throughout the world, probably ran to more than a million specimens, and perhaps several millions."

Dr. Dampf died of cancer in Mexico City on March 17, 1948.

JOHN JUNE DAVIS (1885–1965)

"J. J.," as he was known to his friends, was born in Centralia, Illinois, on October 9, 1885, and was brought up in a small farm community. An uncle of his, who collected butterflies for Dr. William Barnes, helped to develop Davis' interest in insects.

J. J. attended local schools and then enrolled in the University of Illinois, where he came under the influence of the entomologist J. W.

Figure 156. J. J. Davis

Folsom, a "marvelous" teacher who stimulated his interest in insects and natural history. Another man who made a great impression on him was the Illinois State Entomologist, S. A. Forbes. Davis received his B.S. in 1907, graduating with honors in entomology.

For his bachelor's thesis Davis worked on the corn-root aphid, and after graduation he became an assistant to Forbes at the University of Illinois and studied corn, truck-garden, shade-tree, garden insects, and the then ubiquitous bedbug and house fly. He travelled widely throughout Illinois, giving lectures to a variety of audiences, and, in his spare moments, working on the taxonomy of aphids.

From 1911 to 1919 Davis was in charge of the U.S. Cereal and Forage Crops Insect Laboratory at Lafayette, Indiana, where he devoted considerable time to the biology and control of May beetles. F. M. Webster, the noted authority on cereal and forage pests, was an associate of his until 1916. It was Webster who instilled in Davis a great respect for the efficacy of good farm practices in insect control. During 1919 and 1920 Davis was the head of the U.S.D.A. Japanese Beetle Laboratory at Riverton, New Jersey. Then, in 1920, he was appointed head of the then small Department of Entomology at Purdue University and held this position until his retirement in 1956 at the age of seventy. There, as he says in autobiographical notes made at the request of some of his colleagues (Davis, 1958), "I taught, carried on research and extension and spent a lot of time traveling from one end of the state to the other. As Dean Skinner once said, I would be in Evansville in the morning and in Gary that evening for a meeting. It was true. Even so, those were wonderful experiences . . ."

According to Deay *et al.* (1966): "Professor Davis was an inspirational teacher. Of the courses he taught, that of introductory entomology, which he taught each semester for 36 years, was his favorite. He inspired many of his students to choose entomology as a career. These former students now hold responsible positions in all the broad areas of entomology, not only in the United States but throughout the world. His interests in his students did not cease upon their graduation; he kept in touch with them throughout his life."

Davis can truly be called the Father of Scientific Pest Control in the United States, for it was he who introduced pest control operators to the University and the University to pest control operators. He writes of his introduction to the pest control industry as follows:

"I stuck my neck out and I'll never regret it. Entomologists considered pest exterminators questionable characters, and in general they were. In 1935 they met for their second annual convention at Detroit, Michigan. They had organized as the National Association of Exterminators and Fumigators (now the National Pest Control Association) because of requirements of the National Recovery Act (N.R.A.) under the leadership of that dynamic man William O. Buettner. I attended the Detroit convention on my own to find out what a pest control operator looked like. . . .

"A Remarkable Man—Bill Buettner. I really became acquainted with Bill at the third convention of the N.P.C.A. in Cleveland. I don't know how. It was just a coincidence. But from that time until his death in 1953, we were close friends. He knew I was a friend of the industry and I knew I had a friend in Bill. He and I were much alike in that we enjoyed life. We played together and we worked together. When it was a matter of relaxation we took it, when it was a problem to be worked out we did our job as best we could. I never enjoyed [better] companionship or received better advice than with and from Bill."

In 1937 Davis organized the first Purdue Pest Control Conference, which has been an invaluable source of information to the pest control industry and which definitely has greatly helped raise its status. Other colleges followed Purdue's lead in this respect and also instituted short courses for the pest control operator. In 1961 Davis wrote a history of the early years of this industry.

I first met J. J. in California around 1936, but became well acquainted with him only after he retired. A thin, slightly-built man, he had a remarkable sense of humor and a very agile mind. He was a man "in love with life," who made living a pleasure for all who knew him. It was easy to see why his colleagues elected him president of the Entomological Society of America in 1932 and of the American Association of Economic Entomologists in 1938.

Alert in mind and memory until the end, he died in Bethel Park near Pittsburgh, Pennsylvania, on July 13, 1965. The same year the J. J. Davis Research Fund was organized in his honor at Purdue University to aid worthy students in research on pest control.

GEORGE ADAMS DEAN (1873–1956)

Dean was born in Topeka, Kansas, on April 19, 1873. In 1895 he received his B.S. and in 1905 his M.S. from Kansas State College and obtained a teacher's certificate from Kansas State Normal College, Emporia, in 1898.

He was Assistant in Entomology at Kansas State College from 1902 until 1912, when he was appointed head of the newly organized entomology department and Entomologist in the Kansas Entomological Commission, both of which positions he held until he retired in 1943. Under his leadership important studies were undertaken on stored-product

Figure 157. G. A. Dean

pests infesting flour mills. He also pioneered with county-wide campaigns in the use of poisoned bran mash to control grasshoppers, cutworms, and army worms. He was on leave of absence from 1923 to 1925 to head the Division of Cereal and Forage insect investigations in the Bureau of Entomology in Washington. From 1928 to 1930 he was a member of the Mediterranean fruit fly Commission, which supervised the eradication of this pest in Florida.

He was president of the American Association of Economic Entomologists in 1921 and of the Entomological Society of America in 1925.

Dean died on April 23, 1956, at the age of eighty-three. Smith (1956) wrote his obituary.

HARRY FREDERIC DIETZ (1890–1954)

H. F. Dietz was born in Indianapolis, Indiana, on December 8, 1890. He did undergraduate work at Wabash College and Montana State College, serving for a time as a student assistant in the Montana Experiment Station at Bozeman in the investigation of the vectors of Rocky Mountain spotted fever. He later continued his studies at Butler College (now University) receiving a B.A. in zoology in 1914. The next two years he spent as a deputy state entomologist in Indiana. There he began taxonomic studies on scale insects, the results of which he and Harold Morrison published in their book, *The Coccidae or Scale Insects of Indiana.* This work was beautifully illustrated by R. E. Snodgrass.

Figure 158. H. F. Dietz

In 1916 Dietz became associated with the Federal Horticultural Board in Panama and worked on insect pests of ornamental plants and subtropical fruits. He and James Zetek studied the biology of the citrus black fly and published an important article on "The Black Fly of Citrus and Other Subtropical Plants." Returning to the United States in 1919, Dietz first did plant quarantine work for the U.S.D.A. and then became Assistant State Entomologist of Indiana, a position he enjoyed for ten years. In 1929, at the age of thirty-nine, he entered Ohio State University and began graduate work, receiving his M.A. in 1930 and his Ph.D. in 1931.

In 1932 Dietz accepted a field research position with the Du Pont Company and in 1936 transferred to the Du Pont Experimental Station in Wilmington, Delaware. According to Sharp (1954), "His wide knowledge of field conditions, plant ecology, the life history of injurious insects and plant diseases and plant culture made him invaluable in the development of new agricultural chemicals." Dietz was also well-versed in botany, plant pathology, weed control, and related fields.

He died suddenly of a heart attack on September 4, 1954, leaving a widow and four sons.

WESLEY PILLSBURY FLINT (1883–1943)

Flint was born at Southampton, New Hampshire, on May 4, 1883. Since his father was a truck farmer, Wesley was exposed to farmers'

Figure 159. W. P. Flint

problems at an early age. In 1904 he received a certificate from the University of New Hampshire for completing a two-year course in agriculture. In 1906 he worked for the Massachusetts Gypsy Moth Commission, and was privately employed to study forest insects. In 1907 he became an assistant to Dr. S. A. Forbes, Director of the Illinois State Laboratory of Natural History (Illinois State Natural History Survey) at Urbana and for thirty-five years he was associated with this organization, becoming Chief Entomologist of the Illinois Agricultural Experiment Station in 1930, and Chief Entomologist of the Survey in 1935.

Professor Flint was an authority on orchard and field-crop insects, and with C. L. Metcalf wrote the well-known texts *Destructive and Useful Insects* (1928) and *Fundamentals of Insect Life* (1932). He was also coauthor with C. L. Metcalf of *Insects, Man's Chief Competitors* (1932). In 1932 he served as president of the American Association of Economic Entomologists.

Flint died suddenly at his desk in Urbana, Illinois, on June 3, 1943, of a heart attack. His obituaries were written by Davis (1943) and Metcalf (1944).

JUSTUS WATSON FOLSOM (1871–1936)

Folsom was born in Cambridge, Massachusetts, on September 2, 1871. Harvard granted him a B.S. in 1895 and a D.S. in 1899. His early entomological work was largely on anatomy, physiology, and embryology. In 1899–1900 he taught at Antioch College, and in 1900 he became an

Figure 160. J. W. Folsom

instructor in entomology at the University of Illinois in Urbana. He was an assistant professor there from 1908 until he left in 1923. Students found him an excellent teacher, with the ability to imbue them with his own enthusiasm for his subject, though quiet and unassuming.

In his early years at Illinois he wrote his well-known *Entomology with Special Reference to Its Biological and Economic Aspects* (1906). It was revised in 1934 by R. A. Wardle, of the University of Minnesota. Folsom also made notable contributions to the taxonomy of the Collembola and Thysanura.

In 1925 he became an associate entomologist at the U.S. Bureau of Entomology and studied cotton insects in Tallulah, Louisiana, until his final illness. He was active in a number of entomological societies, including the Entomological Society of America, of which he was president in 1931.

Folsom died from a series of heart attacks on September 24, 1936, in Vicksburg, Mississippi. Hinds (1937) wrote his obituary.

HENRY JAMES FRANKLIN (1883–1958)

Franklin was born in Guilford, Vermont, on February 10, 1883. He commenced his undergraduate work at Massachusetts Agricultural College in Amherst in 1899, and obtained his Ph.D. in 1908.

While in college he was outstanding in public speaking and played on the varsity football team for three years. His scholastic achievements

Figure 161. H. J. Franklin

put him at the head of his graduating class, where he was class orator.

Franklin began the study of cranberries as a summer assistant at Wareham, Massachusetts, in 1906 and 1907, and in 1909 was appointed head of the newly established Cranberry Station at East Wareham of the Massachusetts Agricultural Experiment Station. In time he became an authority on cranberry insects and cranberry culture. He wrote numerous publications on other insects, including pioneering studies on North American thrips, and was widely recognized for his *Monograph of the Bumble Bees of the New World*. When Franklin retired in 1952, the cranberry growers of Massachusetts built an addition to the Cranberry Station in his honor named the Doctor Franklin Room.

Franklin died at Buzzards Bay, Massachusetts, on April 16, 1958. Tomlinson and Bourne (1958) prepared his obituary.

ROGER BOYNTON FRIEND (1896–1962)

Friend was born in Hyde Park, Massachusetts, on November 2, 1896. His education in Massachusetts Agricultural College was interrupted by military service in World War I. According to Turner (1962), "He was wounded in action and contracted an infection which later caused Parkinson's disease." He completed his undergraduate studies at the University of Massachusetts in 1923 and in 1927 received his Ph.D. in zoology from Yale University.

As an undergraduate he worked on vegetable insects and later studied

Figure 162. R. B. Friend

the biology of the Asiatic beetle and the birch leaf miner. He taught economic and forest entomology. In 1939 he became head of the Department of Entomology in the Connecticut Agricultural Experiment Station in New Haven and State Entomologist, and a year later was made Vice-Director of the Experiment Stations in New Haven and Storrs. He continued studies which he had begun at Yale on the European pine shoot moth, white pine weevil, and gypsy moth. Because of physical disability he surrendered some of his many responsibilities after 1947 and retired in 1958.

Friend was highly regarded as a teacher and administrator, and he was quite active in state organizations of vegetable and fruit growers, nurserymen, pest control operators and shade-tree workers.

He died after a long illness on February 1, 1962. Turner (1962) wrote his obituary.

THEODORE HENRY FRISON (1895–1945)

Although Frison died at a relatively early age he had already achieved recognition as Chief of the Illinois Natural History Survey, as an accomplished taxonomist, and a notable student of conservation.

He was born in Champaign, Illinois, on January 17, 1895. As chance would have it, J. W. Folsom, Professor of Entomology at the University of Illinois, Urbana, lived not far from him, and Mills (1958) reports that "Not long after the turn of the century, Dr. J. W. Folsom of the Uni-

Figure 163. T. H. Frison

versity of Illinois Department of Entomology was walking down a street in Urbana when he discovered a youngster who was engrossed in observing a colony of ants. Folsom engaged the boy in conversation and was impressed with his interest and knowledge. Thus began a close and personal relationship between Dr. Folsom and young Theodore Henry Frison." Professor Folsom introduced him to Dr. S. A. Forbes, who permitted the young high school student to audit courses in entomology at the University. He was later admitted to the University as a student of entomology, but in April 1918, when in his senior year, he was commissioned a second lieutenant in the infantry. Shortly after the end of the war he returned to the University to complete his studies, receiving an M.S. in 1920 and a doctorate in 1923. Between 1920 and 1923 he worked as Assistant State Entomologist of Wisconsin and then on the staff of the United States Bureau of Entomology and Plant Quarantine at Moorestown, New Jersey. Upon completion of his doctorate in 1923 he became Systematic Entomologist in the Illinois Natural History Survey at Urbana. After Forbes died in 1930, Frison became acting head and in 1931 Chief of the Survey. He was an outstanding administrator of the Illinois Natural History Survey, and it throve under his aegis. Largely through his efforts the Survey became well known as a center for conservation studies of wildlife in Illinois.

A tremendous worker, Frison became well known through such taxonomic monographs as *The Plant Lice, or Aphididae, of Illinois* (with F. C. Holles, 1931), *The Stoneflies, or Plecoptera, of Illinois* (1935), and *Studies of North American Plecoptera* (1942). He also studied the taxonomy of bumblebees.

He was editor of the *Journal of Economic Entomology* from 1935 until 1939 and an active member of numerous societies, including the American Association of Economic Entomologists, of which he was first vice president in 1945.

He had an attractive personality and a fine sense of humor. He was a capable violinist as well as an enthusiastic sportsman—he played tennis and golf and fished and hunted.

Frison died at Urbana after a long illness on December 9, 1945, leaving a wife, a son, and a daughter. Davis (1945), Gibson (1946), Campbell (1946) and Ross (1946) wrote obituaries of him.

HENRY SHEPARD FULLER (1917–1964)

The following sketch is based on an obituary by Traub (1964).

Fuller was born in Washington, D.C., on June 17, 1917. He earned a B.S. in chemical engineering at the Worcester Polytechnic Institute in 1937 and then attended Harvard Medical School, from which he graduated *cum laude* with an M.D. in 1941. Approximately one year later he was commissioned in the Army Medical Corps and served two of his three years' tour of duty in the India-Burma theater.

After his service he was awarded a Harvard traveling fellowship to the London School of Hygiene and Tropical Medicine and won a medal as the most outstanding student. Upon returning to the United States he became an assistant professor of Preventive Medicine at the Bowman

Figure 164. H. S. Fuller

Gray School of Medicine in Winston-Salem, North Carolina. In 1948, as a recipient of a Guggenheim Fellowship, he studied chiggers and ticks at the U.S.P.H.S. Rocky Mountain Laboratory in Hamilton, Montana, and the U.S. National Museum.

In 1949 he joined the faculty of the Harvard School of Public Health as a research associate and an assistant professor of microbiology and proved to be an excellent teacher with a keen sense of humor. In 1953 he became a civilian scientist in the U.S. Army and in 1955 was appointed chief of the Department of Entomology at the Walter Reed Army Institute of Research. A few months later he was made the first chief of the Institute's Department of Rickettsial Diseases. In 1963 he departed for Japan to serve as chief of the Department of Viral and Rickettsial Diseases there.

Though he never had a formal course in entomology, he became an outstanding entomologist through his intelligence, application, and love of natural history. At Harvard, Dr. J. C. Bequaert, a zoologist and entomologist, had encouraged him in the private pursuit of medical entomology. While in Burma, through his tenacious and brilliant research, he identified a mysterious disease as scrub typhus of rickettsial origin and chigger-borne, for which he was later awarded the Typhus Commission Medal. He also studied rickettsial pox, Korean hemorrhagic fever, spotted fever, trench fever, and other arthropod-borne diseases. He did important work on the taxonomy of fleas and chiggers and with G. W. Wharton prepared *The Manual of Chiggers,* which was published in 1952.

Traub writes that Fuller was virtually unique as a physician "interested in and qualified in clinical medicine, preventive medicine, virology, and rickettsiology," with a thorough knowledge of ecology, medical entomology, and the taxonomy of fleas, and trombiculid mites.

Fuller's sudden death on February 3, 1964, in Japan, was caused by a coronary attack.

BENTLEY BALL FULTON (1889–1960)

Fulton was born on August 29, 1889, at Newark, Ohio. He obtained his B.A. from Ohio State in 1912. Professor Herbert Osborn, the well-known entomologist at Ohio State, was largely responsible for influencing Fulton in his junior year to major in entomology. From 1912 to 1919 Fulton was a staff entomologist at the New York Agricultural Experiment Station in Geneva and managed to obtain an M.S. in 1916 from the University of Chicago. From 1919 to 1924 he was a member of the staff of the Oregon Agricultural Experiment Station in Corvallis,

Figure 165. B. B. Fulton

Oregon, where he continued his New York studies of the insect pests of fruit and also studied the biology and control of earwigs and grasshopper control.

Fulton joined the entomology department at Iowa State College (Iowa State University) in Ames, Iowa in 1924. He obtained his Ph.D. there in 1926 and in 1928 moved to North Carolina State College (North Carolina State University) in Raleigh as Professor of Entomology. In 1954, because of ill health, he retired as emeritus professor.

Fulton's biological studies, with their superb illustrations, are famous in entomological circles. He published notable works on the tree crickets, the European earwig, the cheese skipper, and other economic insects. Gurney (1964) states that "the principal groundwork for a satisfactory classification of North American crickets was laid between 1915 and 1956, by the late Bentley B. Fulton." He made significant contributions to the biology, behavior, and the songs of crickets.

Fulton was primarily a research entomologist and did little teaching. Gurney says he "had a dry wit and a keen sense of humor." According to Smith (1961), "He is one of the truly great biologists; however, due to his quiet, unassuming attitude, he was appreciated most by those who had the pleasure of close association with him. He was always most patient and sincere in everything he undertook, attributes of a great biologist."

He died in Raleigh, North Carolina, on December 8, 1960, after a long bout of Parkinson's disease.

C. F. Smith (1961) and A. B. Gurney (1964) wrote obituaries of Fulton.

ARTHUR GIBSON (1875–1959)

Gibson was Dominion Entomologist of Canada in Ottawa for more than two decades. The following information is based on a biographical sketch by C. R. Twinn (1956).

Gibson was born in Toronto December 23, 1875. In 1899 he became an assistant to James Fletcher, the first Dominion Entomologist. He then rose to the position of Assistant Entomologist in 1905 and Chief Assistant in 1908. When Fletcher died in 1908 he was succeeded by C. Gordon Hewitt, who became Dominion Entomologist in 1909. Under Hewitt's leadership entomology became an important separate branch of the Canada Department of Agriculture.

In 1914 Gibson became Chief of the Division of Field Crop and Garden Insects, and when Hewitt died in 1920 he succeeded him as Dominion Entomologist. In 1938 the Entomological Branch became a division of the newly organized Science Service; he was then made Associate Director of the Service and Chief of the Division of Entomology. He retired in 1942 after forty-two years of distinguished service.

What is remarkable about Arthur Gibson's career is that he achieved his eminence in entomology without any formal training in it. However, he had two expert tutors in entomology in the persons of James Fletcher and C. Gordon Hewitt. Gibson's extensive knowledge of entomology combined with his marked administrative ability were responsible for his

Figure 166. Arthur Gibson

leadership in the field. He published more than two hundred scientific and popular articles, dealing with such pests as cutworms, flea beetles, grasshoppers, root maggots, the Colorado potato beetles, greenhouse insects, insects of the flower garden, and household insects.

Gibson enjoyed an international reputation in the field of entomology. He served as president of the American Association of Economic Entomologists in 1926, of the Entomological Society of America in 1922, and at some point, of the Entomological Society of Ontario.

He died on April 18, 1959, at the age of eighty-four, in Brockville, Ontario.

FREDERICK ZELLER HARTZELL (1879–1958)

Hartzell was born on a farm near Easton, Pennsylvania, on December 11, 1879, and received a classical education at Lafayette College, which gave him a B.S. in 1905. For two years he worked in several entomological capacities in Pennsylvania and then went on to Cornell to earn his M.A., which he received in 1909. He was appointed the same year to the New York Agricultural Experiment Station at Fredonia, where he remained until he moved to Geneva, New York, in 1920. He retired in December of 1948 and continued his research on fruit pests.

He was an authority, highly regarded in agricultural circles, on the biology, ecology, and control of grape, pear, apple, and cherry pests.

Figure 167. F. Z. Hartzell

According to Chapman (1960), Professor Hartzell was "well informed" in the fields of ecology, ornithology, biometry, geology, astronomy, botany, meteorology, and in several branches of mathematics and was "a true naturalist, who was courteous and thoughtful, and a great reader and lover of books." He pioneered in the introduction of statistics and biometrics in field experimentation.

He died in Geneva on June 13, 1958.

THOMAS JEFFERSON HEADLEE (1877–1946)

Headlee was born on February 13, 1877, at Headlee, Indiana, and was brought up on a farm. He acquired much of his early education without benefit of schooling. After attending the Indiana State Normal School, he received his A.B. in 1903 and his A.M. in 1906 from Indiana University. In 1906 he was awarded a Ph.D. by Cornell, where he had majored in entomology under Professor J. H. Comstock. From 1906 to 1907 he was Associate Entomologist at the New Hampshire Agricultural Experiment Station in Durham, New Hampshire, and from 1907 to 1912 he was head of the Department of Entomology and Zoology at Kansas State Agricultural College (Kansas State University) in Manhattan.

In 1912 he succeeded Professor J. B. Smith, who was well known as an economic entomologist and a lepidopterist at Rutgers University, as

Figure 168. T. J. Headlee. Courtesy of the College of Agriculture and Environmental Science, Rutgers University

Professor of Entomology and as Entomologist in the New Jersey Agricultural Experiment Station in New Brunswick, and State Entomologist. He held these positions until his retirement, except for that of State Entomologist, which was abolished in 1933.

Headlee's most important achievements were in the field of mosquito control. According to Peairs and Driggers (1946), "He was instrumental in the formation of the New Jersey Mosquito Extermination Association in 1914 and through his qualities of leadership, made it an important and continuously active factor in the control of mosquitoes, a primary insect problem in New Jersey. His more than thirty years' work on mosquitoes was climaxed by the publication in 1945 of *The Mosquitoes of New Jersey and Their Control* which summarizes his laboratory and field work on these insects."

Dr. Headlee taught economic entomology and other entomological subjects. He was especially interested in insect physiology and such problems as temperature in relation to insect development, the effect of radio waves on insects, and the development of mosquito traps. He also actively participated in research on cranberry pests, the biology of sewage disposal, orchard pests, the gypsy moth, and the Japanese beetle, and in many other phases of economic entomology.

He promoted industrial fellowships at Rutgers, helped organize, in 1920, the Northeastern Entomologists, a group which later became the Eastern branch of the American Association of Economic Entomologists, and in 1929 was elected president of the latter.

He "was a man of positive opinions, forthright in speech and manner, impatient with sham and intolerant of injustice," write Peairs and Driggers. His neighbors found him friendly and amiable, but reserved. They were permitted to help themselves to the excellent fruit grown in his beautifully tended orchard. His students could ruefully admire his sense of humor: "When I used to suggest that he accompany me to some tide-gate installaton about three miles out in the meadow," wrote one of them, Leslie (1963), "he would say, 'I am the pedagogue. You, my boy, are the pedestrian.' And then he would promptly proceed to literally walk my feet off. And he was not a little man. He weighed over 200 pounds!"

Headlee died on June 14, 1946, in the Muhlenberg Hospital in Plainfield, New Jersey, from complications following an operation. Two years before his death, the T. J. Headlee Fellowship in Entomology was established in his honor by the Rutgers College of Agriculture. This is a fellowship for fundamental studies in insect physiology, toxicology, and biochemistry.

SAMUEL HENSHAW (1852–1941)

According to Wade and Hyslop (1941), Henshaw was born in Boston on January 29, 1852, a descendant of an old Boston family. He attended local schools in Boston and early evinced an interest in insects and natural history. He was long an active member of the Boston Society of Natural History. He was one of the founders of the Cambridge Entomological Society in 1874 and editor of *Psyche* for many years. From 1876 to 1891 he was an assistant to Professor Alpheus Hyatt, a science instructor at Lowell Technological Institute; he worked on both vertebrates and invertebrates and also prepared material for Professor Hyatt. Henshaw served the Institute as secretary and librarian from 1892 to 1901. He was awarded an honorary M.S. by Harvard in 1903, his only degree. His work as a librarian increased his knowledge of scientific literature, an experience that was invaluable in his preparation of entomological bibliographies at a later period. It no doubt helped him also in accumulating a fine personal library.

In 1891 he began to work for the Museum of Comparative Zoology in Cambridge as an assistant to Dr. H. A. Hagen, whom he succeeded. In 1911 he was appointed Director of the Museum, becoming Director Emeritus in 1927.

He published his *List of the Coleoptera of America North of Mexico* in 1885, with supplements in 1887, 1889, and 1895. This was an in-

Figure 169. Samuel Henshaw. Courtesy of the Museum of Comparative Zoology, Harvard University

dispensable work to all students of Coleoptera until the appearance in 1920 of C. W. Leng's *Catalogue of Coleoptera of America North of Mexico.*

Following in Hagen's footsteps, Henshaw compiled *Bibliography of American Economic Entomology,* parts 1 to 5, 1860–1889, published by the U.S.D.A. between 1890 and 1896. While making these compilations, he served as a field agent for the U.S.D.A. His *Bibliography* was later continued by Nathan Banks and Mabel Colcord.

Jackson (1941) describes Henshaw "as short in stature; but very strong. He had wonderful eyesight so that he could read the finest print and he never wore glasses, even in his old age. He had a very keen and critical mind in affairs and in his relations with men. This gave his judgement great weight with his associates. The keynotes of his character were his sensitiveness, his sense of fun, his personal integrity and his devotion to his work and his friends."

According to Brues (1942), he was an indefatigable worker and disapproved of Colonel T. L. Casey's school of taxonomic "splitters." Brues adds that he would not delegate responsibility to others, so that he was swamped with work. "The various rooms of the museum required many separate keys and Henshaw had a long chain to which these were attached in seemingly endless variety. This chain he carried about with him, selecting one key or another as he personally admitted visiting scientists to study the collections or went about his daily routine through the several collections of the museum."

Although genial with his friends he appeared gruff or "crusty" to others. Children were very fond of him and he took them to circuses and other affairs. At heart a kindly man, he was very solicitous towards Hagen and Samuel Scudder when they were invalids.

Henshaw never fully recovered from the death of his wife, nee Annie Stanwood, in 1900. His latter years were marred by a crippling arthritis. He died on February 5, 1941, at the age of eighty-nine.

WILLIAM BRODBECK HERMS (1876–1949)

The first man to introduce malaria control in the West, Herms was born at Portsmouth, Ohio, on September 22, 1876. Forced to work for a number of years before he was financially able to attend college at Baldwin-Wallace, in Berea, Ohio, his prime interest there was medical entomology, particularly the mosquito vectors of malaria. He won his B.S. in 1902 and got his M.A. at Ohio State in 1906. He also did graduate work at Western Reserve and Harvard.

As a graduate student he made some early and important investigations

Figure 170. W. B. Herms

on the reactions of insects to light. In 1908 he was chosen from many prominent applicants to become Assistant Professor of Parasitology at the University of California.

Shortly after going to California, Professor Herms launched, in the Placer foothill country near Penryn, the first anti-malaria mosquito control campaign in the West. His efforts were so successful that representatives of many other areas of the state approached him for help with their mosquito abatement programs. In many cases it was an uphill fight, since some local Chambers of Commerce bitterly resented the implication that their communities were centers of malarial infection. With true missionary zeal, however, he fought on, even in the face of personal violence in some cases, and lived to see malaria almost eradicated from his adopted state. Work such as this established his reputation as an authority on the control of malaria.

During World War I he volunteered his services in the Army Sanitary Corps, rising from the rank of captain to that of major. His efforts to control insect vectors of disease in Newport News and elsewhere in Virginia were highly successful. After the war he returned to the University of California and became head of the Division of Entomology and Parasitology in 1919, a position he held until shortly before his retirement in 1946. His text *Medical Entomology,* which he revised four times, was revised by M. T. James in its fifth edition. This book served as the backbone of many courses in medical entomology. *Mosquito Control,* published in 1944, was a joint effort by Herms and H. F. Gray.

In the nineteen thirties I attended Professor Herms's classes in medical entomology and ecology and found him an interesting and stimulating teacher, a kind and friendly man. I still remember him walking onto the podium of the classroom, juggling a stack of books with paper slips sticking out as place markers, constantly shuffling the books to read pertinent material to us.

One of Herms's students, D. E. Howell, drew this sketch for inclusion here:

> Biographers have stressed the scientific and educational accomplishments of Professor Herms where indeed he was outstanding, but perhaps his wide range of activities were equally outstanding. His tremendous interest in Boy Scouts left an impact on generations of boys. This interest was infectious and many graduate students in Entomology also became interested in Scouting.
>
> Many of his students were introduced to community responsibilities and the value of service clubs by Professor Herms where his counsel, guidance and active participation were constantly in demand. Church activities also claimed an extensive share of his time where he served as a college age group Sunday School teacher for many years.
>
> Perhaps no one will know how many students were helped financially from Professor Herms' personal funds. These were loaned or given with a request to "keep it quiet" and repayment was never asked. The meals in the Herms' home still stand out in the memories of many of Professor Herms' students. It was a real privilege to be invited to dinner and an evening of stimulating, thought provoking conversation, particularly for those of us away from home who were living on beans and hotdogs (if we could afford the latter).
>
> Perhaps his humility and willingness to admit he was wrong stand out in my memory. Very few people have been willing to apologize for reprimanding that on second thought seemed to be too severe or perhaps unwarranted, yet Professor Herms would do this whenever the opportunity presented itself. . . .
>
> His influence on many generations of students is still very great.

Herms's ability was recognized by his colleagues and he served as president of the American Association of Economic Entomologists in 1928 and the Entomological Society of America in 1941. Baldwin-Wallace College conferred an honorary doctorate on him in 1935.

He died suddenly of a heart attack on May 9, 1949. A very lifelike portrait of him, with his inevitable pipe, is to be seen in his fourth edition of *Medical Entomology,* which appeared posthumously. Furman (1949) and Freeborn (1950) wrote obituaries of him.

GLENN WASHINGTON HERRICK (1870–1965)

Herrick was born on a farm near Otto, New York. He was a cousin of Mrs. John Henry Comstock's, and she probably encouraged him to enter Cornell University in 1892, where he received a B.S.A. in 1896. He did graduate work at Harvard and then taught biology at the State

Figure 171. G. W. Herrick

College of Mississippi (now Mississippi State University) and later entomology at Texas Agricultural and Mechanical University.

In 1909 he joined the entomology staff of Cornell University and there taught beginning courses in entomology as well as economic entomology until his retirement in 1935. His students found him a lively instructor. He was coauthor with J. H. Comstock of the *Manual for the Study of of Insects* and sole author of *Insects Injurious to the Household and Annoying to Man.* In 1915 he was elected president of the American Association of Economic Entomologists.

Herrick died in Ithaca, New York, on February 12, 1965, at the age of ninety-five. Rawlins *et al.* (1965) wrote his obituary.

WARREN ELMER HINDS (1876–1936)

Hinds was born at Townsend, Massachusetts, on September 20, 1876, and raised on a farm. He attended Massachusetts Agricultural College, where he studied under C. H. and H. T. Fernald, father and son, and received the first Ph.D. granted by that institution. His thesis, published as a *Contribution to a Monograph of the Insects of the Order Thysanoptera Inhabiting North America,* set an example for the classification of this order.

In 1902, U.S. Chief Entomologist L. O. Howard appointed him as a field agent to aid W. D. Hunter in the study of the boll weevil in Texas, and his contributions to the knowledge of this pest were outstanding. In 1907 he became head of the department of zoology and entomology at the Alabama Polytechnic Institute at Auburn, and entomologist in

Figure 172. W. E. Hinds

the agricultural experiment station; he held both posts until 1924. There, says Thomas (1936), his "pleasing personality, tact, and consideration for others soon won him a warm place in the hearts of the many friends he made." In 1924 he became an entomologist in the Louisiana Experiment Station and Extension Service. At the request of the Peruvian government he took a leave of absence in 1926 to help Peru control its cotton pests.

Besides thrips and cotton insects, he published on fumigation, stored-product pests, truck-crop, and sugar cane insects. He was active in civic affairs and a number of entomological organizations and was president of the American Association of Economic Entomologists in 1933.

Dr. and Mrs. Hinds were known for their hospitality, and they could count "their student friends by the hundred." (Thomas)

Dr. Hinds died of a heart attack following a bout of influenza on January 11, 1936, in his home in Baton Rouge, Louisiana. He was survived by his wife. They had no children.

A chime was presented to Massachusetts State College in Hinds's honor on May 1, 1937. His collection of thrips is in the Department of Entomology at the University of Massachusetts. His obituaries were written by Thomas (1936) and Bailey (1938, 1939).

JAMES KEEVER HOLLOWAY (1900–1964)

Holloway was born in Aspinwall, Pennsylvania, on December 13, 1900. He received his B.S. from Mississippi Agricultural and Mechanical

Figure 173. J. K. Holloway

College in 1926 and did graduate work at Ohio State in 1926–1927. In 1927 he was employed by the U.S.D.A. in the Japanese Beetle Laboratory in Riverton, New Jersey, where he was concerned with the biological control of insect pests. In 1935 and 1936 he studied the biological control of the Mediterranean fruit fly in Hawaii. From 1937 to 1943 he worked on entomogenous fungi in Florida and sprays against scale insects and red spiders in California.

Through a cooperative U.S.D.A. and University of California program in 1944, he began to study the biological control of the Klamath weed in California, and in 1946 his headquarters was transferred from Riverside to Albany, California. The control of the Klamath weed by *Chrysolina quadrigemina* (Suffr.) and other weeds by insect pests was perfected under his sponsorship and direction. For his efforts along this line he received in 1959 a Superior Service Award from the U.S.D.A.

Holloway died on June 13, 1964. Haeussler (1965) wrote his obituary.

HARRY EDWIN JAQUES (1880–1963)

Jaques is perhaps most widely known as the editor-in-chief of the popular *Pictured Key to Nature Series,* a series which includes his own *How to Know the Insects, How to Know the Beetles, How to Know the Trees,* and *How to Know the Land Birds,* among others.

He was born on July 24, 1880, on a farm near Danville, Iowa. He served as a country schoolmaster in the neighborhood before attending

Figure 174. H. E. Jaques

Howe's Academy. There he was "principal" of the Commercial Department and at the same time attended Iowa Wesleyan University. He received a B.A. from the University in 1911 and in 1912 was appointed Professor of Biology there. Then in 1917 he enrolled at Ohio State University to study entomology; the following year he became an extension entomologist with the U.S. Department of Agriculture and resumed teaching at Iowa Wesleyan. The Iowa Insect Survey Collection was started by him. His *Pictured Key* series consisted, at the time of his death, of twenty-five very useful teaching and reference manuals. Iowa Wesleyan conferred on him an honorary doctorate in 1931.

According to Milspaugh (1964), his home was "a veritable mecca for alumni and his many friends . . . Few left without a gift of one or more of his *How to Know* books, bearing a friendly greeting and his distinguished autograph inside the cover. His hobbies included travel, color photography, insect collecting, stamp collecting . . ."

He died September 18, 1963 at Niagara Falls in Ontario, Canada, and was survived by his wife and five children.

CHARLES WILLISON JOHNSON (1863–1932)

Johnson was born on October 26, 1863, in Morris Plains, New Jersey, and lived there until he was eighteen, when his family moved to St. Augustine, Florida. During his eight years there he developed his first

Figure 175. C. W. Johnson

serious interest in insects and molluscs. In 1888 he became curator of the Wagner Free Institute of Science in Philadelphia and remained in that position until 1903.

On his years in Philadelphia, J. A. G. Rehn (1932) writes: "The Wagner Free Institute of Science, in the northwestern part of Philadelphia, furnished in its then Museum Curator, Charles W. Johnson, a man who could interest boys in nature and hold that interest through the kaleidoscopic changes of youth. To know him as we did was to respect and love him, for unbounded patience, good temper, keen enthusiasm, kindly help and inspiring generosity. That he subsequently removed from Philadelphia, and in the environment of Boston became a national figure in entomology, has never caused us to feel other than that we were the original group of 'Johnson's boys.'" This "original group" included H. L. Viereck, a hymenopterist, and C. T. Greene, a dipterist.

As curator for the Boston Society of Natural History from 1903 until his death in 1932, Johnson built up a notable collection of insects and New England shells. Brooks (1932) states that he had "a keen sense of curatorial duties. By this I mean his methodical arrangement of specimens, their protection from dust and infestation, and an ability to find *what* was wanted *when* it was wanted."

Johnson had an excellent knowledge of the taxonomy of several families of the Diptera. He prepared a number of faunal lists, including *The Diptera of Jamaica, The Diptera of New Jersey,* and *The Diptera of New England.* Most of the specimens in this order were collected by him on his numerous field trips. His work on molluscs included an extensive monograph of the molluscs of New England, done in collaboration with W. J. Clench. From 1890 until his death he served as associate editor and manager of *Nautilus,* a malacological magazine.

Brooks, commenting on the pleasure of doing field work with him, says that Johnson knew flies and the more common species of other orders "as an ornithologist knows his local birds." He worked on the Diptera in his spare time, often until midnight.

Ever ready in his cheerful and patient way to impart his knowledge of insects and molluscs to young or old, Johnson was active in the Cambridge Entomological Club and served as its president and on its committees. "When he was present," writes Brues, "interest in the Club meetings never lagged, as he was always ready with a small box of flies about which an interesting discourse could be woven. On such occasions his marvelous memory and ever vigorous youthful enthusiasm was displayed to best advantage." He was a member of a number of other entomological and natural history societies and was president of the Entomological Society of America in 1924. He was an ardent prohibitionist.

He died on July 19, 1932, after a short illness, his wife having preceded him in death. They had no children. Melander (1932) prepared a list of his publications and Brooks (1932) and Brues (1933) wrote his obituaries.

VERNON LYMAN KELLOGG (1867–1937)

A man of renown not only in entomology and evolution, but also as an administrator, humanitarian, and author, Kellogg was Herbert Hoover's principal assistant in the American relief efforts in Europe during and after World War I and at the same time was largely responsible for the organization and success of the National Research Council as a link between government and science. The most moving memorial of the man is the volume entitled *Vernon Kellogg, 1867–1937,* published by the Belgian American Educational Foundation in 1939. It contains tributes to him by Dorothy Canfield Fisher, William Allen White, Robert A. Millikan, and many other eminent people.

Kellogg was born in Emporia, Kansas, on December 1, 1867, where, as William Allen White puts it, "Vernon and his brother, Fred, in their childhood had the tremendous advantage of wise parents who built for the boys an unusual and beautiful home. His father [Lyman Beecher Kellogg] had been first president of the Kansas State Normal School—a pioneer Teachers College—for many years was probate judge, state

Figure 176. V. L. Kellogg

senator and later attorney general of Kansas. His mother [Abigail Homer Kellogg] died in his infancy and his father married Jennie Mitchell and the two of them surrounded the boys with intelligent love. . . .

"His was a happy boyhood. It was busy and purposeful. It foreshadowed his life. Few boys who have grown up in this town have got so much out of the first years as he did.

"They lived such lives as boys now know only in envious dreams. They skated and swam, trapped and hunted and fished and studied wild life until the whole annual panorama of nature with the going and coming of plants and birds and flowers and the passing colors of grass and trees became a part of their life.

"Is it a wonder that such a boy became a scientist? How could he help it? When he left this town to go to the University of Kansas in 1885 at 18, his fate was written inexorably in the blood and environment of childhood. A college professor's son, Vernon had learned casually to love the outer manifestations of nature. He yearned secretly to study the inner sources of things."

While at the University of Kansas he came under the influence of Dr. F. H. Snow, entomologist, naturalist, and university administrator, and accompanied him on many of his summer camping and collecting trips in Colorado and elsewhere. A Phi Beta Kappa, Kellogg helped edit the college paper, and was a pitcher on the baseball team. According to White, he was "a good dancer and a leader in his college fraternity, Phi Delta Theta. Always a soft-spoken, gentle, diplomatic person, he had his way more by festive intrigue than by force."

Dorothy Canfield Fisher, whose father was a professor at the University of Kansas, testifies that Vernon had "one of the finest, and most admirably controlled disciplined minds" her father had ever had in his classes. He was "sensitive, alert, responsive, with a gay smile hovering in the corner of his mobile mouth, a light in his eye—gallant, young and charming."

He studied science and especially entomology under Snow and S. W. Williston. He received his A.B. in 1889, his M.S. in 1892, and served as assistant professor at the University from 1890 to 1893, during which time he became Secretary to Chancellor Snow. In 1891–1892 Kellogg did graduate work at Cornell University, where he came under the influence of John Henry Comstock. He pursued his studies at the University of Leipzig in 1893–1894 and 1897–1898 and at the University of Paris in 1904–1905.

David Starr Jordan, president of Stanford University and an authority on fish and evolution, was so impressed with Kellogg's accomplishments at the University of Kansas that in 1893 he offered Kellogg a position on his faculty as assistant professor of entomology, which he held until his war-relief activities required him to resign in 1920. At Stanford, Kellogg became an authority on the Mallophaga, biting bird lice, and the Anoplura, sucking lice. He was not only interested in the taxonomy of these groups, but also in the range of variation and their evolution. He also conducted extensive experiments on the silkworm. He collaborated with Jordan in various publications and in presenting a course on evolution. Professor Comstock of Cornell, who taught several winter terms at Stanford during the nineties, often joined Kellogg in his entomological pursuits.

At Stanford special students shared the same room with Kellogg and began to put out papers bearing both his name and theirs. "Such association kindled in them an enthusiasm that went with them throughout their lives," says Doane (1940). Many a student completed his education with the aid of small loans from Kellogg. G. F. Ferris, the noted taxonomist and teacher, was a student of Kellogg's.

In 1908 Kellogg married Charlotte Hoffman of Berkeley, California. Their one child, Jean, a girl, was born in 1910.

According to W. E. Colby in *Vernon Kellogg,* he loved to hike in the High Sierras, where he scaled many formidable peaks. He was a director of the Sierra Club of California, and he and his wife were close friends of John Muir's, the famous California naturalist, who was president of the Club.

Kellogg's works vary from popular articles and reviews to formidable scientific treatises, most of which he wrote in a cabin in the pine woods of Carmel Bay. His numerous publications are listed by McClung (1939).

Among them are *The Elements of Insect Anatomy* (with John Henry Comstock, 1895), *Animal Life* (with David Starr Jordan, 1900), *Elementary Zoology* (1901), *American Insects* (1905), *Darwinism Today* (with David Starr Jordan, 1907), *Evolution and Animal Life* (1907), *Economic Zoology and Entomology* (with R. W. Doane, 1915), and *Herbert Hoover, the Man and His Work* (1920).

McClung (1939) says that when the United States entered World War I, Kellogg was brought to Washington to "aid in the organization of science in support of the Government. The first formal result of these efforts was the organization of the National Research Council in which he became chairman of the Divisions of Agriculture, Botany and Zoology. When the Council was later made a continuing body he was named the Permanent Secretary, in which office he continued active [1919–1931] and emeritus, until his death."

When, in 1915, Kellogg became Director of the American Committee for the Relief of the Belgians, his energies were diverted from entomology and natural history to war relief activities, although he maintained his interest in his former pursuits. From 1917 to 1919 he was Assistant to the U.S. Food Administrator, and from 1918 to 1921 he helped Herbert Hoover administer war relief to Poland and Russia. Kellogg received many national and international honors for his scientific and humanitarian activities.

In 1930 Kellogg was stricken with Parkinson's disease, and after years of illness he died on August 8, 1937, in Hartford, Connecticut, but was buried near Carmel Bay, California; "in the oak-shaded cemetary ceded by Spain to the pueblo of Monterey," says Colby, "a granite boulder from a Pacific beach—symbolic of his quiet and enduring faith—marks the resting place of this lover of California's mountains and shores."

CLARENCE HAMILTON KENNEDY (1879–1952)

Kennedy was born at Rockport, Indiana, on June 25, 1879. He received an A.B. in 1902 and an A.M. in 1903 from Indiana State University and another A.M. from Stanford University in 1915; he obtained his Ph.D. from Cornell in 1919.

Kennedy describes his year at Stanford as follows: "There I interviewed Vernon Kellogg with whom the most delightful year of my life ensued. He looked over my manuscripts on Washington and Oregon dragonflies and remarked, 'You are an entomologist already. I am signing you up for a Master's degree. Your paper on the dragonflies of Washington and Oregon, already finished, will be your M.A. thesis.' What courses? 'Keep on with your research on dragonflies.' Barton Warren Everman, Director

Figure 177. C. H. Kennedy

of the Museum of the California Academy of Science, procured passes for me over the railroads of California and Nevada where I collected natural history specimens and was footloose for dragonflies. When low on money I fell back on my ability at illustrations of fish, birds and insects."

For a time Kennedy was a scientific illustrator of sea fishes in the Bureau of Fisheries and he also worked briefly as a botanical collector for the Mt. Holyoke College Herbarium and later as an instructor in zoology at North Carolina State Agricultural College. In 1919 he joined the staff of Ohio State University as an instructor in entomology. He taught there for thirty years, becoming a full professor in 1933. His courses were on insect morphology and biology.

R. C. Osburn (1952) writes, "It is probably not known to many that Kennedy was severely handicapped by illness during his youth and early manhood and was unable to complete the work for the doctorate at Cornell until the age of forty. Even when he came to this university [Ohio State] in 1920 he was able to undertake only a half-time position. Later, as his health improved, he was rapidly advanced in rank, but even in later years his health was marred and his research slowed-up by recurrent illness. . . .

"As a teacher his work with advanced and graduate students was outstanding and all of those who were privileged to undertake their research under his direction will never forget the care and sincerity with which he supervised their work. But woe to the student who tried to get away with anything. For carelessness or sloppy work more than one student was raked over the coals though he eventually profited by the experience."

Kennedy's mimeographed manual, *Methods for the Study of the Internal Anatomy of Insects,* which appeared in 1932, became an essential part of the undergraduate course in many schools of entomology. He was an authority on dragonflies and Mid-Western ants. He served as managing editor of the *Annals of the Entomological Society of America* from 1929 to 1945 and president of the Society in 1935.

He retired from active teaching in 1949 as professor emeritus and died in Columbus, Ohio, on June 6, 1952.

WILLIAM H. WOOD KOMP (1893–1955)

Komp was born of American parents in Yokohama, Japan, on March 16, 1893. He returned to the United States with his parents and attended school in New York and New Jersey. At Rutgers University he pursued his special interest in entomology by studying under Dr. T. J. Headlee, head of the Department of Entomology, from whom he learned about mosquitoes and mosquito control. He received a B.S. in 1916 and an M.S. in 1917, both from Rutgers, and was working on a doctorate at Cornell from 1917 to 1918, when his studies were interrupted by World War I.

Commissioned in the Regular Corps of the U.S. Public Health Service in 1918, he was assigned to mosquito control in Mississippi and elsewhere. In 1921 he became an assistant to Dr. M. A. Barber and studied malaria

Figure 178. W. H. W. Komp

and malaria control in the Southern states. There he helped develop Paris green as a larvicide and developed a technique of preparing thick bloodstains for use in large-scale malaria surveys. He worked in Honduras and Haiti in 1924, where he became interested in the taxonomy of mosquitoes, which led him to pursue the subject under the dipterist Dr. H. G. Dyar during his annual vacations. The same year Komp investigated and instituted control of malaria in Georgia and among the Pueblo Indians in New Mexico, and did the same for yellow fever in West Africa.

In 1931 he began a sixteen-year association with the Gorgas Memorial Laboratory in Panama, during which he developed basic steps in his microdissection and staining techniques for the study of the male terminalia of mosquitoes.

In 1935 he became an assistant to Dr. Fred Soper of the Rockefeller Foundation and studied yellow fever and malaria in the jungles of Colombia, Ecuador, and the West Indies. This resulted in the publication of *The Anopheline Mosquitoes of the Caribbean Region.* In 1937 he was travelling representative of the Pan-American Sanitary Bureau. In 1942 he was made Consultant in Malariology for the Division of Health and Sanitation of the Institute of Inter-American Affairs, a job which took him all through South America. He also worked with the U.S. Army on the application of DDT by airplane.

Komp resumed his taxonomic studies for the Gorgas Memorial Laboratory in 1947. From 1949 to 1951 he continued his travels to Guatemala, British Honduras, and Costa Rica. His publications include numerous papers on malaria, malaria-carrying mosquitoes, the effect of drugs on the treatment of malaria and on the taxonomy of mosquitoes.

In 1955 Rutgers conferred on him an honorary Doctor of Science. He died on December 7 of that same year; he was survived by his wife and one daughter. Woke *et al.* (1956) wrote his obituary.

SOL FELTY LIGHT (1886–1947)

S. F. Light (he disliked Sol Felty) was born in Elm Mills, Kansas, on May 5, 1886, the son of a Presbyterian minister. S. F. received his A. B. from Park College, Parkville, Missouri, in 1908, and then taught for a time at a high school in Manila. He became an instructor at the University of the Philippines in 1912, where he received an M.S. the following year. From 1914 to 1915 he was a fellow in zoology at Princeton University, which gave him another M.S. in 1915. The next year he returned to the University of the Philippines and eventually became a full professor and chairman of the Department of Zoology.

Figure 179. S. F. Light. Courtesy of F. W. Lechleitner

From 1922 until 1924 he taught at the University of Amoy in China and there also was chairman of the zoology department. In 1924 he set off for the University of California at Berkeley to study with Dr. C. A. Kofoid, a well-known protozoologist; he joined the faculty the next year and was granted a Ph.D. for his work on the flagellates in termites the year after that. He became a full professor in 1929.

At the University, Light studied parthenogenesis, caste determination, and the systematics and biology of termites. He became an authority on Western termites. He helped interest a number of competent students in termites, and they continued his studies after his death. He was considered a remarkable teacher and cooperated with the Division of Entomology and Parasitology in training graduate students at the University, some of whom contributed important sections to *Termites and Termite Control* (1934), edited by C. A. Kofoid.

Part of his large collection of termites went to A. E. Emerson of the University of Chicago, and part to the Department of Entomology of the California Academy of Sciences.

Light was accidentally drowned in Clear Lake, California, on June 21, 1947. Essig (1948) wrote his obituary.

FRANK EUGENE LUTZ (1879–1943)

Lutz was born in Bloomsburg, Pennsylvania, on September 15, 1879. He received an A.B. from Haverford College in 1900. During his first

Figure 180. F. E. Lutz

two years in college he specialized in mathematics, as Weiss (1944) says, "upon the advice of his father, who being an insurance agent, was impressed by the large earnings of life insurance actuaries." Later he took biology because of his interest in medicine. After graduation Lutz was attracted to biometry and in between waiting on tables at the Carnegie Institution's summer biological laboratory at Cold Spring Harbor, New York, he counted the grooves on scallop shells. The results of this study were the basis for his first paper. He then attended the University of Chicago and received an A.M. in 1902. His earnings as a mosquito specialist in Long Island in the summer of that year enabled him to study in London and Berlin in 1902 and 1903. From 1904 to 1909 he was a resident investigator in studies on heredity for the Carnegie Institution at Cold Spring Harbor, where he did some pioneering work on *Drosophila* and was the discoverer of the first white-eyed mutants. In the interim, in 1907, he received his Ph.D. from the University of Chicago with the thesis *The Variation and Correlations of Certain Taxonomic Characters of Gryllus.*

In 1909 he was appointed an assistant curator of invertebrate zoology in the American Museum of Natural History in New York City; when, in 1921, the Department of Insects and Spiders was created, he became its curator. He remained curator until his death at the age of sixty-four. While at the Museum he wrote on geographic distribution, insect sounds, ultraviolet and flower-visitation, wind direction and insect flight, and many other subjects. He was not interested in economic entomology. Although Lutz boasted he had never described a "new species" he was always a strong backer of systematics, and was responsible for the acquisition of many fine collections. He made many field trips throughout the United States as well as in Mexico, British Guiana, Cuba, Puerto Rico, and the Canal Zone.

In the summer of 1926 he started the first trailside museum in the Bear Mountains of New York. He was so enthusiastic about this project that, as Weiss puts it, "he would not go with us for lunch, preferring to dine quickly upon some pieces of bread over which he had broken a raw egg, a nutritious if not appetizing mixture."

A very active member of the New York Entomological Society, Lutz was its president for 1925 and 1926. "For a long period the meetings were held in Dr. Lutz's room on the third floor of the Museum and there, surrounded by preserved spiderwebs, Dr. Lutz's zoo of living insects and entomological books and paraphernalia, many interesting entomological discussions took place, in which he always participated." (Weiss)

The royalties from his *Field Book of Insects* (1st ed., 1918) put four of his children through college. He also wrote for popular consumption *A Lot of Insects* in 1941.

Gertsch (1961) describes him as "a man of penetrating mind who loved nothing more than to shock friend and foe with piercing barbs," and that he "won people by his quiet candor, his complete sincerity, and not least by his apt and at times tart and roguish humor."

Lutz died on November 27, 1943. His obituaries were written by Weiss, Schwarz, Gertsch, and Emerson, all in 1944.

NORMAN EUGENE McINDOO (1881–1956)

McIndoo was born in Lyons, Indiana, on April 11, 1881. He received his A.B. in 1906 and his M.A. in 1909 from the University of Indiana, and a Ph.D. in zoology from the University of Pennsylvania in 1911. His Ph.D. thesis was entitled *The Lyriform Organs and Tactile Hairs of Araneads.* This led to his later studies on the senses of insects and communication among insects.

He served in the U.S. Bureau of Entomology in various capacities from 1911 through 1945, when he retired as a senior entomologist. He was famous for his studies on the olfactory senses of insects and developed the olfactometer, which he used in studies of insect attractants and repellents. He also conducted extensive research on such plant-derived insecticides as derris, cubé, and rotenone.

He died of heart failure in a sanatorium in Takoma Park, Maryland, on September 7, 1956. Siegler (1957) wrote his obituary.

Figure 181. N. E. McIndoo

LEONARD SEPTIMUS McLAINE (1887–1943)

McLaine was born in Manchester, England, on June 27, 1887. His father brought him and his sister to New York in 1897. Leonard attended Massachusetts State College at Amherst and received a B.S. there in 1910 and an M.S. in entomology in 1912.

From 1913 to 1919 he served the Entomological Branch of the Canada

Figure 182. L. S. McLaine

Department of Agriculture as a field officer in charge of the brown-tail moth survey. He then became Chief of the Division of Forest Pests Suppression and Executive Assistant to C. Gordon Hewitt, Dominion Entomologist. In 1938 McLaine was made Chief of the Plant Protection Division and in 1942 succeeded Arthur Gibson as Dominion Entomologist. One of his major concerns was to protect Canadian agriculture from native and imported pests. Highly regarded by entomologists everywhere, McLaine was chosen president of the American Association of Economic Entomologists in 1936, and president of the Entomological Society of Ontario in 1936–1937.

He died on July 20, 1943, in his home at Lake Barnard, Quebec, only one and one-half years after becoming Dominion Entomologist. Fernald (1943) wrote his obituary.

ROBERT MATHESON (1881–1958)

Matheson was born in West River, Nova Scotia, on December 20, 1881. He studied entomology at Cornell, where he obtained a B.S. in 1906, an M.S. in 1907, and a Ph.D. in 1911. After graduation he worked for a time at Cornell, and then at South Dakota State College and Nova Scotia Agricultural College at Truro. He returned to Cornell as Assistant

Figure 183. Robert Matheson and M. D. Leonard.
Courtesy of J. J. Davis

Professor of Entomology in 1914, became Professor of Entomology in 1922, and continued in this position until his retirement in 1949.

His principal field of teaching and research was in medical entomology. He was an authority on mosquitoes and malaria and did important work on the role of mosquitoes as transmitters of plasmodia. As a consultant to the Tennessee Valley Authority he was responsible for many of the measures that eliminated mosquitoes and malaria from that area.

Among his publications are a *Handbook of Mosquitoes of North America* (1929), *Medical Entomology* (1932), and *Entomology for Introductory Courses* (1944). He also wrote a monograph on the North American beetles of the family Haliplidae and published extensively on apple and other crop pests. Schwardt *et al.* (1957–1958) thought that "Although his published works will provide an enduring monument, it is probable that Professor Matheson's greatest contribution to the field of medical entomology was in the training of a long succession of graduate students, many of whom became distinguished leaders in medical entomology. . . .

"An intellectual with strong convictions, and capable of forceful and convincing expression, Professor Matheson's voice was heard in any company of which he became a party. If slightly inclined toward pessimism in his appraisal of mankind, it must be admitted that in most aspects he was right."

Matheson died on December 14, 1958, in Princeton, New Jersey.

AXEL LEONARD MELANDER (1878–1962)

Melander was born in Chicago, Illinois, on June 3, 1878, to parents of Swedish descent. He and his life-long friend C. T. Brues were encouraged to pursue their interest in insects by two stimulating high school teachers. During the summers the boys were able to indulge their interest in natural history at the summer home of Melander's parents in northern Illinois.

Both Melander and Brues began to contribute to scientific publications while they were in high school. Upon completing high school they enrolled in the University of Chicago in order to study with William Morton Wheeler, a biologist who became an authority on ants. But when Wheeler moved to the University of Texas to head the Department of Zoology, the two boys followed him. Melander was an assistant in the Department of Zoology there from 1898 to 1902 and received his M.S. in 1902. The next year he married a childhood sweetheart, Mabel Evans; she was as enthusiastic a field naturalist as he and collected with him

Figure 184. A. L. Melander

on many of his field trips. The Melanders had two sons; one became an engineer and the other a musician.

Melander taught at Washington State College in Pullman from 1904 to 1926 and was Professor and Head of the Department of Entomology for twenty years. During this period, Wheeler shifted from Texas to Harvard, and Melander went there to take his Sc.D. in 1914. After Washington State he joined the faculty of the College of the City of New York as Professor of Biology and Head of the Department until his retirement as professor emeritus in 1943. The Melanders then made their home in Riverside, California, where he served as Research Associate at the Citrus Research Center.

Melander was equally at home in both applied science and basic research. In 1914 while at Washington State University he showed that lime-sulfur had become less effective against the San Jose scale in some localities than it was in others. This was the first demonstration of the development of insect resistance to insecticides. He also worked on the Coulee cricket, elm leaf beetle, asparagus beetle, the squash bug, and the codling moth. He was an authority on numerous families of flies, some in obscure genera. With C. T. Brues he wrote the *Key to Families of North American Insects* (1915) and *Classification of Insects* (1932), both of which went through several revisions; in 1937 Frank Carpenter, a paleoentomologist at Harvard, added a chapter on fossil insects to a revision of the latter book. Melander received many honors for his work, and was president of the Entomological Society of America in 1938.

He gathered together one of the world's largest collections of Diptera: it consisted of 250,000 specimens with 12,000 named species. This collection and an equally famous library on the Diptera are now in the Smithsonian Institution.

Melander and his wife collected throughout the United States and many parts of Canada and Mexico. He was skillful in photographing insects, both in motion and still. Spieth (1966), from whom these notes are taken, says that he used these films and slides "to inform and excite the interest of students in his classroom and audiences in lecture halls across the United States, including the American Museum of Natural History."

Melander died in Riverside, California, on August 14, 1962, at the age of eighty-four. According to Spieth's obituary he was "a strong and yet a very gentle man who won the affection and respect of students, colleagues, and friends for his integrity, his subtle humor, his never-failing courtesy, and his indomitable courage in hewing to the line of his convictions."

CLELL LEE METCALF (1888–1948)

Metcalf was born in Lakeville, Ohio, on March 25, 1888, and became interested in natural history as a youngster on his father's farm. He had four brothers, of whom one named Zeno also became a prominent entomologist.

Figure 185. C. L. Metcalf

From 1903 to 1906 Metcalf took high school correspondence courses and then entered a high school in Wooster, Ohio, from which he graduated in 1907. From there he went to Ohio State University to study entomology, receiving his B.S. in 1911 and M.S. in 1912. Later, on a scholarship, he continued his studies at Harvard, where he obtained his doctorate in 1919. He held a variety of positions in entomology in several states, including Ohio, North Carolina, Maine, and New York, and was Professor of Entomology and head of the entomology department at the University of Illinois from 1921 to 1947, when ill health forced him to retire.

At the University Metcalf applied himself zealously to teaching, administration, and research. Much of the latter was done far into the night and resulted in lengthy articles and important books. Three of these books he wrote with W. P. Flint: *Destructive and Useful Insects* (1928), which is now in its fifth edition (the fourth and fifth editions were revised by his son, R. L. Metcalf), *Fundamentals of Insect Life* (1932), and *Insects, Man's Chief Competitor* (1932). Metcalf was also an authority on the Syrphids and published numerous papers on these flies and other insects.

"In his twenty-five years in Illinois," writes Balduf (1948), "Doctor Metcalf gave courses of instruction in beginning entomology, insect control, medical entomology, and the bionomics of insects, and directed the investigations of many advanced students. As a consequence of his reputation as a teacher, graduate students from many states and Canada, as well as Illinois, came to work under his direction. He truly loved to teach . . ." Balduf describes him as being "slight of physique, with flashing dark eyes and seemingly inexhaustible energy."

He was in great demand as a speaker and was heavily involved with many civic affairs and organizations. He was president of the Entomological Society of America in 1934.

Metcalf died after a prolonged illness on August 21, 1948.

AUSTIN WINFIELD MORRILL (1880–1954)

Morrill was born in Tewksbury, Massachusetts, on September 11, 1880. He studied entomology under the Fernalds at Massachusetts Agricultural College and received his bachelor's degree in 1900. He continued his studies there and did original work on greenhouse white flies, pentatomid bugs, and the effects of fumigation upon plant growth, earning his Ph.D. in 1903.

For the next three years he worked with W. D. Hunter of the Bureau of Entomology on the boll weevil and other cotton pests in Victoria,

Figure 186. A. W. Morrill

Texas. He was then transferred to Florida to study aleyrodid pests of citrus fruits and established a laboratory at Orlando. There he was assisted by E. A. Back, R. S. Woglum, and W. W. Yothers in the investigation of the biology of citrus white flies and their control by fungi, spraying, and fumigation.

From 1909 to 1919 Morrill was State Entomologist for the Arizona State Department of Agriculture in Tucson and Entomologist of the Arizona Agricultural Experiment Station. Through his "ability, enthusiasm and skill," according to Back and Reed (1955) he developed excellent facilities for research, teaching and extension in that state.

In 1920 Morrill began his long career as a consulting entomologist on truck-crop and cotton pests. His California Insectaries, Inc., in Glendale supplied the great demand from all parts of the United States for *Trichogramma* parasites for cotton pests. He also investigated soil treatment for subterranean termites and fumigation for dry-wood termites and assisted in the development of a code of ethics for pest control operators in California. He contributed extensively to the literature on entomology in a great variety of publications.

He died in Arcadia, California, on September 28, 1954.

WILMON NEWELL (1878–1943)

Newell was born in Hull, Iowa, on March 4, 1878. He earned a B.S. in 1897 and an M.S. in 1899 from Iowa State College. In 1920 Iowa

Figure 187. Wilmon Newell

State College and in 1937 Clemson Agricultural College conferred honorary doctorates upon him.

In 1898 he was Deputy Entomologist of the Iowa State Experiment Station in Ames under Herbert Osborn and from 1899 to 1902 he was associated with F. M. Webster at the Ohio Experiment Station in Columbus. Then in 1902 he became an assistant entomologist in the Texas Experiment Station in College Station, in 1903 State Entomologist of Georgia, and in 1905 Entomologist of the Louisiana Experiment Station in Baton Rouge. In 1910 he joined the faculty of the Texas Agricultural and Mechanical College as Professor of Entomology and Experiment Station Entomologist. Then, in 1915, he moved to Florida, where he became the first Plant Commissioner of the newly created State Plant Board. He served in this position until the end of his life. In 1920 he was elected president of the American Association of Economic Entomologists. In 1921 he was appointed Dean of the College of Agriculture, Director of the Florida Experiment Station, and Director of the Agricultural Extension Division, and, in 1938, Provost for Agriculture at the University of Florida in Gainesville.

During his long career he published technical papers on apiculture, cotton insects, scale insects, quarantine, and insect eradication. Some of his finest research was on control methods for the cotton boll weevil, the Argentine ant, and the eradication of American foul brood. He was largely responsible for the campaign in Florida to eradicate citrus canker in 1915 and the Mediterranean fruit fly in 1929.

Newell's tremendous zest for work and executive ability were responsible for the number of administrative positions he held and in great part accountable for the present prominence of the Florida Experiment Station.

Creighton (1943) writes: "Those of us who were privileged to know this great man intimately, encountered in him striking qualities of human warmth, and a cleanness of spirit which resulted in a response of affection. He had a deep interest in the development of young scientific workers and served as their invaluable adviser. He had a tender regard for the unfortunate. In the words of Dr. Dwight Sanderson, 'I remember his personal generosity. There were poor families around College Station, Texas, which Newell helped—often with cash loans—in ways which only later were revealed.' "

Dr. Newell died in Gainesville on October 25, 1943, and the Newell Entomological Society and Newell Hall at the University were both named in his honor. Creighton (1943) and Mowry (1944) wrote obituaries of him.

LEONARD MARION PEAIRS (1886–1956)

Peairs' father was a teacher and supervisor of Indians in Lawrence, Kansas, and L. M. Peairs was born there on June 5, 1886. He graduated from Kansas State Agricultural College at Manhattan in 1905, and received his master's degree in 1907. For a number of years, as a graduate student in entomology, he was associated in various capacities with the University of Illinois, Maryland Agricultural College, College Park, and

Figure 188. L. M. Peairs

Kansas State Agricultural College. In 1912 he became professor of entomology at the West Virginia University, a position he held until his retirement in 1952. During 1916 and 1917 he was a Fellow at the University of Chicago, and in 1925 he received a doctorate in zoology from that institution. His thesis was on *The Relation of Temperature to the Development of Insects,* a subject in which he was interested for many years.

Peairs aided E. D. Sanderson in the revision of his text *Insect Pests of Farm, Garden and Orchard* in 1921 and 1931, and a third revision in 1941 carried Peairs's name as sole author. R. H. Davidson is responsible for some recent revisions of this book.

Cory (1956) writes: "Peairs served as editor of the *Journal of Economic Entomology* from 1940 to 1953. For this task he was especially qualified as he was conversant with mathematics, physics and chemistry, in addition to his extensive knowledge of entomology and his exact knowledge of rhetoric. He gave of his knowledge, time and energy without stint. He drove himself relentlessly in this work, which was a 'labor of love' as he felt he had a service to perform to the American Association of Economic Entomologists. In fact, he exhausted his slim physical resources, demonstrating to all who knew the facts his outstanding unselfishness towards his fellow scientists."

Peairs died on January 29, 1956.

EVERETT FRANKLIN PHILLIPS (1878–1951)

Phillips, who was regarded as "the greatest scientific apiculturist in our time" (Anon., 1951–1952), was born in Ohio on November 14, 1878. He graduated from Allegheny College in 1899 and studied zoology at the University of Pennsylvania, where he obtained his Ph.D. in 1904. In 1905 he began to work for the U.S.D.A. in Washington, D.C., and became head of its Division of Bee Culture in 1907.

Phillips had become interested in beekeeping upon completion of his studies for his doctorate on the compound eye of the honeybee. Among the many subjects he studied at the U.S.D.A. were the behavior of bees in winter and the wintering of bees, the eradication of European foulbrood, the physical properties of honey, the bee louse, and others. He also worked on legislation to protect bees from introduced diseases, on pollination, the effect of insecticides on pollinating insects, and the relation of bees to the spread of fireblight.

For many years he was in charge of research and instruction in apiculture at Cornell University. Many of his students became authorities on beekeeping in their own right. Students flocked to him from all over

Figure 189. E. F. Phillips

the United States and all over the world. He was the author of *Beekeeping,* published by Macmillan, and he wrote more than six hundred bulletins and articles on this subject. The beekeeping library he assembled at Cornell became known as the "Phillips Beekeeping Library." He was an enthusiastic and stimulating teacher and was active in the American Association of Economic Entomologists and in a variety of community affairs in Ithaca.

Phillips died at his home in Ithaca on August 21, 1951. Anon. (1951–1952) wrote his obituary.

ALTUS LUCIUS QUAINTANCE (1870–1958)

Porter (1959) says of Quaintance that "To the younger generation of entomologists he is a shadowy figure in the early history of the profession. Those of us who knew him think of him as an extremely able scientist and administrator who for about thirty years exerted a potent influence on the development of our profession."

Quaintance was born in Iowa on December 19, 1870. He received his B.S. from the Florida Agricultural College (University of Florida) in 1893, an M.S. from Alabama Polytechnic Institute in 1894, and a Sc.D. from the same institution in 1915. "From 1894 through 1902 he served successively as instructor in biology and professor of entomology at the University of Florida; biologist and horticulturist of the Georgia Agricultural Experiment Station; and as zoologist, entomologist at [the

Figure 190. A. L. Quaintance

University of] Maryland. In 1903 he received an appointment in the then Division of Entomology, U.S.D.A. Shortly thereafter he became entomologist in charge of the Division of Deciduous Fruit Insect Investigations in the newly formed Bureau of Entomology, a position that he held until his retirement in 1930. He also served as associate chief of the Bureau from 1924 to 1930." (Porter, 1959)

Quaintance worked chiefly on insects affecting deciduous fruits; in taxonomy, he was a recognized authority on the Aleyrodidae or white flies, a few species being serious pests of greenhouse and ornamental plants and citrus fruits.

"Dr. Quaintance, or 'Q,' as he was known to his associates," writes Porter, "had an uncanny understanding of human nature and behavior, and an unusual ability to look ahead and visualize the probable effects of any given course of action. In spite of the importance of the various positions that he held, he was a modest, unassuming individual who never worried about personal credit. When he started his career it was customary for the head of an organization to be senior author of publications reporting research carried on by members of his staff. However, 'Q' would rarely permit his name to appear as an author of a research paper unless he had actually participated in the work, or on a popular paper unless he had had some part in its preparation.

"At the close of 1930 Dr. Quaintance retired to his farm north of Washington, to which he had given the unique name 'Quaint Acres.' Here he devoted his time to the production of fruit and ornamental nursery stock. Hundreds of Washingtonians stopped at his place on their

Sunday afternoon drives to buy peaches or apples if in season, ornamental plants, or some of the high quality apple cider which was locally quite famous."

Howard (1930) said of him, "He has built up modern field laboratories and has directed plans of research that have brought very important results. His keen appreciation of the value of fundamental studies has led him to initiate work through ably trained assistants in directions that had not been pushed by other workers, and many of the strong ideas adopted in practice by the Bureau have originated with him."

Quaintance was a charter member of the erstwhile Entomological Society of America and in 1904 he was elected president of the American Association of Economic Entomologists.

He died at his home near Silver Spring, Maryland, on August 7, 1958.

HENRY JOSEF QUAYLE (1876–1951)

Quayle was born on the Isle of Man on April 29, 1876, but moved to the United States with his parents when he was a youngster. He graduated from the University of Illinois in Urbana in 1903 and then served as an assistant in entomology at the University of California. From 1905 to 1906 he was an instructor in zoology at Iowa State College (now University). He then returned to the University of California in 1906 as assistant professor of entomology and rose to the rank of Professor

Figure 191. H. J. Quayle

of Entomology there and Chairman of the Division of Entomology at the Citrus Experiment Station in Riverside.

Quayle became an authority on citrus scales and their control. Smith *et al.* (1952) claim that he was "the first to detect the development of resistance in populations of certain scale insects to treatment by fumigation with hydrocyanic acid gas." He was also an expert on the Mediterranean fruit fly, early recognizing its danger to the American fruit industry, and on several occasions co-operated with the U.S.D.A. on problems arising from this pest. In 1938 he published his *Insects of Citrus and Other Subtropical Fruits.*

He died on October 24, 1951, in Riverside, eight years after his retirement.

PHIL RAU (1885–1948)

I am indebted to Dr. E. P. Meiners' published (1949) and unpublished notes (the latter made available to me through the kindness of Dr. L. F. Pinkus of Washington University, St. Louis) for the information in this sketch.

Phil Rau was born in St. Louis on April 12, 1885. His schooling ended shortly after he was orphaned at the age of ten. Fortunately he had an inquiring mind and the ability to teach himself and in time he overcame his lack of formal education.

According to Meiners, when Rau was twenty-three he enrolled in the biology department of Washington University in St. Louis as a special

Figure 192. Phil Rau

student. For two years he was under the influence of Dr. James F. Abbott, head of the department of biology and a student of insect life.

As a young man Rau collected beetles and studied their classification, but soon lost his interest in the taxonomy of Coleoptera. Shortly thereafter he began to conduct experiments on the Cecropia moth, and his first paper was entitled "Observations on the Duration of Life, on Copulation, and on Oviposition in *Samia cecropia* L." Rau was so fascinated by the life history and behavior of this insect that he published nine additional papers on it.

At about this time he came under the influence of Charles H. Turner, head of the biology department of Sumner Teachers College in St. Louis. Turner was a noted investigator of insect behavior and Rau became fascinated by the subject.

There can be no doubt that his wife's interest in natural history stimulated Rau in his investigations. Meiners reports that he married Nellie S. Harris, a librarian employed at the Missouri Botanical Garden, in 1911; that she was a graduate of the University of Kansas and very much interested in nature study. He adds that she accompanied Rau on many of his field excursions and collaborated with him in many of his published articles.

Rau's articles appeared in numerous publications and, although they were primarily concerned with bees and wasps, they also encompassed the behavior of cockroaches, moths, and other insects. In 1918 he and his wife published *Wasp Studies Afield;* it was well received by entomologists and naturalists, being written in an entertaining style that attracted both the professional and the layman.

In his unpublished notes Meiners writes: "Rau was a keen and careful observer of insect life. Once, on his rambles, he found a clay bank under a club-house porch on the banks of the Meramec River that was teeming with bees, wasps and other insects. He studied this bank for one entire summer, which resulted in the publication of a paper, 'The Ecology of a Sheltered Clay Bank; A Study in Insect Sociology,' in the *Transactions of the Academy of Science of St. Louis* in 1926. At the end of the season he undertook to transport this clay bank, section by section, to the back yard of his home in Kirkwood, where he maintained it under observation for many years."

After spending five weeks in 1928 in the humid, tropical rain forest on Barro Colorado Island in the Canal Zone he wrote another book, *Jungle Bees and Wasps of Barro Colorado Island,* which appeared in 1933.

Meiners says that although Rau would speak informally at meetings he did not care to present formal lectures. "The local entomological societies did not appeal to him very much since the nature of the discussions were uninteresting to him. He felt even more keenly about this later in life. On one occasion when an effort was made to create an organization on a higher level he wrote, 'Thanks for your kind letter. I can appreciate your efforts to give St. Louis a high brow institution. We have been thriving on this "milk-and-water food for babes" and "ain't nature grand" stuff long enough. I no longer wish to be guilty of perpetuating it. What we need is something along the line of a seminar, not necessarily entomological, but biological, not necessarily a large group but one of mature mentalities.' "

In 1939 and 1940 Rau visited Mexico, where he continued his studies on insects and published some of his observations. He was finishing a "long protracted work" which summed up his observations on wasps when he died after a long illness on May 30, 1948, in Kirkwood.

Meiners (1949) writes: "Phil Rau was not only an entomologist but he found an interest in all phases of natural history. He was always attentive toward young and budding naturalists and did everything within his ability to foster their interests. His home was the mecca of the entomologists and other naturalists of the vicinity."

WILLIAM ALBERT RILEY (1876–1963)

Riley was born at Mankato, Minnesota, on January 10, 1876. He did his undergraduate work at DePauw University and received his Ph.D. from Cornell in 1903. He taught at DePauw University, Columbia University, Cornell Medical College, and New York University.

Figure 193. W. A. Riley

In 1899 he became an instructor at Cornell University and in 1912 was made Professor of Entomology and Parasitology. At Cornell he was associated with such entomologists as J. H. Comstock, J. G. Needham, M. V. Slingerland, and O. A. Johannsen.

In 1918 Riley was made Professor and Chief of the Department of Entomology and Economic Zoology at the University of Minnesota on the St. Paul campus. In 1925 he became Chairman of the Department of Zoology on the Minneapolis campus. He returned to the St. Paul campus in 1930 and remained there until his retirement in 1944.

During the early years of his career he studied insect anatomy; later he became interested in parasitology. He investigated hookworm in Puerto Rico and Panama. In 1915 he published a *Handbook of Medical Entomology* and in 1932 there appeared *Medical Entomology,* a text written in collaboration with O. A. Johannsen, also of Cornell. Riley was well known as a teacher, and many prominent entomologists did their graduate work under his guidance.

In 1917 he was president of the Entomological Society of America and in 1931 president of the American Society of Parasitologists. In 1931–1932 he was a visiting professor at Lingnan University in Canton, China.

Riley died on October 2, 1963, at the age of eighty-seven and was survived by a daughter. Hodson (1964) wrote his obituary.

RURIC CREEGAN ROARK (1887–1962)

Roark was born in Glasgow, Kentucky, on March 13, 1887. He earned a B.A. from the University of Cincinnati in 1907, an M.A. from the University of Illinois in 1908, and a Ph.D. from George Washington University in 1917. His doctoral thesis was on a chemical study of pyrethrum.

In 1908 he was appointed Assistant Chemist in the U.S.D.A. Bureau of Chemistry, and in 1910 he transferred to the Insecticide and Fungicide Laboratory "to work on the development of chemical methods of analysis for insecticides, fungicides, and disinfectants, which the Insecticide Act of 1910 brought under federal inspection."

Roark left the Federal service in 1918 to become Chief Chemist, U.S. Sanitary Specialties Corporation in Chicago. From 1918 to 1923, he was also a research chemist with the General Chemical Corporation in Baltimore. He patented the PDB (paradichlorobenzene) deodorizing block which found such wide usage in lavatories. He returned to the U.S.D.A. in 1923 as an associate chemist and worked in Texas with F. C. Bishopp and E. W. Laake investigating repellents for screwworm fly.

Figure 194. R. C. Roark

Later he collaborated with R. T. Cotton of the Bureau of Entomology in developing fumigants such as ethylene dichloride and ethylene oxide.

In 1927 Roark was appointed chief of the new Insecticide Division in the Bureau of Chemistry and Soils of the U.S.D.A. This group under varying titles made many advances in insecticidal research. For his contributions and leadership, Roark received the Superior Service Award from the U.S.D.A. in 1956. Roark helped industry organize the National Association of Insecticide and Disinfectant Manufacturers, the predecessor of the Chemical Specialties Manufacturers Association.

According to Ruth L. Busbey (1963): "Dr. Roark's even temperament, unfailing courtesy, and keen but kindly sense of humor won him many friends. His respect for the individual and his encouragement of his coworkers to develop their abilities and work out their own ideas were characteristic of the bigness of his nature."

Dr. Roark died suddenly on May 9, 1962, at his home in Washington, D.C.

SIEVERT ALLEN ROHWER (1888–1951)

The son of a mining engineer, Rohwer was born at Telluride, Colorado, on December 22, 1888. He attended the University of Colorado from 1907 to 1909, where he studied under Professor T. D. A. Cockerell and absorbed some of his enthusiasm for taxonomy and became a specialist on the sawflies even before leaving college, according to

Figure 195. S. A. Rohwer

Muesebeck *et al.* (1951). He entered the U.S. Bureau of Entomology in 1909 as a specialist on sawflies without completing his college course, and received his first academic degree in 1948, when the University of Colorado conferred an honorary doctorate on him.

Rohwer worked on sawflies, wasps, and parasitic Hymenoptera between 1908 and 1928 and published about one hundred and eighty papers, including descriptions of eighty new genera and some eleven hundred species. In 1923 the Division of Taxonomic Investigations was created by the Bureau of Entomology and Rohwer was chosen to organize, develop, and direct this division.

Rohwer became Business Manager of the Bureau in 1927; his ability in this position resulted in his appointment as Assistant Chief of the Plant Quarantine and Control Administration in 1928. After this his involvement in such projects as the control of the Mediterranean fruit fly in Florida, the Mexican fruit fly in Texas, and the pink bollworm in Florida left him with no time to devote to his taxonomic studies.

In 1941 Rohwer was assigned as liaison between the Government and the pesticide industry. He was also responsible for the destribution of insecticides and equipment to farmers during this critical period.

Rohwer served as Assistant Chief in charge of Regulatory Activities in the Agricultural Research Administration under the U.S.D.A. through 1950, and in 1951 he assumed greater responsibilities (we are not told what these were).

"Briefly, Sievert Rohwer might be characterized as an old-line, highly individualistic, unorthodox government official, with an intense loyalty to his convictions and an indefatigable zest for work." (Bishopp *et al.*, 1951.)

He was president of the American Association of Economic Entomologists in 1948 and was considered a "godfather" of the National Pest Control Association. In recognition of his services during World War II the U.S.D.A. awarded him the Superior Service Certificate and Medal in 1947.

Rohwer died suddenly of a heart attack at work on February 12, 1951. He was survived by his wife, four sons, and a daughter.

DWIGHT SANDERSON (1878–1944)

Sanderson was born in Clio, Michigan, on September 25, 1878. He earned two B.S. degrees, one from Michigan Agricultural College in East Lansing in 1897 and another from Cornell University in 1898. After graduating, he worked in various capacities in entomology and zoology for colleges in Maryland and Delaware, at Texas Agricultural and Mechanical at College Station and at the University of New

Figure 196. Dwight Sanderson

Hampshire. After several years he became Director of the Agricultural Experiment Station at Durham, New Hampshire, and in 1910 was elected president of the American Association of Economic Entomologists. In that same year he moved on to become Dean of the College of Agriculture at West Virginia University and from 1912 to 1915 was also Director of the West Virginia Agricultural Experiment Station at Morgantown.

Sanderson wrote many articles on entomology and was especially interested in the effect of temperature on insects. He also wrote bulletins on important insect pests and their control. He was the author of *Insects Injurious to Staple Crops* in 1902 and *Insect Pests of Farm, Garden and Orchard* in 1911 and was one of the prime founders of the *Journal of Economic Entomology,* having helped to get it financed. It was by means of his guidance that the Federal Insecticide Act in 1910 was founded on sound principles.

In 1915 he entered the University of Chicago to take graduate work in sociology and received a doctorate in this field in 1921. In 1918 he became associated with the Rural Social Organization at Cornell, where he remained until his retirement in 1943. He became an authority on rural sociology and wrote four well-known books on the subject.

He died at his home in Ithaca, New York, on September 27, 1944. Phillips (1944) and Anderson (1946) wrote his obituaries.

ERNEST RALPH SASSCER (1883–1955)

Sasscer was born on October 25, 1883, at Waldorf, Maryland. He received his B.S. in 1904 from Maryland Agricultural College in College Park and then became a Scientific Aide in the Federal Bureau of Entomology. He gradually advanced to the rank of Chief Inspector of the Federal Horticultural Board. In the early years of his association with the U.S.D.A., he studied the taxonomy of scale insects, which eventually led to his interest in the development of vacuum fumigation for plant

Figure 197. E. R. Sasscer

pests. This permitted the entry of some plant materials in the United States that otherwise would have been barred.

In 1928 he was appointed Entomologist in Charge of the Division of Foreign Plant Quarantines of the Bureau of Entomology and Plant Quarantines, as the Bureau came to be designated in 1934. In the course of his work, he traveled widely, studying plant quarantine regulations in Europe, Mexico, and Cuba.

Sasscer was said to have been a congenial companion with a keen sense of humor, skilled at repartee and the use of apt anecdotes. He was impatient with red tape, knew all his workers personally, and was interested in them as individuals.

Sasscer retired in 1953 but continued to travel widely until he died of a heart attack in Minneapolis, Minnesota, on July 7, 1955. Cory (1955) wrote his obituary.

HERBERT HENRY SCHWARDT (1903–1962)

Schwardt was born at Savonburg, Kansas, May 14, 1903. He received his B.S., M.S., and Ph.D. degrees from the University of Kansas. His early interests were in chemistry but later he changed to entomology and zoology. For a time he worked for the U.S.D.A. on fruit-tree insects in Arkansas and later was appointed to the faculty of the University of Arkansas in Fayetteville. There he became an authority on the biology,

Figure 198. H. H. Schwardt

ecology, and control of bloodsucking flies on livestock. He was also concerned with the pests of ricelands, stored rice, mosquitoes, and gnats.

In 1938 Schwardt became a member of the Department of Entomology and Limnology at Cornell, and in 1957 became head of the Department.

Holland *et al.* (1962) write that "Professor Schwardt had a boundless enthusiasm for his work and a fund of patience and humor that endeared him to everyone. At Cornell he served in many capacities: first, he worked on the control of insect pests of forage crops, then on control of pests of stored products, and finally on control of livestock insects." He was a stimulating teacher and was held in high esteem by his students, entomological colleagues, farmers, and manufacturers of agricultural chemicals.

He died on May 14, 1962, on his fifty-ninth birthday.

RAYMOND CORBETT SHANNON (1894–1945)

Shannon was born in Washington, D.C., on October 4, 1894. Orphaned as a child, he was fortunate to have two outstanding entomologists, John R. Malloch and Frederick Knab, lodged in the home of his foster mother, Mrs. Susan McCormick. They helped direct his interests towards entomology. "Not much is known of his early life, but he left school in 1912 and entered government work as a student assistant in Knab's office [U.S. Bureau of Entomology]. He was a born naturalist, bright, eager, and appreciative, and everyone was glad to aid him." (McAtee and Wade, 1951)

Although Shannon specialized on the Diptera, two of his most helpful friends were the coleopterists E. A. Schwarz and H. S. Barber. Shannon often accompanied them to Plummers Island, Maryland, and became an adept student of entomology and collector of insects. Shannon gradually advanced in the Bureau until his departure for college in 1916. All his friends in the Bureau helped him prepare for Cornell. His schooling was interrupted by service in the Sanitary Corps in World War I. He returned to Government service in 1921 and received his B.S. from Cornell in 1923. During the next few years he made expeditions to Panama to study mosquitoes, did graduate work at George Washington University, and studied insects of importance to public health in Argentina.

In 1927 he was employed by the Rockefeller Foundation as a special member of the International Health Division to study the mosquitoes of Brazil, jungle yellow fever, and malaria transmitted by *Anopheles gambiae*. From Brazil he wrote in 1930 that "Since returning I made another trip and now have amoebiasis again, also hookworm, schizostomiasis, round worms, and Giardia. This business of wading around for *gambiae* larvae gets into your blood in more ways than one." In March 1930 he collected African *A. gambiae* in Natal, Brazil. His recommendations for exterminating these mosquitoes went unheeded at the time; when some years later the campaign was undertaken, it was at a much greater cost because of the extent of the area infested.

In his early years he was interested in the taxonomy of syrphid flies, botflies, and blowflies; later he became more involved in the ecology of biting flies as vectors of disease. His work on mosquitoes carried him from Brazil to Argentina, Peru, and Trinidad; he also went to Greece to study the Tabanidae. In some of these South American countries he investigated verruga fever. In 1938 he demonstrated the transmission of yellow fever by *Haemagogus, Aedes leucocelaenus,* and other mosquitoes in South America. Suffering from mosquito-borne dengue fever, from pain caused by a tumor on his arm, and from insomnia, Shannon died from an overdose of drugs on March 7, 1945, at the age of fifty, and was buried in Port-of-Spain, Trinidad. McAtee and Wade (1951) say he died "a martyr to the hazards of medical entomology."

MARK VERNON SLINGERLAND (1864–1909)

Slingerland was born in Otto, New York, on October 3, 1864. John Henry Comstock (1914) writes that Slingerland's "call to his life work came to him suddenly and with irresistible force. When he came to

Figure 199. M. V. Slingerland

Cornell he knew nothing of entomology. In speaking of this fact afterward he said that when he entered the University he did not know that a butterfly was developed from a caterpillar. During his freshman year he listened to a lecture on the transformation and habits of insects, and the wonders of the insect world took such a deep hold on his imagination that he could not sleep during the following night. From that moment there was no doubt in his mind of what his life work should be."

As an undergraduate at Cornell he took care of the insectary and did such a good job that he was made Assistant Entomologist of the Experiment Station there. In 1892 he received his B.S. from the University, which then gave him the post of Instructor in Entomology. He became Assistant Professor of Economic Entomology there in 1899.

Comstock and Slingerland began publishing on fruit insects and wireworms in 1890. Thereafter the work of the Experiment Station was conducted almost exclusively by Slingerland. He wrote monographs on many of the important insect pests in the state, and his photographic illustrations were far ahead of the times. J. G. Needham (1946), a fellow student at Cornell, wrote that Slingerland "took to insect photography like a duck to water. When I knew him he almost lived in the insectary with that big, old long-bellows camera. He went to the new insectary with it, stayed with it, all but slept with it; and it became the instrument of his chief contribution to economic entomology." His writings were scientifically accurate and comprehensible. He drew up the first "Spray Calendar" of dates for spraying various crops.

Slingerland was president of the American Association of Economic Entomologists in 1903.

Comstock reports that "As a teacher Professor Slingerland was clear, direct, and painstaking. He had the keenest interest in the needs of each individual student. In the last conversation that the writer had with him, only a few hours before his death, he discussed the work of several of his students. Even at that hour, when it was evident to others that the end was near, his thought was not of himself but of his students."

His death, on March 10, 1909, in Ithaca, was caused by Bright's disease. He was only forty-five.

HARRY SCOTT SMITH (1883–1957)

Smith was born in Aurora, Nebraska, on November 29, 1883, and was brought up on a farm. He obtained his A.B. in 1907 and his M.S. in 1908 from the University of Nebraska, where he studied under Lawrence Bruner, a famous orthopterist. From about 1909 to 1913 he was employed

Figure 200. H. S. Smith

by the U.S. Bureau of Entomology to study the natural enemies of the cotton boll weevil, the gypsy moth, and the alfalfa weevil. While studying the alfalfa weevil he traveled to Italy, where he struck up a friendship with Filippo Silvestri, an eminent Italian entomologist.

In 1913 Smith was made Superintendent of the California State Insectary in Sacramento, where he organized the State's researches in biological control of insect pests on a scientific basis. When in 1923 the research activities in biological control were transferred to the Department of Biological Control at the Citrus Experiment Station in Riverside, California, Smith was put in charge of the Department and remained its head until his retirement in 1951.

During his many years in California he was largely responsible for the large-scale use of *Cryptolaemus montrouzieri* in the control of destructive mealybugs on citrus and of *Macrocentrus ancylivorus* against the Oriental fruit moth as well as other successful biological controls of insect pests. He was the first to use insects in the United States for the control of the infamous Klamath weed. He was an authority on plant quarantines, insect resistance before the advent of DDT, ecology, and population dynamics. Many of our foremost students of biological control first learned the subject under his benevolent eye.

One of Smith's students, Paul DeBach, has been kind enough to prepare this profile of him:

Prof. Harry S. Smith ("Prof. Harry" to all who knew him) was universally liked, admired, and respected. He was a gentleman in the true sense of the word; kindly, modest, and considerate. He was easy to talk to on a wide variety of subjects and always found the time to do so. Visiting scientists, former students, and colleagues regularly sought him out, often just to chat, but also frequently to have some problem critically analyzed or to have some idea scrutinized.

Prof. Smith was "down to earth" in his philosophy of life. He never put on a front and had no social ambitions, although he would often be the center of attention at gatherings. He loved the out-of-doors and in his earlier years did considerable hiking and camping. He was a compulsive sports fan, especially with football, and was virtually a walking encyclopedia on the subject. He went to every game whenever possible, particularly those played by the University of California (Berkeley), his adopted alma mater. He knew each coach personally and for many years was an authority on game statistics.

Prof. Smith was an "idea" man. Any research problem presented to him for discussion or suggestions would be sure to benefit from his thinking. This was an inspiration to his graduate students and a great help to colleagues consulting him. He could nearly always come up with some new idea or novel approach, not only in biological control but in widely diverse areas of entomology. His ideas and interests were extremely broad. Prof. Smith was one of the first modern biologists to understand and propound an adequate theory of animal population regulation and his ideas are still among the most important ones currently accepted by population ecologists. Needless to say, he was unsurpassed in the field of biological control of insect pests and for many years was considered to be *the* world authority on this subject.

Smith was elected president of the American Association of Economic Entomologists in 1940; in 1953 an honorary doctorate was conferred on him by the University of Nebraska.

He died suddenly at his home in Riverside on November 28, 1957. He was survived by his wife, Psyche Bruner Smith, and five children. Clausen and Flanders (1958) wrote his obituary.

GEORGE JOHNSTON SPENCER (1888–1966)

Spencer was born of misionary parents in southern India in 1888. After an early education in India, he attended the University of Manchester from 1906 to 1908. He then emigrated to Canada, and received a B.S.A. from the Ontario Agricultural College in Guelph.

During World War I he served with distinction on the Western Front as an officer in the Canadian Expeditionary Force. Following the war he returned briefly to the University of Manchester, and later lectured at Ontario Agricultural College. During the summers he studied the European corn borer for the Ontario Department of Agriculture. In 1924 he received an M.Sc. from the University of Illinois, whereupon he joined the University of British Columbia as Assistant Professor of

Figure 201. G. J. Spencer

Zoology. He taught not only general zoology but histology and entomology for thirty years or more to "generations of grateful students," as Holland (1967) puts it.

According to Holland, "George Spencer always considered himself an 'applied entomologist' and a large part of his effort over the years was devoted to practical investigations on ants, earwigs, carpet beetles, stored-products insects, and household and garden pests of the Vancouver area. His summers were mainly spent in the dry interior of British Columbia except for that of 1926, which was devoted to investigations on commercial crabs on the west coast of Vancouver Island. A famous story is told of a protracted journey in an open rowboat, with a female assistant, collecting plankton. As the hours went by, a personal problem in hydrostatics became unbearable, but he eventually solved it after seeking and receiving her firm assurance that she was, indeed, a professional biologist!"

During the summers he studied grasshopper ecology and parasites for the Canada Department of Agriculture. He also made important collections of insects for the University of British Columbia, being especially interested in the Diptera and such ectoparasites of mammals and birds as lice and fleas.

Spencer "opened the door" of entomology to many of his students, says Holland. "The student found himself sharing the boundless enthusiasm and complete absorption of the teacher and he left the university with his entomological batteries fully charged. Contact with the master was never lost and Spencer students generally, wherever they became established, maintained a close correspondence with their mentor, who took an undiminishing interest in their careers and in their families."

Although he became professor emeritus in 1954, he continued to lecture and work in the Spencer Entomological Museum until shortly before his death. He died after a lingering illness on July 24, 1966 in his home in Vancouver. Canadian and American entomologists will long remember him as a kindly man with a sense of humor, who was a devoted and highly competent student of insects.

JAMES MALCOLM SWAINE (1879–1955)

Swaine was born on October 14, 1879, in Truro, Nova Scotia, attended public school there, and then entered the Nova Scotia Agricultural College. Later he attended Cornell University, where he studied entomology and received a B.S. in 1905 and an M.S. in 1906. In 1915 he completed his thesis on Canadian bark beetles and received his Ph.D. from Cornell.

From 1907 to 1912 he lectured on entomology and zoology at Macdonald College. Because of his studies on bark beetles, C. Gordon Hewitt, the Dominion Entomologist, in 1912 invited him to head the Forest Insect Investigations in Ottawa. In 1919 he became Chief of the Division of Forest Insects, and in 1923 Associate Dominion Entomologist.

Trueman (1955) writes: "During the quarter of a century following his appointment in 1912, Dr. Swaine laid the foundations of forest en-

Figure 202. J. M. Swaine

tomology in Canada. He was that happy combination of scientist, naturalist, and woodsman [who] sees a problem, understands what must be done, and does it. His work on the control of the outbreak of the European spruce sawfly is a classic example."

In 1937 he was appointed Director of the newly formed Science Service, and this organization grew and prospered under his administration. He retired from active duty in 1945.

Swaine was a man of broad interests, as Trueman suggests in his obituary of him: "To sit in his study and read archaeology and ancient history, to make pastel paintings of old sailing ships (and he knows their rigging from stem to stern), and to talk with old friends not only of the past but of the future of the Department are his wintertime hobbies." With the coming of summer, his garden and silver streams with hungry trout were two attractions that lured him from the confines of his study.

Swaine died in Ottawa, Canada, on November 11, 1955.

OTTO HERMAN SWEZEY (1869–1959)

One of Hawaii's most distinguished and deeply respected scientists, Swezey was born on June 7, 1869, on a backwoods farm six miles east of Rockford, Illinois. His early schooling was in a one-room schoolhouse. He attended Lake Forest College in Illinois, where he received his B.A. in 1896. From there he went to Northwestern University,

Figure 203. O. H. Swezey

earned his M.A. in 1897, and served as a biology instructor from 1898 to 1902. From 1902 to 1903 he studied under Professor Herbert Osborn at Ohio State University, where he became very much interested in Hemiptera-Heteroptera. By 1904 he had published several excellent papers on Ohio leafhoppers and their parasites. These caught the eye of L. O. Howard, Chief of the Bureau of Entomology who recommended him to the Hawaiian Sugar Planters Association (H.S.P.A.). He was employed as an Assistant Entomologist by the H.S.P.A. around June 1904, and this was the beginning of an association that lasted for forty-eight years and brought distinction to him and the organization. In July of 1904 he married Mary H. Walsh and they went to Hawaii in August of that year. Swezey was the first American entomologist to be employed by a private company.

As soon as he became established in Hawaii, says Pemberton (1960), he "launched into the work of receiving, breeding and distributing parasites of the sugarcane leafhopper, then being shipped to Hawaii from Australia, where they were discovered by Albert Koebele and R. C. L. Perkins." Pemberton adds that Swezey "played a major role in the successful biological control of insects attacking sugar cane in Hawaii and the sugar industry must always remain indebted to him."

In 1915 H.S.P.A. promoted him to the position of Entomologist, a post he held until his retirement in 1933. He then became Consulting Entomologist and continued his work on the insects of Hawaii for another eighteen fruitful years.

He was a founder of the entomological section of the Bernice P. Bishop Museum in Honolulu and was Honorary Curator of the insect collections there from 1907 to 1919. Swezey had been an ardent student of the biology of Lepidoptera since boyhood and he constantly pursued this study in Hawaii. Although he was more interested in studying the life histories of insects than in describing them, he nevertheless named 107 Lepidoptera. E. C. Zimmerman, who himself had served as a curator of the Bishop Museum, and other workers on Hawaiian insects have acknowledged their great debt to the observations made by Swezey. See the Swezey Memorial Number of the *Proceedings of The Hawaiian Entomological Society* (Anon., 1960).

He edited *Proceedings of The Hawaiian Entomological Society* for thirty-nine years and was president of the Society for four years.

At various times Swezey went on important collecting expeditions to some of the islands of the Pacific such as Guam and Samoa. The insects were then sent to various specialists for further study. For many years he studied the native insects and plants of Hawaiian forests and became an authority on both, which resulted in the publication in 1954 of *Forest Entomology in Hawaii.*

According to Usinger and Zimmerman (1960), Swezey "was a completely unique personality—modest to a fault and of unusually calm disposition. He was much interested in the teachings of the Bahai faith. He was a vegetarian and abstained from the use of tobacco and alcoholic beverages. A patron of the arts, he rarely missed attending symphony concerts and other musical events."

Pemberton (1960) reports that Swezey "was greatly revered by all who knew him. His wide interest and his enthusiasm in studying the Hawaiian biota was stimulating to his colleagues. He was tolerant of all conditions and people that confronted him and was prone to overlook or excuse the faults or mistakes of others, rather than criticize. And finally though rich in achievements and honors in his chosen field of science, he always remained a modest and unassuming man."

In 1944 he received a much deserved honorary doctorate from the University of Hawaii.

Swezey died in San Jose, California, on November 3, 1959, at the age of ninety.

CHARLES AUBREY THOMAS (1895–1962)

Thomas was born at Kennett Square, Pennsylvania, on May 29, 1895. He served as a private in the U.S. Sanitary Corps in World War I. He then attended Pennsylvania State University at University Park and graduated in entomology in 1921. From 1921 to 1925 he did graduate work in entomology at the University of Pennsylvania, during which

Figure 204. C. A. Thomas

time he was also employed by the U.S.D.A. Japanese Beetle Laboratory in Moorestown, New Jersey.

In 1925 he was employed by Pennsylvania State College (now University) to study the animal pests of mushrooms and greenhouse ornamentals. When he retired in 1960, he served as a consulting entomologist for the famous Longwood Gardens in Kennett Square.

Snetsinger (1962) notes that "During his thirty-five years of work on mushroom and greenhouse pests he published some eighty papers. His papers on the life histories of mushroom pests are classical studies which are referred to by all workers on these pests." He adds that "his love of natural history resulted in a home museum for an office and a virtual botanical garden for a yard." A witty and friendly man, he was well liked by all who knew him.

He died in the city of his birth on April 7, 1962.

CHARLES HENRY TURNER (1867–1923)

Turner was born in Cincinnati, Ohio, on February 3, 1867.

He received a B.S. in 1891 and an M.S. in 1892 from the University of Ohio and graduated from the University of Chicago in 1907 with a Ph.D., *magna cum laude*. He taught at Clark University in Georgia from 1892–

Figure 205. C. H. Turner

1905, at the Haynes Normal School in Tennessee in 1907–1908, and was Professor of Biology and Psychology at Sumner Teacher's College in St. Louis from 1908 until the time of his death.

He is known for his studies in the behavior and comparative psychology of invertebrates. He worked on the crayfish and was coauthor of a book on the smaller crustaceans, or Entomostraca, of Minnesota. Later he studied the behavior and tropisms of spiders, ants, bees, moths, cockroaches, and other insects. Some of his studies are in the *Journal of Animal Behavior, Transactions of The St. Louis Academy of Sciences* and similar periodicals. Turner's studies were quoted by many authorities. He also devoted himself to the sociological problems of the Negro race, of which he was a member. Ray (1923) says that Turner worked under unusual handicaps, with no library and no money to purchase laboratory equipment.

Professor Turner died in Chicago on February 14, 1923, at the age of fifty-six.

WILLIAM HUNTER VOLCK (1879–1943)

Volck was born in Riverside, California, on September 24, 1879, and was raised on a citrus farm. He attended public schools in southern California and entered the College of Agriculture at the University of California in 1901.

C. W. Woodworth, Professor of Entomology at the University of California, found him so capable that he recommended his appointment by the Los Angeles County Board of Horticultural Commissioners as an inspector of citrus pests in 1902, and Volck never completed his uni-

Figure 206. W. H. Volck

versity training. He made careful studies of red spider mites in Los Angeles County and later in Sutter County, California.

In 1902 he was also appointed an assistant to Woodworth and began a thorough study of the oils used against scale insects. In time he became an authority on oil emulsions, and "Volck" oil was one of his contributions. He also investigated the use of arsenicals and had a great deal to do with the introduction and use of arsenicals such as basic lead arsenate which were low in phytotoxicity. As a field assistant at the University of California from 1908 to 1913 and as Entomologist and Horticultural Commissioner for several counties in central California, he did important research on many apple and pear pests.

In 1920 Volck became a research director of the California Spray Chemical Company in Berkeley. His efforts contributed significantly to the growth of this company, and he was granted many important patents on insecticides developed in his research. Until his death in Watsonville, California, on January 12, 1943, Volck was active in the development of new products and the direction of technical workers in research of commercial pesticides. The raising of tropical plants in greenhouses was one of his hobbies in later years.

Essig (1943) wrote his obituary.

JOSEPH SANFORD WADE (1880–1961)

Wade was born in Cumberland County, Kentucky, on July 20, 1880, and spent his boyhood in rural surroundings that filled him with a

Figure 207. J. S. Wade

love for nature. When he was fifteen his family moved to a farm in Kansas. He attended Fairmount College from 1905 to 1906 and then returned to his farm, where he wrote articles on farming and historical subjects. In 1913 F. M. Webster, a noted Federal entomologist, read his article "The Grasshopper Year of 1874" in the Wellington, Kansas, *Daily Journal* and later that year appointed him a field assistant in the U.S. Bureau of Entomology. Wade studied the false wireworms in Kansas and wrote several important papers on them.

In 1917 he was transferred to Washington, D.C., as an assistant to the Federal entomologist, W. R. Walton. He lived there until his death of a heart attack on January 1, 1961. The intellectual life of Washington with its scientific societies and its libraries was a heady stimulant to him and over the years he wrote hundreds of articles, reviews, and reprints. He prepared bibliographies on the Hessian fly, immature stages of North American Coleoptera, and sugar-cane predators and parasites. He published articles on Kentucky history and the activities of naturalists in Kentucky. He was president of the Entomological Society of Washington in 1934. He was known not only as an entomologist and a bibliographer, but as a biographer and lover of art.

Wade never married and lived in Washington with his older sister Mary. A deeply religious man, he was a regular attendant at a Presbyterian church in Washington.

Leonard and Larrimer (1961) describe him as a "cultured gentleman of the old school, a distinguished writer and historian, and above all else a thoughtful and helpful friend to many more than we can well count."

CLAUDE WAKELAND (1888–1960)

Wakeland was born at La Jara, Colorado, on August 2, 1888, and worked on his father's ranch as a boy. He earned a B.S. from Colorado Agricultural College (Colorado State University) in 1914 and an M.S. in 1924. In 1934 he obtained his Ph.D. from Ohio State University. From 1914 to 1920 he worked for the Colorado Agricultural Experiment Station, where he was especially concerned with the alfalfa weevil and its control. In 1920 he became the Idaho Extension Entomologist with the Department of Entomology at the University of Idaho and in 1928 became Head of the Department of Entomology and Entomologist in the Extension Division. There he studied a variety of agricultural pests, including the pea weevil and false wireworms.

In 1938 he joined the U.S. Bureau of Entomology and in 1942 he became the first Chief of its Division of Grasshopper Control and directed the control of grasshoppers, Mormon crickets, and chinch bugs in twenty-

Figure 208. Claude Wakeland

seven co-operating states. Two important Government papers published by him were *The High Plains Grasshopper* in 1958 and *Mormon Crickets in North America* in 1959. He retired from active service in 1958.

Shockley (1963) describes Wakeland as "a musician (mandolin and piano), skilled cabinet worker, carpenter, and amateur photographer."

He died of a heart attack in a hospital in Denver, Colorado, on November 9, 1960.

WILLIAM RANDOLPH WALTON (1873–1952)

Walton was born in Brooklyn, New York, on September 23, 1873. He attended school at Middletown, New York. He combined a lively interest in natural history with, in the words of Wade *et al.* (1953), "manual, mechanical, and mathematical proficiency, and noteworthy skill in drawing." After his father died in 1890 he became a telegraph operator on the Erie Railway. During his "long nocturnal vigils" he observed, collected, and reared insects. Dr. J. A. Lintner, State Entomologist of New York, corresponded with him and helped him in his studies. From 1890 to 1906 Walton worked for the railroad in several engineering capacities and for a time served as an engineering draftsman.

In 1905–1906 he attended an art school in Allegheny (now part of Pittsburgh) and proved to be an outstanding student. In 1906 he joined the Pennsylvania State Department of Agriculture in Harrisburg, where he worked for four years as an entomological artist, botanist, photographer,

Figure 209. W. R. Walton

and maker of museum models. He also began to study and publish papers on the Diptera, especially on the Syrphidae, Asilidae, and Tachinidae. Walton's *An Illustrated Glossary of Chaetotaxy and Anatomical Terms Used in Describing Diptera* appeared in 1909.

The following year he began his long association with the U.S. Bureau of Entomology as an entomological illustrator. He also worked and wrote on predacious and parasitic Diptera. In 1917, a year after F. M. Webster's death, Walton succeeded him as head of the Division of Cereal and Forage Insect Investigations in the Bureau. He was made Senior Entomologist in 1923 and worked on a variety of insect problems, including the European corn borer, grasshoppers, white-fringed beetles, Hessian flies, greenbugs, and chinch bugs. He retired from the Bureau in 1943.

A great fisherman, Walton wrote on outdoor life for a number of publications. For many years he was a very active member of the Entomological Society of Washington, of which he served as president in 1920 and 1921. He edited its *Proceedings* from 1927 to 1943. After a long illness he died on October 20, 1952, and was survived by his wife and five children.

FRANKLIN GERSHOM WHITE (1873–1937)

White was born at Hooksburg, Ohio, on December 22, 1873. He obtained a B.S. from Ohio University in 1901, a Ph.D. from Cornell in 1905, and an M.D. from George Washington University in 1909.

Figure 210. F. G. White

From 1903 to 1906 he served as an instructor in bacteriology at Cornell, where he began a study of the diseases of honeybees. In 1906 he was employed by the U.S.D.A. as an expert in animal bacteriology in the Bureau of Animal Industry. That same year the Bureau of Entomology published his Ph.D. thesis, a technical paper on bee diseases. This paper showed that foul brood actually consisted of two distinct bacterial diseases, American foul brood and European foul brood, and their causative organisms were *Bacillus larvae* and *Bacillus pluton*, respectively.

In 1907 he joined the Bureau of Entomology and continued his work on such bee diseases as American foul brood, European foul brood, sacbrood, and *Nosema*. According to Bishopp and Burnside (1937) other diseases described by White "include a neosporidian infection of flour beetles; a protozooan and a bacterial disease of the Mediterranean flour moth; septicemia of the Colorado potato beetle; pink bollworm septicemia, and polyhedral diseases of insects."

During his long career White also made important contributions to the use of sterile maggots in the treatment of osteomyelitis, "creeping eruption," milky spore disease of Japanese beetle, and other diseases of insects. He was one of the pioneers in the study of insect diseases, a subject of immense interest to present-day students of the biological control of insect pests.

White died at Moorestown, New Jersey, on April 27, 1937. A bachelor, he was survived by a brother and three sisters.

Bishopp and Burnside (1937) wrote his obituary.

CHARLES WILLIAM WOODWORTH (1865–1940)

Woodworth was born in Champaign, Illinois, on April 28, 1865. He studied under Dr. S. A. Forbes, the State Entomologist, at the University of Illinois and received his B.S. there in 1885 and his M.S. in 1886. He continued his studies at the University as a graduate student and completed his paper "The Wing Veins of Insects," which was published in 1906. He studied with Dr. H. Hagen at Harvard University from 1886 to 1888 and again in 1900–1901. From 1888 to 1891 he was an entomologist and botanist at the Arkansas Agricultural Experiment Station in Fayetteville. He then joined the University of California as an assistant professor of entomology, becoming head of the Division of Entomology in 1920. In 1918 and from 1922–1924, he organized and taught entomology in various cities in China. Woodworth retired from the University of California in 1930 as professor emeritus.

His interests in entomology were both taxonomic and economic, and he prepared lists of butterflies, grasshoppers, and other insects in California. In the economic field he worked on such agricultural pests as the peach twig borer, the California peach tree borer, the potato tuber moth, the grape leafhopper, the codling moth, the red spider, and the Argentine ant. He made important contributions in his studies of arsenicals, HCN, and petroleum oils as insecticides.

Essig (1941) writes that Woodworth "was a man of many sterling qualities and the very personification of honesty, kindness, modesty, and

Figure 211. C. W. Woodworth

tolerance." Essig adds that he was like a father to many of his students and "gave them more than the credit they earned and deserved. I do not recall that he ever added his name to the publication by a single student or by an assistant. He was extremely tenacious in what he believed to be right."

Woodworth was said to have kept a California tarantula as a pet under a bell-jar, and when lecturing would remove the spider from the bell-jar and stroke the back of his hairy and good-natured companion.

His interest in science extended to biology, ecology, mathematics, physics, and chemistry, and he was, says Essig, "an inventor of considerable ability": he developed new insecticides and methods of applying and manufacturing them.

He was a member of many scientific organizations and a founder of the Pacific Slope Branch of the American Association of Economic Entomologists. He received many honors.

Woodworth died at his home in Berkeley, California, on November 19, 1940.

References

CHAPTER I

Allen, E. G. 1951. The History of American Ornithology before Audubon. *Trans. Amer. Phil. Soc.* 41:463–478.

Anon. 1843. Obituary Notice of Professor Peck. *Mass. Hist. Soc. Coll.* 2:161–170.

Bassett, A. S. 1938. Some Georgia Records of John Abbot. *Auk* 55:244–254.

Brinton, D. G. 1882. Memoir of S. S. Haldeman. *Proc. Amer. Phil. Soc.* 19:279–285.

Doubleday, E. 1869. *Entomological Correspondence: Thaddeus William Harris, M.D.* Edited by S. H. Scudder. *Occasional Papers of the Boston Soc. Nat. Hist.*

Dow, R. P. 1913. The Work and Times of Dr. Harris. *Bull. Bklyn. Entomol. Soc.* 8:106–118.

Dow, R. P. 1914. John Abbot, of Georgia. *J. N. Y. Entomol. Soc.* 22:65–72.

Frost, S. W. 1937. Frederick Valentine Melsheimer. *Pap. Lancaster Co. Hist. Soc.* 41:164–168.

Grote, A. R. 1889. The Rise of Practical Entomology in America. *20th Annu. Rep. Entomol. Soc. Ont.,* pp. 75–82.

Hagen, H. A. 1884. The Melsheimer Family and the Melsheimer Collection. *Can. Entomol.* 16:191–197.

Harris, E. 1882. Memoir of Thaddeus William Harris, M.D. *Mass. Hist. Soc. Proc.* 19:313–322.

Hart, C. H. 1881. Samuel Stehman Haldeman. *Pa. Mo.* 12:584–601.

Heisey, M. L. 1937. Frederick Valentine Melsheimer, Entomologist. *Pap. Lancaster Co. Hist. Soc.* 41:103–111, 162–163.

Higginson, T. W. 1869. *Entomological Correspondence of Thaddeus William Harris, M.D.* Edited by S. H. Scudder. *Occasional Papers of the Boston Soc. Nat. Hist.*

Holland, W. J. 1929. The First Picture of an American Butterfly. *Sci. Mo.* 38:44–48.

Howard, L. O. 1930. *A History of Applied Entomology*. *Smithsonian Misc. Coll.* Vol. 84.

Lesley, J. P. 1886. Samuel Stehman Haldeman. *Nat. Acad. Sci. Biog. Memoirs* 2:139–172.

Marx, G. 1891. A List of the Araneae of the District of Columbia. *Proc. Entomol. Soc. Wash.* 2:148–161.

Ord, G. 1869. A Memoir of Thomas Say. *A Description of the Insects of North America by Thomas Say*. Edited by John L. LeConte, M.D. Vol. I. New York, J. W. Bouton, pp. vii–xxi.

Remington, C. L. 1948. Notes on My Life (John Abbot). *Lepidopterists' News* 2:28–30.

Schwarz, E. A. 1890. Annual Address of the President. *Proc. Entomol. Soc. Wash.* 2:5–23.

Scudder, S. H. 1869. *Entomological Correspondence of Thaddeus William Harris, M.D.* Edited by S. H. Scudder. *Occasional Papers of the Boston Soc. Nat. Hist.*

Scudder, S. H. 1888. John Abbot, the Aurelian. *Can. Entomol.* 20:150–154.

Webster, F. M. 1895. Thomas Say. *Entomol. News* 6:1–4, 33–34, 80–81, 101–103.

Weiss, H. B., and G. M. Ziegler. 1931. *Thomas Say, Early American Naturalist.* Springfield, Ill., Charles C. Thomas.

CHAPTER II

Anon. 1894. Benjamin Dann Walsh. *Entomol. News* 5:269–270.

Collins, D. L. 1954. The "Bug Catcher of Salem." *Bull. of the Schools* (New York) 40:193–196.

Comstock, A. B. 1953. *The Comstocks of Cornell.* Ithaca, N. Y., Comstock Publ. Associates.

Felt, E. P. 1899. Memorial of Life and Entomology Work of Joseph Albert Lintner. *Bull. N. Y. State Mus.* 5:303–611.

Fitch, A. 1856–1872. *Report on the Noxious, Beneficial, and Other Insects of the State of New York.* Albany, N. Y., C. Van Benthuysen.

Forbes, S. A. 1907. Grierson's Cavalry Raid. *Ill. State Hist. Soc. 8th Annu. Meet.,* Springfield, Jan. 24.

Forbes, S. A. 1910. Dr. Cyrus Thomas. *J. Econ. Entomol.* 3:383–384.

Forbes, S. A. 1930. Autobiographical Sketch Written in 1923. *Sci. Mo.* 30:475–476.

Goding, F. W. 1885. Biographical Sketch of William LeBaron. *Entomologica Americana* 1: 122–125.

Goding, F. W. 1888. A Pen Sketch of Cyrus Thomas, Third State Entomologist. *Ill. Hort. Soc. Trans. for 1888* n. s. 22:106–108.

Howard, L. O. 1930. *A History of Applied Entomology. Smithsonian Misc. Coll.* Vol. 84.

Howard, L. O. 1932. Biographical Memoir of Stephen Alfred Forbes. 1844–1930. *Nat. Acad. Sci. Biog. Memoir* 15:1–54.

Lintner, J. A., and F. G. Sanborn. 1879. An Account of the Collections Which Illustrate the Labors of Dr. Asa Fitch. *Psyche* 2:273–276.

Mallis, A. 1963. The Diaries of Asa Fitch, M.D. *Bull. Ent. Soc. Amer.* 9:262–265.

Metcalf, C. L. 1930. Stephen Alfred Forbes, May 29, 1844–March 13, 1930. *Entomol. News* 41:175–178.

Mills, H. B. 1964. Stephen Alfred Forbes. *Systematic Zoology* 13:208–214.

Osten Sacken, C. R. 1903. *Record of My Life Work in Entomology*. Cambridge, Mass., Cambridge Univ. Press.

Rezneck, S. 1961. Diary of a New York Doctor in Illinois—1830–1831. *J. Ill. State Hist. Soc.* 54:25–50.

Riley, C. V. 1869–1870. In Memoriam—B. D. Walsh. *Amer. Entomol.* 2:65–68.

Riley, C. V. 1870. The Walsh Entomological Collection. *Amer. Entomol.* 2:95.

Riley, C. V. 1880. Dr. Asa Fitch. *Amer. Entomol.* 3:121–123.

Slingerland, M. V. 1898. Dr. Joseph Albert Lintner. *Can. Entomol.* 30:165–166.

Thurston, E. P. 1879. Sketch of Dr. Asa Fitch. *Pop. Sci. Mo.* 16:116–120.

Tucker, E. A. 1920. Benjamin D. Walsh—First State Entomologist of Illinois. *Trans. Ill. State Hist. Soc.*, pp. 54–61.

Walsh, B. D. 1869. Universal Remedies. *Amer. Entomol.* 2:33–35.

Ward, H. B. 1930. S. A. Forbes. *Science* 71(1841):378–381.

Weiss, H. B. 1936. *The Pioneer Century of American Entomology*. New Brunswick, N. J. Published by the Author.

CHAPTER III

Anon. 1871. Destruction of the Walsh Cabinet in the Chicago Fire. *Can. Entomol.* 3:196.

Anon. 1950. Leland Ossian Howard, 1857–1950. *J. Econ. Entomol.* 43:958–962.

Bishopp, F. C. 1957. Leland Ossian Howard Centennial. *Bull. Entomol. Soc. Amer.* 3:1–3.

Bissell, T. L. 1960. History of Entomology at the University of Maryland. *Bull. Entomol. Soc. Amer.* 6:80–85.

Blake, D. H. 1951/1952. Two Old Coleopterists. *Coleopterists' Bull.* 5:49–54, 65–72, 6:3–9, 19–26.

Brooks, F. E. 1925. West Virginia Scientists. *W. Va. Review* 2:308–345.

Clark, A. H. 1952. Leland Ossian Howard (1857–1950). *Cosmos Club Bull.* 5:2–4.

Comstock, A. B. 1953. *The Comstocks of Cornell*. Ithaca, N. Y., Cornell Publ. Associates.

Cory, E. N., W. D. Reed, and E. R. Sasscer. 1955. Charles Lester Marlatt, 1863–1954. *Proc. Entomol. Soc. Wash.* 57:37–43.

Dodge, C. R. 1883. Townend Glover. *Psyche* 4:115–116.

Dodge, C. R. 1886. T. E. Glover. *Proc. Entomol. Soc. Wash.* 1:60.

Dodge, C. R. 1888. The Life and Entomological Work of the Late Townend Glover, First Entomologist of the U. S. Dept. of Agriculture. *U. S. D. A. Div. Entomol. Bull.* No. 18.

Doutt, R. L. 1958. Vice, Virtue, and the Vedalia. *Bull. Entomol. Soc. Amer.* 4:119–123.

Forbes, S. A. 1916. Francis Marion Webster. *J. Econ. Entomol.* 9:239–241.

Finch, K. 1930. Mrs. Anna Botsford Comstock. *Entomol. News* 41:277–279.

Finch, K. 1931. Professor John Henry Comstock. *Entomol. News* 42:153–157.

Gahan, A. B., G. J. Haeussler, E. R. Sasscer, and J. S. Wade. 1950. Leland Ossian Howard, 1857–1950. *Proc. Entomol. Soc. Wash.* 52:224–233.

Goode, G. B. 1896. A Memorial Appreciation of Charles Valentine Riley. *Science* 3:217–225.

Howard, L. O. 1909. The Entomological Society of Washington. *Proc. Entomol. Soc. Wash.* 11:8–18.

Howard, L. O. 1916. Francis Marion Webster. *Proc. Entomol. Soc. Wash.* 18:79–83.

Howard, L. O. 1925A. Walter David Hunter, LL.D. *Proc. Entomol. Soc. Wash.* 27:170–181.

Howard, L. O. 1925B. Walter David Hunter. *J. Econ. Entomol.* 18:844–848.

Howard, L. O. 1929. Frank Hurlbut Chittenden. *J. Econ. Entomol.* 22:989–990.

Howard, L. O. 1930. *A History of Applied Entomology. Smithsonian Misc. Coll.* Vol. 84.

Howard, L. O. 1933. *Fighting the Insects.* New York, Macmillan Co.

Howard, L. O., H. S. Barber, and A. Busck. 1928. Dr. E. A. Schwarz. *Proc. Entomol. Soc. Wash.* 30:154–164.

Howard, L. O., E. A. Schwarz, and H. G. Hubbard. 1895. Charles V. Riley, Ph.D. *Proc. Entomol. Soc. Wash.* 3:293–298.

Marlatt, C. L. 1953. *An Entomologist's Quest: The Story of the San Jose Scale.* Baltimore, Md., Monumental Printing Co.

Meiners, E. P. 1943. Charles Valentine Riley, *C. V. Riley Entomol. Soc.* Columbia, Mo., Univ. of Missouri (Mimeo.).

Osborn, H. 1937. *Fragments of Entomological History.* Columbus, Ohio. Published by the Author.

Packard, A. S. 1895. Charles Valentine Riley. *Science* 2:745–751.

Packett, F. S. In Howard, L. O., 1925A. Walter David Hunter, LL.D. *Proc. Entomol. Soc. Wash.* 27:170–181.

Rohwer, S. A. 1947. Greetings to L. O. Howard. *Proc. Entomol. Soc. Wash.* 49:146–147.

Rohwer, S. A. 1950. Andrew Delmar Hopkins, 1857–1948. *Proc. Entomol. Soc. Wash.* 52:21–26.

Schwarz, E. A. 1929. Letters. *J. N. Y. Entomol. Soc.* 37:312.

Schwarz, E. A., L. O. Howard, and O. F. Cook. 1901. Henry Guernsey Hubbard. *Proc. Entomol. Soc. Wash.* 4:350–360.

Skinner, H. 1895. Prof. C. V. Riley, M.A., Ph.D. *Entomol. News* 8:241–243.

Smith, J. B. 1899. Obituary of Henry Guernsey Hubbard. *Entomol. News* 10:80–82.

Snyder, T. E., and J. M. Miller. 1949. Andrew Delmar Hopkins, 1857–1948. *J. Econ. Entomol.* 42:868–869.

Stemple, R. M. 1966. Andrew Delmar Hopkins; Pioneer in Forest Entomology: A Bibliography. *Bull. Entomol. Soc. Amer.* 12:25–28.

Torre-Bueno, J. R. de la. 1948. Some More Entomologists. *Bull. Bklyn. Entomol. Soc.* 43:150–153.

Walton, W. R. 1916. Francis Marion Webster. *Science* 43:162–164.

Walton, W. R., and F. C. Bishopp. 1937. Dr. L. O. Howard and the Entomological Society of Washington. *Proc. Entomol. Soc. Wash.* 39:121–132.

CHAPTER IV

Anon. 1932. Charles James Stewart Bethune, M.A., D.C.L. *Can. Entomol.* 64:98–99.

Bethune, C. J. S. 1908. James Fletcher. *Can. Entomol.* 40:433–437.

Bethune, C. J. S. 1914. Dr. William Saunders, C.M.G. *Can. Entomol.* 46:333–336.

Dearness, J. 1939. Reminiscences of the Early Days of the Society. *Can. Entomol.* 71:21.

Essig, E. O. 1931. *A History of Entomology.* New York, Macmillan Co.

Gibson, A. 1909. James Fletcher. *Ottawa Naturalist* 22:189–234.

Gibson, A., and J. M. Swaine. 1920. Charles Gordon Hewitt. *Can. Entomol.* 52:97–105.

Goding, F. W. 1894. A Pen Sketch of Prof. William Saunders, F.R.S.C., F.L.S. *Annu. Rep. Entomol. Soc. Ont.* 25:120–121.

Harrington, W. H. 1892. The Abbé Provancher. *Can. Entomol.* 24:130–131.

Harrington, W. H. 1909. James Fletcher. *Ottawa Naturalist* 22:196–205.

Hewitt, C. G. 1921. *The Conservation of the Wild Life of Canada.* New York, Charles Scribner's Sons.

Howard, L. O. 1930. *A History of Applied Entomology. Smithsonian Misc. Coll.* Vol. 84.

Lyman, H. H. 1910. The Rev. Charles James Stewart Bethune, M.A., D.C.L., F.R.S.C. *Can. Entomol.* 42:2–3.

Macoun, J. 1909. James Fletcher. *Ottawa Naturalist* 22:212–214.

Maheux, G. 1923. L'Abbé Provancher. *Fifty-third Annu. Rep. Entomol. Soc. Ontario,* pp. 28–30.

Ormerod, E. 1904. *Eleanor Ormerod, LL.D., Economic Entomologist. Autobiography and Correspondence.* New York, E. P. Dutton & Co.

Osborn, H. 1937. *Fragments of Entomological History.* Columbus, Ohio. Published by the Author.

Saunders, W. 1909. James Fletcher. *Ottawa Naturalist* 22:192–196.

Saunders, W. E. 1939. Entomological Memories. *Can. Entomol.* 71:20–24.

Shutt, F. T. 1909. James Fletcher. *Ottawa Naturalist* 22:220–222.

Spencer, G. J. 1964. A Century of Entomology in Canada. *Can. Entomol.* 96:33–59.

CHAPTER V

Alexander, C. P. 1943. Henry Torsey Fernald (1886–1952). *Entomol. News* 64:85–88.

Braun, A. F. 1921. Charles Henry Fernald. *Entomol. News* 32:129–133.

Calvert, P. P. 1893. Hermann August Hagen. *Entomol. News* 4:313–317.

Comstock, A. B. 1953. *The Comstocks of Cornell.* Ithaca, N. Y., Comstock Publ. Associates.

Essig, E. O. 1931. *A History of Entomology*. New York, Macmillan Co.

Felt, E. P. 1920. Maria E. Fernald. *J. Econ. Entomol.* 15:153.

Foote, R. H. 1962. R. E. Snodgrass. *Proc. Entomol. Soc. Wash.* 64:210.

Green, J. W., *et al.* 1908. Francis Huntington Snow. *Graduate Mag.,* Univ. Kansas. 7:121–127.

Henshaw, S. 1894. Hermann August Hagen. *Proc. Amer. Acad. Arts & Sci.* 29:419–423.

Herrick, G. W. 1931. Professor John Henry Comstock. *Ann. Entomol. Soc. Amer.* 24:199–204.

Hinds, W. E. 1933. *Fernald Club Yearbook.* Amherst, Univ. of Mass.

Howard, L. O. 1910. C. H. Fernald. *Science* 32:769–775.

Howard, L. O. 1930. *A History of Applied Entomology. Smithsonian Misc. Coll.* Vol. 84.

Howard, L. O. 1933. *Fighting the Insects.* New York, Macmillan Co.

Hyder, C. K. 1953. *Snow of Kansas.* Lawrence, Univ. Kansas Press.

Kellogg, V. L. 1909. Francis Huntington Snow. *J. Econ. Entomol.* 2:83–85.

Knull, D. J., and J. N. Knull. 1954. H. Osborn. *J. Econ. Entomol.* 47:1164–1165.

Leech, H. B. 1959. Bibliography of Gordon F. Ferris. *Pan-Pac. Entomol.* 35:29–50.

Loew, H., in Osten Sacken, C. R. 1903.

McKenzie, H. L. 1959. Gordon Floyd Ferris as a Student of the Scale Insects. *Pan-Pac. Entomol.* 35:25–28.

Michelbacher, A. E. 1965. Edward Oliver Essig, 1884–1964. *Pan-Pac. Entomol.* 41:207–234.

Needham, J. G. 1931. John Henry Comstock. *Science* 73 (1894):409–410.

Needham, J. G. 1946. The Lengthened Shadow of a Man and His Wife. *Sci. Mo.* 62:140–150, 219–229.

Osborn, H. 1937. *Fragments of Entomological History.* Columbus, Ohio. Published by the Author.

Osten Sacken, C. R. 1903. *Record of My Life Work in Entomology.* Cambridge, Mass., Cambridge Univ. Press.

Schmitt, J. B. 1962. R. E. Snodgrass. *Proc. Entomol. Soc. Wash.* 64:211–213.

Schmitt, J. B. 1963. Robert Evans Snodgrass. *Entomol. News* 74:141–142.

Thurman, E. B. 1959. Robert Evans Snodgrass, Insect Anatomist and Morphologist. *Smithsonian Misc. Coll.* 137:1–17.

Thurman, E. B. 1962. R. E. Snodgrass. *Proc. Entomol. Soc. Wash.* 64:213–216.

Usinger, R. L. 1959. Gordon Floyd Ferris, 1893–1958. *Pan-Pac. Entomol.* 35:1–12.

Wiggins, I. L. 1959. Gordon Floyd Ferris, the Teacher. *Pan-Pac. Entomol.* 35:13–24.

Williston, S. W. 1908. Frank H. Snow, the Man and the Scientist. *Graduate Mag.,* Univ. Kansas, 7:128–134.

CHAPTER VI

Calvert, P. P. 1935. Edward Bruce Williamson. *Entomol. News* 46:1–13.

Carpenter, F. M., and P. J. Darlington. Jr. 1954. Nathan Banks: A Biographic Sketch and List of Publications. *Psyche* 61:81–110.

Gunder, J. D. 1930. North American Institutions Featuring Lepidoptera. XIV. Museum of Comparative Zoology, Cambridge, Mass. *Entomol. News* 41:147–152.

Rehn, J. A. G. 1962. Philip Powell Calvert (1871–1961). *Entomol. News* 73:113–121.

Schwardt, H. H. 1959. James George Needham. *Ann. Entomol. Soc. Amer.* 52:338–339.

CHAPTER VII

Cockerell, T. D. A. 1911. Samuel Hubbard Scudder. *Science* 34:338–342.

Cockerell, T. D. A. 1911. Scudder's Work on Fossil Insects. *Psyche* 18:179–180.

Dow, R. 1937. The Scientific Work of Albert Pitts Morse. *Psyche* 44:1–11.

Ewing, H. E. 1936. Obituary—Andrew Nelson Caudell, 1872–1936. *J. Econ. Entomol.* 29:471–472.

Fay, C. E. 1911. Obituary of S. H. Scudder. *Appalachia* 12:276–279.

Field, W. L. W. 1911. Doctor Scudder's Work on the Lepidoptera. *Psyche* 18:179–180.

Gurney, A. B. 1965. James Abram Garfield Rehn, 1881–1965. *J. Econ. Entomol.* 58:805–807.

Howard, L. O. 1901. Otto Lugger. *Entomol. News* 12:222–224.

Howard, L. O. 1930. *A History of Applied Entomology. Smithsonian Misc. Coll.* Vol. 84.

Howard, L. O., and A. Busck. 1936. Andrew Nelson Caudell. *Proc. Entomol. Soc. Wash.* 38:34–37.

Kingsley, J. S. 1911. Samuel Hubbard Scudder. *Psyche* 18:175.

Mayor, A. G. 1924. Samuel Hubbard Scudder. *Nat. Acad. Biog. Sci. Memoir* 17:81–104.

Mickel, C. E. 1937. Andrew Nelson Caudell. *Ann. Entomol. Soc. Amer.* 30:181–182.

Morse, A. P. 1911. The Orthopterological Work of Mr. S. H. Scudder with Personal Reminiscences. *Psyche* 18:187–192.

Phillips, M. E. 1950. James A. G. Rehn Completes Fifty Years of Research. *Entomol. News* 61:85–88.

Phillips, M. E. 1965. James Abram Garfield Rehn, 1881–1965. *Entomol. News* 76:57–61.

Rehn, J. A. G. 1948. Morgan Hebard (1887–1946). *Entomol. News* 59:57–69.

Scudder, S. H. 1899. *Everyday Butterflies.* Boston, Mass., Houghton Mifflin Co.

Scudder, S. H. 1903. Hunting for Fossil Insects. *34th Rep. Entomol. Soc. Ont.*, pp. 101–103.

Skinner, H. 1911. Samuel Hubbard Scudder. *Entomol. News* 22:289–292.

Swenk, M. H. 1937. In Memoriam—Lawrence Bruner. *Nebr. Bird Review* 5:35–48.

CHAPTER VIII

Abbott, M. 1949. *The Life of William T. Davis.* Ithaca, N. Y., Cornell Univ. Press.

Adams, J. B., and G. W. Simpson. 1955. Edith Marion Patch. *Ann. Entomol. Soc. Amer.* 48:313–314.

Anon. 1916. Obituary of Theodore Pergande. *Entomol. News* 27:240.

Cleaves, H. 1942. W. T. D. *Bull. Bklyn. Entomol. Soc.* 37:132–138.

Comstock, A. B. 1953. *The Comstocks of Cornell.* Ithaca, N. Y., Comstock Publ. Associates.

Essig, E. O. 1959. Charles Fuller Baker, Entomologist, Botanist, Teacher. 1872–1927. *Men and Moments in the History of Science.* Seattle, Wash., Univ. Washington Press.

Essig, E. O., and R. L. Usinger. 1940. The Life and Works of Edward Payson Van Duzee. *Pan-Pac. Entomol.* 16:147–177.

Froeschner, R. C. 1966. Carl John Drake, 1886–1965. *Ann. Entomol. Soc. Amer.* 59:1028–1029.

Gurney, A. B., J. P. Kramer, and W. W. Wirth. 1966. Carl John Drake. *Proc. Entomol. Soc. Wash.* 68:63–71.

Howard, L. O. 1913. Philip Reese Uhler, LL.D. *Entomol. News* 24:433–439.

Howard, L. O. 1933. *Fighting the Insects: The Story of an Entomologist.* New York, Macmillan Co.

Howard, L. O., E. A. Schwarz, and A. Busck. 1916. A Biographical and Bibliographical Sketch of Otto Heidemann. *Proc. Entomol. Soc. Wash.* 18:203–205.

Hungerford, H. B. 1948. Notice—The Torre-Bueno Collection of Hemiptera. *Bull. Bklyn. Entomol. Soc.* 43:148.

Hungerford, H. B. 1958. Raymond Hill Beamer. *J. Kan. Entomol. Soc.* 31:58–66.

Hungerford, H. B., and P. W. Oman. 1958. Raymond Hill Beamer, 1889–1957. *Ann. Entomol. Soc. Amer.* 51:410.

Leonard, M. D., and R. I. Sailer. 1960. Harry Gardner Barber, 1871–1960. *Proc. Entomol. Soc. Wash.* 62:125–138.

List, G. M. 1942. Clarence Preston Gillette. *Ann. Entomol. Soc. Amer.* 35:122.

Needham, J. G. 1946. The Lengthened Shadow of a Man and His Wife. *Sci. Mo.* 62:140–150, 219–229.

Olsen, C. E. 1948. Memories of Early Visits to J. R. de la Torre-Bueno and His Bug Sanctuary. *Bull. Bklyn. Entomol. Soc.* 43:135–137.

Osborn, H. 1937. *Fragments of Entomological History.* Columbus, Ohio. Published by the Author.

Parshley, H. M. 1951. On the Life of William T. Davis. *Entomol. News* 62:84–86.

Riley, W. A. 1924. Alexander Dyer MacGillivray. *Entomol. News* 35:224–228.

Russell, L. M. 1963. Harold Morrison. *Proc. Entomol. Soc. Wash.* 65:311–313.

Schwarz, E. A., O. Heidemann, and N. Banks. 1914. Life and Writings of Philip Reese Uhler. *Proc. Entomol. Soc. Wash.* 16:1–7.

Sherman, J. W., Jr. 1948. J. R. de la Torre-Bueno. *Bull. Bklyn. Entomol. Soc.* 43:154–156.

Smith, C. F. 1956. Zeno Payne Metcalf. *Proc. Entomol. Soc. Wash.* 58:121–122.

Teale, E. W. 1942. William T. Davis—An Appreciation. *Bull. Bklyn. Entomol. Soc.* 37:118–126.

Torre-Bueno, J. R. 1941. Edward Payson Van Duzee—An Appreciation. *Bull. Bklyn. Entomol. Soc.* 36:80–81.

Torre-Bueno, J. R. 1948. Entomology in the United States. *Bull. Bklyn. Entomol. Soc.* 43:141–148.

Usinger, R. L. 1954. Howard Madison Parshley. *Pan-Pac. Entomol.* 30:1–4.

Van Duzee, E. P. 1934. Millard Carr Van Duzee. *Pan-Pac. Entomol.* 10:90–96.

Wade, J. S., and H. G. Barber. 1945. William Thompson Davis, 1862–1945. *Proc. Entomol. Soc. Wash.* 47:230–235.

Woodruff, L. C. 1956. Herbert Barker Hungerford. *Kan. Sci. Bull.* 38:ii–v.

Woodruff, L. C. 1963. Obituary of Herbert Barker Hungerford. *J. Kan. Entomol. Soc.* 36:197–199.

CHAPTER IX

Abbott, M. 1949. *The Life of William T. Davis.* Ithaca, N. Y., Cornell Univ. Press.

Anderson, W. H., G. B. Vogt, and A. B. Gurney. 1950. Herbert Spencer Barber, 1882–1950. *Proc. Entomol. Soc. Wash.* 52:259–269.

Banks, N., E. A. Schwarz, and H. L. Viereck. 1910. Henry Ulke. *Proc. Entomol. Soc. Wash.* 12:105–111.

Blackwelder, R. D. 1950. The Casey Room: Memorial to a Coleopterist. *Coleopterists' Bull.* 4:65–80.

Blaisdell, F. E. 1925. Thomas Lincoln Casey. *Pan-Pac. Entomol.* 2:90–91.

Blake, D. H. 1952. Two Old Coleopterists. *Coleopterists' Bull.* 6:3–9.

Blatchley, W. S. 1928. "Quit-claim" Specialists vs. the Making of Manuals. *Bull. Bklyn. Entomol. Soc.* 23:10–18.

Blatchley, W. S. 1930. *Blatchleyana.* Indianapolis, Nature Publ. Co.

Blatchley, W. S. 1932. *In Days Agone.* Indianapolis, Nature Publ. Co.

Blatchley, W. S. 1939. *Blatchleyana II.* Indianapolis, Nature Publ. Co.

Buchanan, L. L. 1934. Henry Frederick Wickham. *Proc. Entomol. Soc. Wash.* 36:60–64.

Buchanan, L. L. 1935. Thomas Lincoln Casey and the Casey Collection of Coleoptera. *Smithsonian Misc. Coll.* 94:1–15.

Calvert, P. P. 1898. A Biographical Notice of George Henry Horn. *Trans. Amer. Entomol. Soc.* 25:i–xxiv.

Caudell, A. N. 1929. Dr. E. A. Schwarz. *Entomol. News* 40:31–32.

Darlington, P. J., Jr. 1940. Henry Clinton Fall (1862–1939). *Psyche* 47:45–54.

Davis, J. J. 1941. Willis Stanley Blatchley. *Ann. Entomol. Soc. Amer.* 34:279–283.

Davis, W. T. 1941. Dr. Willis Stanley Blatchley. *Bull. Bklyn. Entomol. Soc.* 36:18–19.

Davis, W. T. 1941. Charles W. Leng and the New York Entomological Society. *J. N. Y. Entomol. Soc.* 49:189–192.

Davis, W. T. 1941. Charles W. Leng and the Brooklyn Entomological Society. *Bull. Bklyn. Entomol. Soc.* 36:45–49.

Dow, R. P. 1914. The Greatest Coleopterist. *J. N. Y. Entomol. Soc.* 22:185–191.

Essig, E. O. 1953. Edwin Cooper Van Dyke. *Pan-Pac. Entomol.* 29:72–97.

Hatch, M. H. 1926. Thomas Lincoln Casey as a Coleopterist. *Entomol. News* 37:175–179, 198–202.

Horn, G. H. 1883. Memoir of John L. LeConte, M.D. *Proc. Amer. Phil. Soc.* 21:294–299.

Horn, G. H. 1884. John Lawrence LeConte. *Proc. Amer. Acad. Arts & Sci.* 19:511–516.

Horn, G. H. 1890. A Reply to Dr. C. V. Riley. Privately Printed.

Horn, G. H. 1896. A Visit to Cambridge. *Entomol. News* 7:49–50.

Howard, L. O. 1933. *Fighting the Insects: The Story of an Entomologist.* New York, Macmillan Co.

Howard, L. O., H. S. Barber, and A. Busck. 1928. Dr. E. A. Schwarz. *Proc. Entomol. Soc. Wash.* 30:153–183.

Leng, C. W. 1925. Thomas Lincoln Casey. *Entomol. News* 36:97–100.

Lesley, J. P. 1883. J. L. LeConte. *Proc. Amer. Phil. Soc.* 21:291–294.

Linsley, E. G. 1940. Henry Clinton Fall. *Pan-Pac. Entomol.* 16:1–3.

Mann, W. M. 1948. *Ant Hill Odyssey.* Boston, Little, Brown and Co.

Noland, E. J. 1897. Dr. George Henry Horn. *Proc. Acad. Sci. Philadelphia.* 49:515–518.

Riley, C. V. 1883. John Lawrence LeConte. *Psyche* 4:107–110.

Schaupp, F. G. 1883. J. L. LeConte. *Bull. Bklyn. Entomol. Soc.* 6:i–ix.

Schwarz, E. A., and W. M. Mann. 1925. Colonel Thomas Lincoln Casey. *Proc. Entomol. Soc. Wash.* 27:42–43.

Scudder, S. H. 1884. A Biographical Sketch of Dr. John Lawrence LeConte. *Trans. Amer. Entomol. Soc.* 11:i–xxviii.

Scudder, S. H. 1886. John Lawrence LeConte. *Nat. Acad. Sci. Biog. Memoirs* 2:261–293.

Sherman, J. D., Jr. 1929. Letters of E. A. Schwarz. *J. N. Y. Entomol. Soc.* 37:182–392.

Sherman, J. D., Jr. 1940. Henry Clinton Fall. *J. N. Y. Entomol. Soc.* 48:33–36.

Sherman, J. D., Jr. 1941. Charles W. Leng. *J. N. Y. Entomol. Soc.* 49:185–187.

Skinner, H. 1898. Dr. George H. Horn. *Entomol. News* 9:1–3.

Skinner, H. 1910. Henry Ulke. *Entomol. News* 21:99–100.

Smith, J. B. 1885. Review of Casey's "Contributions." *Entomologica Americana* 1:58–59.

Smith, J. B. 1897. Dr. Horn's Contributions to Coleopterology. *Proc. Acad. Sci. Phila.* 49:529–535.

Smith, J. B. 1898. George H. Horn. *Science* 7:73–77.

Smith, J. B. 1910. Insects and Entomologists, Their Relation to the Community at Large. II. *Pop. Sci. Mo.* 76:469–477.

Snyder, T. E. 1950. Herbert Spencer Barber, 1882–1950. *Coleopterists' Bull.* 4:50–54.

Van Dyke, E. C. 1947. The Biography of Frank Ellsworth Blaisdell, Sr. *Pan-Pac. Entomol.* 23:49–58.

Wade, J. S. 1940. Obituary Notice of Willis Stanley Blatchley. *Proc. Entomol. Soc. Wash.* 42:204–208.

Weiss, H. B. 1936. *The Pioneer Century of American Entomology.* New Brunswick, N. J. Published by the Author.

CHAPTER X

Abbott, M. 1949. *The Life of William T. Davis.* Ithaca, N. Y., Cornell Univ. Press.

Alexander, C. P. 1952. *Fernald Club Yearbook.* Vol. 22. Univ. Mass.

Anon. 1890. A Needless Alarm. *Entomol. News* 1:44.

Anon. 1891. (Henry Edwards.) *Entomol. News* 2:138.

Anon. 1895. Rev. J. G. Morris, D.D. *Entomol. News* 6:9–10.

Anon. 1902. Herman Strecker. *Entomol. News* 13:1–4.

Anon. 1904. Dr. Hans Herman Behr. *Entomol. News* 15:142–144.

Avinoff, A. 1928. Testimonial to Dr. Holland on His Eightieth Birthday. *Ann. Carnegie Museum* 19:11–13.

Avinoff, A. 1933. William Jacob Holland. *Entomol. News* 44:141–144.

Barber, H. S. 1912. In Howard *et al.* Dr. John Bernard Smith. *Proc. Entomol. Soc. Wash.* 14:111–117.

Bell, E. L. 1946. Roswell Carter Williams, Jr. *Entomol. News* 57:167–171.

Bethune, C. J. S. 1903. Prof. Augustus Radcliffe Grote. *Annu. Rep. Entomol. Soc. Ont.* 34:109–112.

Bethune, C. J. S. 1909. William Henry Edwards. *Can. Entomol.* 41:245–248.

Beutenmuller, W. 1891. Henry Edwards. *Can. Entomol.* 23:141–142.

Brown, F. M. 1958–1960. The Correspondence between William Henry Edwards and Spencer Fullerton Baird. *J. N. Y. Entomol. Soc.* 66:191–227, 67:107–123, 125–149, 68:157–175.

Calvert, P. P. 1926. The Entomological Work of Henry Skinner. *Entomol. News* 37:225–249.

Cockerell, T. D. A. 1920. Biographical Memoir of Alpheus Spring Packard, 1839–1905. *Nat. Acad. Sci. Biog. Memoir* 9:181–236.

Cockerell, T. D. A. 1943. Alpheus Spring Packard. *Bios* 14:58–63.

Coolidge, K. R., and H. H. Newcomb. 1920. Richard H. Stretch—An Appreciation. *Entomol. News* 31:181–185.

Cottle, J. E. 1926. *Euphydryas quino* Behr. *Pan-Pac. Entomol.* 3:75–76.

Dexter, R. W. 1957. The Development of A. S. Packard, Jr., as a Naturalist and an Entomologist. *Bull. Bklyn. Entomol. Soc.* 52:57–66, 101–112.

Dickerson, E. L. 1911–1913. The Work of Professor John B. Smith in Economic Entomology. *Proc. S. I. Assoc. Arts and Sciences* 4:17–24.

Dos Passos, C. F. 1956. William Phillips Comstock, 1880–1956. *J. N. Y. Entomol. Soc.* 64:1–5.

Dyar, H. G. 1905. A Synonymic Catalogue of the North American Rhopalocera. Supplement No. 1. By Henry Skinner, M.D., Philadelphia. *J. N. Y. Entomol. Soc.* 13:217.

Edwards, W. H. 1879. *The Butterflies of North America.* American Entomol. Soc. 1868–1872. Reprinted by Houghton, Osgood & Co. in 1879.

Engelhardt, G. P. 1930. Dr. William Barnes. *Bull. Bklyn. Entomol. Soc.* 25:143–144.

Essig, E. O. 1931. *A History of Entomology.* New York, Macmillan Co.

Ferguson, D. C. 1962. James Halliday McDunnough (1877–1962). *J. Lepidopterists' Soc.* 16:209–228.

Forbes, W. T. M., and J. M. Aldrich, 1929. Harrison Gray Dyar. *Entomol. News* 40:165–168.

Fox, R. M. 1963. Arthur Ward Lindsey. *Entomol. News* 74:134.

Gibson, A. 1912. John Bernhardt Smith. *Can. Entomol.* 44:97–99.

Graef, E. L. 1914. Some Early Brooklyn Entomologists. *Bull. Bklyn. Entomol. Soc.* 9:47–56.

Grossbeck, J. A. 1912. Professor John Bernhardt Smith, Sc.D. *Entomol. News* 23:193–196.

Grossbeck, J. A. 1911–1913. John B. Smith as a Lepidopterist. *Proc. S. I. Assoc. Arts and Sciences* 4:28–31.

Grote, A. R. 1889. The Rise of Practical Entomology in America. *Annu. Rep. Entomol. Soc. Ont.* 20:75–82.

Gunder, J. D. 1929. The Carnegie Museum, Pittsburgh, Pa. *Entomol. News* 40:205–217.

Gunder, J. D. 1929. North American Institutions Featuring Lepidoptera. *Entomol. News* 40:245–252.

Gutzkow, F., G. Chismore, and A. Eastwood. 1904. Hans Herman Behr. Read before the Calif. Acad. Sci., pp. 1–7.

Hansens, E. J., and H. B. Weiss. 1958. *Entomology in New Jersey.* New Brunswick, N. J., Rutgers Univ. Press.

Heinrich, C., and E. A. Chapin. 1942. William Schaus. *Proc. Entomol. Soc. Wash.* 44:188–195.

Heinrich, C., and U. C. Loftin. 1944. August Busck, 1870–1944. *Proc. Entomol. Soc. Wash.* 46:230–239.

Holland, W. J. 1903. *The Moth Book.* New York, Doubleday, Page & Co.

Holland, W. J. 1926–1927. Dr. Henry Skinner. *Ann. Carnegie Museum* 17:197–198.

Howard, L. O. 1929. Harrison Gray Dyar. *Science* 69:151–152.

Howard, L. O. 1930. *A History of Applied Entomology. Smithsonian Misc. Coll.* Vol. 84.

Howard, L. O., F. M. Webster, and A. D. Hopkins. 1912. Dr. John Bernard Smith. *Proc. Entomol. Soc. Wash.* 14:111–117.

Howland, H. R. 1907. *The Buffalo Soc. of Nat. Sciences* 7:10–14.

Hunt, R. 1963. In Leslie, J. B. Highlights. *Proc. 50th Annu. Meet. N. J. Mosquito Exterm. Assoc.*, pp. 21–30.

Kingsley, J. S. 1888. Sketch of Alpheus Spring Packard. *Pop. Sci. Mo.* 33:260–267.

Leng, C. W. 1911–1913. John B. Smith as a Coleopterist. *Proc. S. I. Assoc. Arts and Sciences* 4:25–27.

Leslie, J. B. 1963. Highlights in the Fifty Years of the New Jersey Mosquito Extermination Assoc. *Proc. 50th Annu. Meet. N. J. Mosquito Exterm. Assoc.*, pp. 21–30.

Mead, A. D. 1905. Alpheus Spring Packard. *Pop. Sci. Mo.* 67:43–48.

Mengel, L. W. 1902. Dr. Herman Strecker. *Can. Entomol.* 34:25–26.

Milburn, J. G. 1913. Recollections of A. R. Grote. *Entomol. News* 24:182–183.

Muesebeck, C. F., and H. Heinrich. 1936. Foster Hendrickson Benjamin. *Proc. Entomol. Soc. Wash.* 38:25–26.

Osborn, H. 1912. John Bernhardt Smith. *J. Econ. Entomol.* 5:234–236.

Osborn, H. 1937. *Fragments of Entomological History*. Columbus, Ohio. Published by the Author.

Ruckes, H., and C. F. dos Passos. 1965. Ernest Layton Bell. *J. N. Y. Entomol. Soc.* 73:49–56.

Schaus, W. 1930. William Barnes. *Proc. Entomol. Soc. Wash.* 32:114.

Skinner, H. S. 1899. The Butterfly Book. *Entomol. News* 10:18–19.

Skinner, H. S. 1909. William H. Edwards. *Entomol. News* 20:193–194.

Slosson, A. T. 1915. A Few Memories. *J. N. Y. Entomol. Soc.* 23:85–91. 1917. 25:93–97.

Smith, J. B. 1894. A New Insecticide. *Entomol. News* 5:223.

Smith, J. B. 1900. Rev. George Duryea Hulst, Ph.D. *Entomol. News* 11:613–615.

Smith, J. B. 1905. The Entomological Work of Dr. A. S. Packard. *Psyche* 12:33–35.

Smith, J. B. 1906. (Letter.) *Entomol. News* 17:69.

Smith, J. B. 1910. Insects and Entomologists, Their Relation to the Community at Large. II. *Pop. Sci. Mo.* 76:469–477.

Snyder, T. E., H. H. Shepard, and J. F. G. Clarke. 1955. Austin Hobart Clarke. *Proc. Entomol. Soc. Wash.* 57:83–88.

Strecker, H. 1885. Letter from Strecker to Geo. A. Ehrman. Jan. 16, 1885. Not published.

Teale, E. W. 1942. George Paul Englehardt. *Bull. Bklyn. Entomol. Soc.* 37:153–157.

Torre-Bueno, J. R. 1948. Entomology in the United States. *Bull. Bklyn. Entomol. Soc.* 43:141–148.

Voss, E. G. 1963. Arthur Ward Lindsey. *J. Lepidopterists' Soc.* 17:181–190.

Wade, J. S. 1930. Henry Edwards. *Sci. Mo.* 30:240–250.

Wade, J. S., and H. W. Capps. 1955. Carl Heinrich. *Proc. Entomol. Soc. Wash.* 57:249–255.

Walton, W. R. 1931. The Butterfly Book. *Proc. Entomol. Soc. Wash.* 33:187.

Weeks, A. C. 1900. In Memoriam: Rev. Dr. George D. Hulst. *J. N. Y. Entomol. Soc.* 8:248–250.

Weeks, A. C. 1906. Letter to the Editor. *Entomol. News* 17:15–21.

Weiss, H. B. 1936. *The Pioneer Century of American Entomology*. New Brunswick, N. J. Published by the Author.

Williams, R. C., Jr. 1931. Dr. Holland's New Butterfly Book. *Entomol. News* 42:291.

CHAPTER XI

Anon. 1926. Ezra Townsend Cresson. *Entomol. News* 37:161–163.

Brues, C. T. 1937. Professor William Morton Wheeler. *Psyche* 44:61–96.

Burks, B. D. 1965. Obituary. Louis Hart Weld. *Ann. Entomol. Soc. Amer.* 58:133–134.

Calvert, P. P. 1928. A Contribution to the History of Entomology in North America. *Trans. Amer. Entomol. Soc.* Supplement to Vol. 52:i–lxiii.

Calvert, P. P. 1926. Ezra Townsend Cresson. *Science* 64(1644):8–9.

Calvert, P. P. 1948. Theodore Dru Alison Cockerell, 1866–1948. *Year Book of the Amer. Philos. Soc. for 1948,* pp. 247–252.

Cockerell, T. D. A. 1935. Recollections of a Naturalist. *Bios* 6:372.

Cockerell, T. D. A. 1935–1938. Recollections of a Naturalist. *Bios* 8:122–127.

Cory, E. N., and C. F. W. Muesebeck. 1960. Arthur Burton Gahan, 1880–1960. *Proc. Entomol. Soc. Wash.* 62:198–204.

Cushman, R. A., and L. M. Russell. 1940. Grace A. Sandhouse. *Proc. Entomol. Soc. Wash.* 42:188–189.

Doutt, R. L. 1958. Vice, Virtue and the Vedalia. *Bull. Entomol. Soc. Amer.* 4:119–123.

Essig, E. O. 1931. *A History of Entomology*. New York, Macmillan Co.

Essig, E. O. 1948. Theodore Dru Alison Cockerell. *Pan-Pac. Entomol.* 24:117–121.

Ewan, J. 1950. *Rocky Mountain Naturalists*. Denver, Univ. Denver Press, pp. 95–116.

Fullaway, D. T. 1922. Albert Koebele. *Proc. Hawaiian Entomol. Soc.* 5:20–28.

Gahan, A. B., L. M. Russell, and C. H. Heinrich. 1951. James Chamberlin Crawford. *Proc. Entomol. Soc. Wash.* 53:107–109.

Gertsch, W. J. 1961. H. F. Schwarz, 1883–1960. *Entomol. News* 72:85–89.

Henderson, L. J., T. Barbour, F. M. Carpenter, and H. Zinsser. 1937. Obituary: William Morton Wheeler. *Science* 85:533–535.

Howard, L. O. 1925. Albert Koebele. *J. Econ. Entomol.* 18:556–562.

Howard, L. O., E. A. Schwarz, and N. Banks. 1908. W. S. Ashmead. *Proc. Entomol. Soc. Wash.* 10:126–156.

Klopsteg, P. E. 1963. Potpourri and Gallimaufry. *Science* 140:594–598.

Krombein, K. V. 1961. V. S. L. Pate, 1903–1958. *Entomol. News* 72:1–5.

Linsley, E. G. 1948. Theodore D. A. Cockerell. *Bull. Bklyn. Entomol. Soc.* 43:116–118.

Mann, W. M. 1934. Stalking Ants, Savage and Civilized. *Nat. Geog. Mag.* 65:171–192.

Mann, W. M. 1948. *Ant Hill Odyssey*. Boston, Little, Brown and Co.

Melander, A. L., and F. M. Carpenter. 1937. William Morton Wheeler. *Ann. Entomol. Soc. Amer.* 30:433–437.
Michener, C. D. 1948. T. D. A. Cockerell. *J. N. Y. Entomol. Soc.* 56:171–174.
Muesebeck, C. F. W. 1942. Alexander Arsene Girault. *Ann. Entomol. Soc. Amer.* 35:122–123.
Muesebeck, C. F. W. 1951. Sievert Allen Rohwer. *Proc. Entomol. Soc. Wash.* 53:227–230.
Muesebeck, C. F. W. 1957. Robert Asa Cushman, 1880–1957. *Proc. Entomol. Soc. Wash.* 59:247–254.
Muttkowski, R. A. 1914. George William Peckham, M.D., LL.D. *Entomol. News* 25:145–148.
Parker, G. H. 1938. Biographical Memoir of William Morton Wheeler. *Nat. Acad. Sci. Biog. Memoir* 19:203–241.
Peckham, G. W., and E. G. Peckham. 1905. *Wasps: Social and Solitary.* Westminster, Eng., Constable and Co., Ltd.
Perkins, R. C. L. 1925. Early Work of Albert Koebele in Hawaii. *Hawaiian Planters' Record* 29:359–364.
Rehn, J. A. G. 1932. Henry Lorenz Viereck. *Entomol. News* 43:141–148.
Rohwer, S. A. 1948. Theodore Dru Alison Cockerell. *Proc. Entomol. Soc. Wash.* 50:103–108.
Schwarz, H. F. 1948. Theodore D. A. Cockerell. *Entomol. News* 59:85–89.
Skinner, H. 1908. Dr. Wm. H. Ashmead. *Entomol. News* 19:397–398.
Snyder, T. E., J. E. Graf, and M. R. Smith. 1961. William M. Mann. *Proc. Entomol. Soc. Wash.* 63:68–73.
Swezey, O. H. 1925. Biographical Sketch of the Work of Albert Koebele in Hawaii. *Hawaiian Planters' Record* 29:364–368.
Weiss, H. B. 1936. *The Pioneer Century of American Entomology.* New Brunswick, N. J. Published by the Author.
Wheeler, W. M. 1927. Carl Akeley's Early Work and Environment. *Natural History* 27:133–141.
Wilson, E. O. 1959. William M. Mann. *Psyche* 66:55–59.

CHAPTER XII

Aldrich, J. M. 1906. Baron Osten Sacken. *Entomol. News* 17:269–272.
Aldrich, J. M. 1918. Samuel Wendell Williston. *Entomol. News* 29:322–327.
Aldrich, J. M. 1919. Samuel Wendell Williston the Entomologist. *Sigma Xi Quart.* 7:19–21.
Aldrich, J. M. 1930. Informal Reminiscences. *J. Wash. Acad. Sci.* 20:495–498.
Allen, T. C. 1959. Charles Lewis Fluke, 1891–1959. *J. Econ. Entomol.* 52:787.
Banks, N. 1911. Daniel William Coquillet. *Proc. Entomol. Soc. Wash.* 13:195–210.
Bromley, S. W. 1944. Ephraim Porter Felt, 1868–1943. *J. N. Y. Entomol. Soc.* 52:223–236.
Burgess, A. F. 1943. Ephraim Porter Felt, 1868–1943. *J. Econ. Entomol.* 36:950–952.

Caudell, A. N., A. Busck, and L. O. Howard. 1919. Frederick Knab. *Proc. Entomol. Soc. Wash.* 21:41–52.

Cresson, E. T., Jr. 1911. Daniel William Coquillet. *Entomol. News* 22:337–338.

Davis, J. J. 1958. Memories of Years Agone—Mostly Entomological. Dept. Entomology, Purdue University.

Doutt, R. L. 1958. Vice, Virtue and the Vedalia. *Bull. Entomol. Soc. Amer.* 4:119–123.

Essig, E. O. 1931. *A History of Entomology*. New York, Macmillan Co.

James, M. T. 1945. C. H. Tyler Townsend. *Ann. Entomol. Soc. Amer.* 38:143–144.

Johnson, C. W. 1906. Charles Robert b.v. Osten Sacken. *Entomol. News* 17:273–275.

Lull, R. S. 1924. Samuel Wendell Williston. *Nat. Acad. Sci. Biog. Memoir* 17:115–141.

McClung, C. E. 1919. A Student Appreciation. *Sigma Xi Quart.* 7:9.

Melander, A. L. 1934. John Merton Aldrich. *Psyche* 41:133–149.

Muesebeck, C. F. W., and C. W. Collins. 1944. Ephraim Porter Felt. *Proc. Entomol. Soc. Wash.* 46:26–29.

Osborn, H. F. 1919. The Man and the Paleontologist. *Sigma Xi Quart.* 7:2–8.

Osten Sacken, C. R. 1903. *Record of My Life Work in Entomology*. Cambridge, Mass., Cambridge Univ. Press.

Sabrosky, C. W. 1963. John Russell Malloch, 1875–1963. *Ann. Entomol. Soc. Amer.* 56:565.

Sherman, F., Jr. 1906. Letter to the Editor. *Entomol. News* 17:32.

Wade, J. S. 1930. John Merton. Aldrich. *J. Wash. Acad. Sci.* 20:495–498.

Walton, W. R. 1914. On the Work of the Late Daniel W. Coquillet and Others. *J. N. Y. Entomol. Soc.* 22:159–164.

Walton, W. R. 1934. John Merton Aldrich, Ph.D. *Proc. Entomol. Soc. Wash.* 36:180–183. 1935. 37:53–59.

Wilson, C. H. 1942. *Ambassadors in White*. New York, Henry Holt & Co.

CHAPTER XIII

Anon. 1895. (George Marx.) *Entomol. News* 6:265–266.

Baker, E. W., and A. B. Gurney. 1951. Henry Ellsworth Ewing. *Proc. Entomol. Soc. Wash.* 53:147–149.

Banks, N. 1931. J. H. Emerton. *Can. Entomol.* 63:23–24.

Banks, N. 1932. J. H. Emerton. *Psyche* 39:1–8.

Bosbyshell, O. C. 1911. Henry C. McCook. *J. Presb. Hist. Soc.* 6:97–150.

Burgess, E. 1875. *A Collection of the Arachnological Writings of Nicholas Marcellus Hentz, M.D.* Edited by Edward Burgess. *Boston Soc. Nat. Hist.*

Calvert, P. P. 1911. Henry Christopher McCook. *Entomol. News* 22:433–438.

Cobb, C. 1932. Nicholas Marcellus Hentz. *J. Elisha Mitchell Sci. Soc.* 47:47–51.

Comstock, A. B. 1953. *The Comstocks of Cornell*. Ithaca, N. Y., Cornell Univ. Press.

Hutchinson, G. E. 1945. Alexander Petrunkevitch: An Appreciation of His Scientific Works and a List of His Published Writings. *Trans. Conn. Acad. Arts & Sci.* 36:9–24.

Johannsen, O. A. 1937. Cyrus R. Crosby, 1879–1937. *J. Econ. Entomol.* 30:221.

Kinkead, E. 1950. Arachnologist I. *New Yorker,* April 22, p. 37; Arachnologist II. *Ibid.* April 29, p. 37.

Riley, C. V. 1896. Dr. George Marx. *Proc. Entomol. Soc. Wash.* 3:195–201.

Skinner, H. 1911. Henry C. McCook. *J. Presb. Hist. Soc.* 6:115–121.

Woodruff, L. L. 1945. Alexander Petrunkevitch, Colleague and Friend. *Trans. Conn. Acad. Arts & Sci.* 36:7–8.

Wadsworth, C., Jr. 1911. Henry C. McCook. *J. Presb. Hist. Soc.* 6:100–102.

Walton, W. R. 1921. Entomological Drawings and Draughtsman. *Proc. Entomol. Soc. Wash.* 23:69–99.

Woodson, W. D. 1950. Our first "spider man." *Nature Mag.* 43:485–487.

CHAPTER XIV

Alexander, C. P. 1948. Dr. Alfonso Dampf Tenson. *Entomol. News* 59:89–91.

Alexander, C. P. 1952. Guy Chester Crampton. *Entomol. News* 63:1–3.

Alexander, C. P. 1957. James Speed Roger. *Entomol. News* 68:85–88.

Anderson, W. A. 1946. Dwight Sanderson, Rural Social Builder. *Rural Sociology* 11:7–14.

Anon. 1951–1952. Everett Franklin Phillips. *Cornell University—Necrology of the Faculty,* pp. 10–13.

Anon. 1960. Otto Herman Swezey Memorial Number. *Proc. Hawaiian Entomol. Soc.* 17:159–310.

Back, E. A., and W. D. Reed. 1955. Austin Winfield Morrill. *J. Econ. Entomol.* 48:230–231.

Baker, W. A., and J. S. Wade. 1953. William Randolph Walton. *J. Econ. Entomol.* 46:532–533.

Bailey, S. F. 1938. Bells in Memory of W. E. Hinds. *Pan-Pac. Entomol.* 14:96.

Bailey, S. F. 1939. The Hinds Collection of Thysanoptera. *Pan-Pac. Entomol.* 15:91–93.

Balduf, W. V. 1948. Clell Lee Metcalf. *J. Econ. Entomol.* 41:997–998.

Bishopp, F. C., and C. E. Burnside. 1937. In Memoriam: Gershon Franklin White. *Proc. Entomol. Soc. Wash.* 39:184–188.

Bishopp, F. C., E. R. Sasscer, and C. F. Muesebeck. 1951. Sievert A. Rohwer. *J. Econ. Entomol.* 44:437–439.

Blake, D. H. 1951. Two Old Coleopterists. *Coleopterists' Bull.* 5:49–54, 65–72. 1952. *Ibid.* 6:3–9, 19–26.

Bourne, A. I. 1953. Alfred Franklin Burgess. *J. Econ. Entomol.* 46:918–920.

Brooks, W. S. 1932. Bull. 65. Boston Soc. Nat. Hist., October.

Brues, C. T. 1933. Charles Willison Johnson. *Entomol. News* 44:113–116.

Brues, C. T. 1942. Samuel Henshaw. *Ann. Entomol. Soc. Amer.* 35:123–124.

Busbey, R. L. 1963. Ruric Creegan Roark. *Proc. Entomol. Soc. Wash.* 65:69–77.

Calvert, P. P. 1941. Samuel Henshaw, 1852–1941. An Appreciation. *Entomol. News* 52:241–242.

Campbell, F. L. 1946. Valediction—Theodore Henry Frison. *Sci. Mo.* 62:91–93.

Cardon, P. V., and W. H. White. 1951. Percy Nicol Annand. *J. Econ. Entomol.* 44:268–270.

Chapman, P. J. 1960. Frederick Zeller Hartzell. *J. Econ. Entomol.* 53:180–181.

Clausen, C. P., and S. E. Flanders. 1958. Harry Scott Smith. *J. Econ. Entomol.* 51:266–267.

Comstock, J. H. 1909. The Late Professor Slingerland. *Entomol. News* 20:217–219.

Comstock, J. H. 1914. Mark Vernon Slingerland. *N. Y. Agr. Expt. Sta. (Cornell) Bull.* 348, pp. 623–624.

Cory, E. N. 1955. Ernest Ralph Sasscer. *Proc. Entomol. Soc. Wash.* 57:309–310.

Cory, E. N. 1956. Leonard Marion Peairs. *J. Econ. Entomol.* 49:430–431.

Craighead, F. C., H. H. Stage, and L. L. Buchanan. 1944. Maulsby Willett Blackman. *Proc. Entomol. Soc. Wash.* 46:14–18, 21.

Creighton, J. T. 1943. Wilmon Newell. *J. Econ. Entomol.* 36:947–949.

Davis, J. J. 1943. Wesley Pillsbury Flint. *J. Econ. Entomol.* 36:644–645.

Davis, J. J. 1945. Theodore Henry Frison. *J. Econ. Entomol.* 38:729–730.

Davis, J. J. 1958. J. W. Folsom. Memories of Years Agone—Mostly Entomological. Dept. Entomology, Purdue University.

Davis, J. J. 1961. A Contribution to the History of Commercial Pest Control. Dept. Entomology, Purdue University.

Deay, H. O., J. V. Osmun, and G. E. Lehker. 1966. John June Davis. *Ann. Entomol. Soc. Amer.* 59:871–872.

DeLong, D. M. 1944. Elmer Darwin Ball. *J. Econ. Entomol.* 37:159.

Doane, R. W. 1940. Vernon L. Kellogg. *Ann. Entomol. Soc. Amer.* 33:599–607.

Eaton, C. B., and F. P. Keen. 1964. Harry E. Burke. *J. Econ. Entomol.* 57:613–614.

Emerson, A. E. 1944. Frank Eugene Lutz. *Science* 99:233–234.

Essig, E. O. 1931. *A History of Entomology*. New York, Macmillan Co.

Essig, E. O. 1941. Charles William Woodworth. *J. Econ. Entomol.* 34:128–129, 34:595–596.

Essig, E. O. 1943. William Hunter Volck. *J. Econ. Entomol.* 36:484–486.

Essig, E. O. 1948. Sol Felty Light. *Pan-Pac. Entomol.* 24:49–55.

Felt, E. P. 1939. Wilton Everett Britton. *J. Econ. Entomol.* 32:350–351.

Fernald, H. T. 1943. Leonard S. McLaine. *J. Econ. Entomol.* 36:946–947.

Fisher, C. C. 1939. *Vernon Kellogg, 1867–1937*. Anderson House, Washington, D. C., The Belgian American Foundation, Inc.

Freeborn, S. B. 1950. W. B. Herms. *Proc. & Papers 18th Annu. Conf. Calif. Mosquito Control Assoc.*, pp. 2–3.

Furman, D. P. 1949. William Brodbeck Herms. *Pan-Pac. Entomol.* 25:192.

Gertsch, W. J. 1944. Frank E. Lutz. *Ann. Entomol. Soc. Amer.* 37:133–135.

Gibson, A. 1946. Obituary. Theodore Henry Frison. *Can. Entomol.* 78:23–24.

Gurney, A. B. 1964. The Entomological Work of Bentley B. Fulton. *Proc. Entomol. Soc. Wash.* 66:151–159.

Haeussler, G. J. 1965. James Keever Holloway. *J. Econ. Entomol.* 58:808–809.

Hinds, W. E. 1937. J. W. Folsom. *Ann. Entomol. Soc. Amer.* 30:182.

Hodson, A. C. 1964. William A. Riley. *Ann. Entomol. Soc. Amer.* 57:266.

Holland, G. P. 1967. George Johnston Spencer. *Ann. Entomol. Soc. Amer.* 60:710–711.

Holland, R. F., G. C. Kent, and C. E. Palm. 1962. Herbert Henry Schwardt. *J. Econ. Entomol.* 55:1027.

Howard, L. O. 1930. *A History of Applied Entomology. Smithsonian Misc. Coll.* Vol. 84.

Jackson, R. T. 1941. Samuel Henshaw. *Science* 93:342–343.

Leonard, M. D., and W. H. Larrimer. 1961. Joseph Sanford Wade. *Proc. Entomol. Soc. Wash.* 63:219–222.

Leslie, J. B. 1963. Highlights in the Fifty Years of the N. J. Mosquito Exterminating Association. *Proc. 50th Annu. Meeting N. J. Mosquito Exterm. Assoc.*, pp. 21–30.

Martin, D. F., M. A. Price, and N. M. Randolph. 1954. Sherman Weaver Bilsing. *J. Econ. Entomol.* 47:1163–1164.

McAtee, W. L., and J. S. Wade. 1951. Raymond Corbett Shannon. *Proc. Entomol. Soc. Wash.* 53:211–222.

McClung, C. E. 1939. Vernon Lyman Kellogg. *Nat. Acad. Sci. Biog. Memoirs* 20:245–247.

Meiners, E. P. 1949. Phil Rau. *Lepidopterists' News* 2:62.

Melander, A. L. 1932. The Entomological Publications of C. W. Johnson. *Psyche* 39:87–99.

Melander, A. L. 1955. Charles Thomas Brues, II. *Ann. Entomol. Soc. Amer.* 48:422–423.

Metcalf, C. L. 1944. Wesley P. Flint. *Ann. Entomol. Soc. Amer.* 37:131–132.

Mills, H. B. 1958. From 1858 to 1958. *Ill. Nat. Hist. Survey Bull.* 27:99–100.

Millspaugh, D. D. 1964. Harry Edwin Jaques. *Ann. Entomol. Soc. Amer.* 57:265–266.

Mowry, H. 1944. Wilmon Newell. *Science* 99:377–378.

Muesebeck, C. F. W., A. B. Gahan, and E. R. Sasscer. 1951. Sievert Allen Rohwer. *Proc. Entomol. Soc. Wash.* 53:227–230.

Muesebeck, C. F. W., J. S. Wade, and W. H. Anderson. 1958. Adam Giede Böving. *Proc. Entomol. Soc. Wash.* 60:33–40.

Needham, J. G. 1946. The Lengthened Shadow of a Man and His Wife. *Sci. Mo.* 62:140–150, 219–229.

Osborn, H. 1937. Vernon L. Kellogg. *J. Econ. Entomol.* 31:325–326.

Osborn, H. 1952. Clarence Hamilton Kennedy. *Ann. Entomol. Soc. Amer.* 45:362–363.

Osborn, H., J. E. Graf, and F. W. Poos. 1944. Elmer Darwin Ball. *Proc. Entomol. Soc. Wash.* 46:21–22.

Osburn, R. C. 1952. Clarence Hamilton Kennedy. *Buckeye—Ohio State Univ.* 11:2.

Packard, C. M., and J. S. Wade. 1940. Charles Nicholas Ainslie. *J. Econ. Entomol.* 33:206–207.

Peairs, L. M., and B. F. Driggers. 1946. Thomas J. Headlee. *J. Econ. Entomol.* 39:681–683.

Pemberton, C. E. 1960. Otto Herman Swezey. *J. Econ. Entomol.* 53:332–333.

Pemberton, C. E. 1960. Otto Herman Swezey. *Proc. Hawaiian Entomol. Soc.* 17:182–193.

Phillips, E. F. 1944. Dwight Sanderson. *J. Econ. Entomol.* 37:858–859.

Porter, B. A. 1959. Altus L. Quaintance. *J. Econ. Entomol.* 52:182.

Rau, P. 1923. Dr. Charles Henry Turner. *Entomol. News* 34:289–292.

Rawlins, W. A., J. G. Franclemont, and H. Dietrich. 1965. Glen Washington Herrick. *J. Econ. Entomol.* 58:809–810.

Reed, W. D., P. Simmons, and A. W. Morrill, Jr. 1959. Ernest Adna Back. *J. Econ. Entomol.* 52:1229–1230.

Rehn, J. A. G. 1932. Henry Lorenz Viereck. *Entomol. News* 43:141–148.

Romer, A. S., I. W. Bailey, and F. M. Carpenter. 1955. Charles Thomas Brues. *Harvard Univ. Gazette,* Dec. 17, pp. 107–108.

Ross, H. H. 1946. Theodore Henry Frison. *Ann. Entomol. Soc. Amer.* 39:345.

Schwardt, H. H., H. Dietrich, and B. V. Travis. 1957–1958. Robert Matheson. *Cornell University–Necrology of the Faculty,* pp. 20–22.

Schwarz, H. F. 1944. Frank E. Lutz. *Entomol. News* 55:29–32.

Sharp, S. S. 1954. Harry Frederic Dietz. *J. Econ. Entomol.* 47:1162–1163.

Shockley, W. 1963. Claude Wakeland. *J. Econ. Entomol.* 56:423.

Siegler, E. H. 1957. Norman Eugene McIndoo. *Proc. Entomol. Soc. Wash.* 59:43–44.

Smith, C. F. 1961. Bentley Ball Fulton. *J. Econ. Entomol.* 54:613–614.

Smith, H. S., A. M. Boyce, and L. D. Batchelor. 1952. Henry Josef Quayle. *J. Econ. Entomol.* 45:559.

Smith, R. C. 1956. George Adam Dean. *J. Econ. Entomol.* 49:573.

Snetsinger, R. 1962. Charles Aubrey Thomas. *J. Econ. Entomol.* 55:576.

Spieth, H. T. 1966. Axel Leonard Melander. *Ann. Entomol. Soc. Amer.* 59:235–237.

Thomas, F. L. 1936. Warren Elmer Hinds. *J. Econ. Entomol.* 29:225–226.

Tomlinson, W. E., Jr., and A. I. Bourne. 1958. Henry James Franklin. *J. Econ. Entomol.* 51:564–565.

Traub, R. 1964. Henry Shepard Fuller. *J. Econ. Entomol.* 57:793–795.

Trueman, H. L. 1955. James Malcolm Swaine. *Can. Sci. Service Entomol. Div. News Letter* 33:2–3.

Turner, N. 1962. Roger B. Friend. *J. Econ. Entomol.* 55:422.

Twinn, C. R. 1956. Arthur Gibson. *Canada Sci. Service Entomol. Div. News Letter* 34:2–3.

Usinger, R. L., and E. C. Zimmerman. 1960. Otto Herman Swezey. *Pan-Pac. Entomol.* 36:151–153.

Vorhies, C. T. 1944. Elmer Darwin Ball. *Ann. Entomol. Soc. Amer.* 37:129–130.

Wade, J. S. 1949. George Ware Barber. *J. Econ. Entomol.* 42:163–165.

Wade, J. S., and J. A. Hyslop. 1941. Obituary Notice of Samuel Henshaw. *Proc. Entomol. Soc. Wash.* 43:108–110.

Wade, J. S., W. A. Baker, and F. W. Poos. 1953. William Randolph Walton. *Proc. Entomol. Soc. Wash.* 55:103–108.

Walton, W. R. 1921. Entomological Drawings and Draughtsman: Their Relation to the Development of Economic Entomology in the United States. *Proc. Entomol. Soc. Wash.* 23:77–78.

Walton, W. R., and D. J. Caffey. 1940. Charles N. Ainslie. Obituary Notice. *Proc. Entomol. Soc. Wash.* 42:27–30.

Weiss, H. B. 1944. Frank Eugene Lutz. *J. N. Y. Entomol. Soc.* 52:63–73.

Weiss, H. B. 1946. Thomas Jefferson Headlee. *Ann. Entomol. Soc. Amer.* 39:347–348.

Woke, P. A., A. Stone, and R. H. Foote. 1956. William H. Wood Komp. *Proc. Entomol. Soc. Wash.* 58:47–55.

Index

O

P

Q

R

S

T

ABOUT THE AUTHOR

Arnold Mallis is currently Assistant Professor of Entomology Extension at Pennsylvania State University and has served for nearly a quarter of a century as Entomologist for the Gulf Research & Development Co. He received his B.S. and M.S. degrees at the University of California, Berkeley. He is the author of numerous papers on household insects and several on early entomologists.

The text of this book was set in Baskerville Linotype and printed by offset on P & S Special GL manufactured by P. H. Glatfelter Co., Spring Grove, Pa. Composed, printed and bound by Quinn & Boden Company, Inc., Rahway, N.J.